ISBN 978-1-332-29658-3
PIBN 10310683

English
Français
Deutsche
Italiano
Español
Português

www.forgottenbooks.com

Mythology Photography **Fiction**
Fishing Christianity **Art** Cooking
Essays Buddhism Freemasonry
Medicine **Biology** Music **Ancient**
Egypt Evolution Carpentry Physics
Dance Geology **Mathematics** Fitness
Shakespeare **Folklore** Yoga Marketing
Confidence Immortality Biographies
Poetry **Psychology** Witchcraft
Electronics Chemistry History **Law**
Accounting **Philosophy** Anthropology
Alchemy Drama Quantum Mechanics
Atheism Sexual Health **Ancient History**
Entrepreneurship Languages Sport
Paleontology Needlework Islam
Metaphysics Investment Archaeology
Parenting Statistics Criminology
Motivational

THE

NATURALIST:

MONTHLY JOURNAL OF

NATURAL HISTORY FOR THE NORTH OF ENGLAND

EDITED BY

THOMAS SHEPPARD, F.G.S.,

CURATOR OF THE MUNICIPAL MUSEUMS, HULL.

AUTHOR OF "GEOLOGICAL RAMBLES IN EAST YORKSHIRE"; "THE MAKING OF EAST YORKSHIRE"
EDITOR OF MORTIMER'S "FORTY YEARS' RESEARCHES"; AND OF THE PUBLICATIONS OF THE
EAST RIDING ANTIQUARIAN SOCIETY, AND HULL SCIENTIFIC AND FIELD NATURALISTS' CLUB.

PRESIDENT OF THE HULL SCIENTIFIC AND FIELD NATURALISTS' CLUB, OF THE HULL
GEOLOGICAL SOCIETY, AND OF THE HULL SHAKESPEARE SOCIETY; HON. SECRETARY OF
THE YORKSHIRE NATURALISTS' UNION; MEMBER OF THE COUNCILS AND COMMITTEES
OF THE BRITISH ASSOCIATION, MUSEUMS' ASSOCIATION, YORKSHIRE ROMAN ANTI-
QUITIES COMMITTEE, THE YORKSHIRE GEOLOGICAL SOCIETY, THE EAST
RIDING NATURE STUDY ASSOCIATION, THE EAST RIDING ANTIQUARIAN
SOCIETY, AND HULL LITERARY CLUB; HON. LIFE MEMBER OF
THE SPALDING GENTLEMEN'S SOCIETY; OF THE DONCASTER
SCIENTIFIC SOCIETY; ETC.

AND

THOMAS WILLIAM WOODHEAD, Ph.D., F.L.S.

LECTURER IN BIOLOGY, TECHNICAL COLLEGE, HUDDERSFIELD;

WITH THE ASSISTANCE AS REFEREES IN SPECIAL DEPARTMENTS OF

J. GILBERT BAKER, F.R.S., F.L.S.   GEORGE T. PORRITT, F.L.S., F.E.S.
PROF. PERCY F. KENDALL, M.SC., F.G.S.   JOHN W. TAYLOR.
T. H. NELSON, M.B.O.U.   WILLIAM WEST, F.L.S.

1907.

LONDON:

A. BROWN & SONS, LTD., 5, FARRINGDON AVENUE, E.C.
AND AT HULL AND YORK.

PRINTED AT BROWNS' SAVILE PRESS,
DOCK STREET, HULL.

LIST OF PLATES.

JANUARY 1907.

No. 600
(No. 378 of current series).

THE NATURALIST.

A MONTHLY ILLUSTRATED JOURNAL OF

NATURAL HISTORY FOR THE NORTH OF ENGLAND.

EDITED BY

T. SHEPPARD, F.G.S.,

THE MUSEUM, HULL;

AND

T. W. WOODHEAD, Ph.D., F.L.S.,

TECHNICAL COLLEGE, HUDDERSFIELD;

WITH THE ASSISTANCE AS REFEREES IN SPECIAL DEPARTMENTS OF

J. GILBERT BAKER, F.R.S. F.L.S.,   GEO. T. PORRITT, F.L.S., F.E.S.,
Prof. P. F. KENDALL, M.Sc., F.G.S.,   JOHN W. TAYLOR,
T. H. NELSON, M.B.O.U.,   WILLIAM WEST, F.L.S.

Contents :—

LONDON:

A. BROWN & SONS, LIMITED, 5, FARRINGDON AVENUE, E.C.

And at HULL AND YORK.

Printers and Publishers to the Y.N.U.

PRICE 6d. NET. BY POST 7d. NET.

POST FREE.

(Of a few of these there are several copies.)

1. **Inaugural Address,** Delivered by the President, Rev. W. FOWLER, M.A in 1877. **6d.**
2. **On the Present State of our knowledge of the Geography of Britis** Plants (Presidential Address). J. GILBERT BAKER, F.R.S. **6d.**
3. **The Fathers of Yorkshire Botany** (Presidential Address). J. GILBER BAKER, F.R.S. **9d.**
4. **Botany of the Cumberland Borderland Marshes.** J. G. BAKER, F.R.S. **6**
5. **The Study of Mosses** (Presidential Address). Dr. R. BRATHWAITE F.L.S. **6d.**
6. **Mosses of the Mersey Province.** J. A. WHELDON. **6d.**
7. **Strasburger's Investigation on the Process of Fertilisation in Phaner** gams. THOMAS HICK, B.A., B.Sc. **6d.**
8. **Additions to the Algæ of West Yorkshire.** W. WEST, F.L.S. **6d.**
9. **Fossil Climates.** A. C. SEWARD, M.A., F.R.S. **6d.**
10. **Henry Thomas Soppitt** (Obituary Notice). C. CROSSLAND, F.L.S. **6**
11. **The Late Lord Bishop of Wakefield** (Obituary Notice). WILLIA WHITWELL, F.L.S. **6d.**
12. **The Flora of Wensleydale.** JOHN PERCIVAL, B.A. **6d.**
13. **Report on Yorkshire Botany for 1880.** F. ARNOLD LEES. **6d.**
14. **Vertebrates of the Wertern Ainsty (Yorkshire).** EDGAR R. WAITE F.L.S. **9d.**
15. **Lincolnshire.** JOHN CORDEAUX, M.B.O.U. **6d.**
16. **Heligoland.** JOHN CORDEAUX, M.B.O.U. **6d.**
17. **Bird-Notes from Heligoland for the Year 1886.** HEINRICH GÄTK C.M.Z.S. **1s.**
18. **Coleoptera of the Liverpool District.** Part IV., Brachelytra. JOH W. ELLIS, L.R.C.P. **6d.**
19. **Coleoptera of the Liverpool District.** Parts V. and VI., Clavicorni and Lamellicornia. JOHN W. ELLIS, L.R.C.P. **6d.**
20. **The Hydradephaga of Lancashire and Cheshire.** W. E. SHARP. **6d.**
21. **The Lepidopterous Fauna of Lancashire and Cheshire.** Part I. Rhophalocera. JOHN W. ELLIS, L.R.C.P. **6d.**
22. **The Lepidopterous Fauna of Lancashire and Cheshire.** Part II. Sphinges and Bombyces. JOHN W. ELLIS, L.R.C.P. **6d.**
23. **Variation in European Lepidoptera.** W. F. DE VISMES KANE, M.A. M.R.I.A. **6d.**
24. **Yorkshire Lepidoptera in 1891.** A. E. HALL, F.E.S. **6d.**
25. **Yorkshire Hymenoptera** (Third List of Species). S. D. BAIRSTOW F.L.S., W. DENISON ROEBUCK, and THOMAS WILSON. **6d.**
26. **List of Land and Freshwater Mollusca of Lancashire.** ROBER STANDEN. **9d.**
27. **Yorkshire Naturalists at Gormire Lake and Thirkleby Park.** **6d.**

WANTED.

Quarterly Journal Geological Society, Vols. 1-22.
Phillip's Life of William Smith.
British Association Reports for 1839 and 1840.
Proceedings of Yorkshire Geological and Polytechnic Society, Vol. I.
Barnsley Naturalist Society's Quarterly Reports, Set.

GOOD PRICES GIVEN.

Apply—Hon. Sec., Y.N.U., Museum, Hull.

On the Edge of the Moor.

(See p. 4).

THE NATURALIST

FOR 1907.

NOTES AND COMMENTS.

LIVERPOOL BIOLOGISTS.

IN a substantial report of 208 pages,* Prof. Herdman, with the assistance of Messrs. A. Scott and J. Johnstone, gives a valuable account of a year's work at Liverpool and Piel. An idea of the

An Over=crowded Mussel Skear.

nature of the contents of the volume may be gathered by the following list of reports, etc., which it contains :—Introduction and general account of the work ; ' Sea-fish Hatchery at Piel ' ; ' Classes, Visitors, etc., at Piel '; ' Report on the Tow-nettings ' ; ' Faunistic Notes ' ; ' Mussel Transplantation ' ; ' Trawling Observations '; ' Marked Fish Experiments '; ' Parasites of Fishes '; ' Ichthyological Notes ' ; ' Sewage Pollution·

* No. XIV. Report for 1905 of the Lancashire Sea-Fisheries Laboratory at the University of Liverpool, and the Sea-fish Hatchery at Piel. Liverpool, 1906.

at Llanfairfechan'; 'Oligodynamic Action of Copper'; and
'Sea Fish Hatchery at Port Erin.'

We regret that the demands upon our space enable us to do
little more than draw the attention of naturalists to this useful
report.   The effect on the growth of mussels by overcrowding
on the skears is admirably shown, and good work was done by
transplanting young mussels in over-crowded areas, to places
where they had better opportunity of flourishing.   Under
favourable conditions, mussels grow fully half-an-inch a year,
but when they are too thick upon the skears progress is not
so rapid.

In the same report a curious instance of arrested metamorphosis in a flounder is referred to.   As will be seen from the

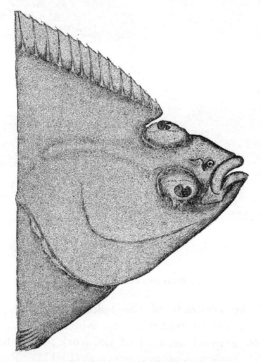

Flounder, showing arrested metamorphosis.

figure, the left eye does not occupy the normal position, but is
very distinctly on the (secondary) dorsal margin of the head,
and, indeed, is easily visible from the 'blind' side of the
specimen.   The fish was also pigmented on both sides.   For
the illustrations we are indebted to Prof. Herdman.

LIVERPOOL GEOLOGISTS.

THAT our geological friends in Liverpool are as enthusiastic as ever, is shown by the regular appearance of the 'Proceedings of the Liverpool Geological Society,' in a brilliantly coloured cover. Part 2 of Vol. X., containing particulars of the work accomplished in the Society's forty-seventh session, is recently to hand. The principal items are :—(1) The President's Address (presumably by Mr. H. C. Beasley), in which he concludes that 'The secrets of the Trias are only to be discovered by the study of the desert regions of the world existing as they do in each of the continents'; (2) 'The Colorado Canyon and its lessons,' by W. M. Davis, of Harvard ; (3) 'The Pleistocene Clays and Sands of the Isle of Man,' by T. Mellard Reade and Joseph Wright. What *would* the Liverpool geologists do without these two veterans, who must be congratulated on their enthusiasm, and on the persistency with which, by means of their lists of foraminifera, they endeavour to drive nails in the coffin of the glacialists. But this coffin is not yet made,* and if ever it is, we doubt whether the numerous lists published by Messrs. Reade and Wright will have had anything to do with it. On the other hand, the day *may* come (and may it be long postponed), when the last of the 'submergers' will be sent to sea, Viking-like, and be heard of no more ; as it is, they 'pass on and on, and go from less to less.' Mr. J. Lomas gives an account of his examination of 'The Dwyka in South Africa,' which is illustrated by two blocks from photographs. Mr. W. D. Brown writes 'On Some Erratics of the Boulder Clay in the Neighbourhood of Burscough,' and gives the results of experiments upon boulders by artificial sand blast. In a lengthy and elaborate paper, Messrs. T. M. Reade and Philip Holland, conclude their useful researches among 'Sands and Sediments.' This is illustrated by two excellent plates.

LEEDS GEOLOGISTS.

Our Leeds Geological friends must be congratulated upon placing on record an account of their doings during the years 1900-5.† This record appears in a pamphlet of 52 pages, and contains particulars of the officers for each year, titles of papers read, etc., from the twenty-seventh to the thirty-first session. There are abstracts of papers on 'The Airedale Glacier,' by

---

* Some think that a coffin will not be required—as the glacialists will be *cremated!*

† 'Transactions of the Leeds Geological Association,' Part XIII., 1900-1, 1901-2, 1902-3, 1903-4, 1904-5. Price 1/6. Jowett & Sowry, Printers, Leeds.

H. B. Muff; 'The Geology of Ingleborough,' by J. H. Howarth;
'The Aims of a Local Geological Society' and 'The Causes of
Volcanic Action,' by D. Forsyth ; 'Geological Photography,'
by Godfrey Bingley ; 'Glacier Lakes of the Cleveland Hills,' by
P. F. Kendall ; 'The relation of the Geology to the Vegetation
of the West Riding of Yorkshire,' by W. G. Smith ; 'Some
Drift Deposits near Leeds, [Re-printed from the 'Proceedings
of the Yorkshire Geological Society,'] by E. Hawkesworth ;
'River Capture in Yorkshire,' by Rev. W. Lower Carter ;
'The Eastern Extension of the Yorkshire Coalfield,' by P. F.
Kendall ; and 'Report of the Magnesian Limestone Committee,' by A. R. Dwerryhouse. Lists of other papers to be
consulted on the various subjects referred to are given, and
add to the value of the abstracts.

Presumably the ' lack of funds,' referred to in the Preface,
which has accounted for the delay in the appearance of these
Transactions, is also responsible for the brevity of some of the
abstracts, and the omission of some papers altogether. The
appearance of the publication would have been improved had the
various lists of officers, etc., been put all together at one end of
the pamphlet, instead of cutting up the reading matter. Mr.
Hawkesworth's map would also have been much clearer if the
'gravel patches' had been shaded. We commend the Editors
on the prominence given to *local* papers—that by Mr. J. H.
Howarth being particularly valuable—and trust that in future
the finances of the Society will enable them to produce details
of their proceedings more promptly and more fully.

NATURE PHOTOGRAPHS.

We have frequently, recently, called attention to the excellence
of the illustrations appearing in books on natural history. A
little volume just issued by Mr. T. N. Foulis * is an example.
It is a companion to that noticed in this Journal for December,
1905, but, if anything, the illustrations are even more attractive
than in that volume. There are seventy in all, after photographs by Charles Reid, and to these letter-press has been
provided by the Rev. C. A. Johns, compiled from 'British
Birds and their Haunts' and other sources. By the courtesy
of the publishers we are able to reproduce one of the most
charming of the illustrations — entitled, 'The Edge of the
Moor.' (Plate I.)

---

* 'I go A-walking Through the Woods and O'er the Moor.' 23, Bedford
Street, London. 79 pp. Price 2/6 net.

John Gilbert Baker,
Aug. 4 1906

PROMINENT YORKSHIRE WORKERS.

II.—JOHN GILBERT BAKER, F.R.S., F.L.S., M.R.I.A., V.M.H.

(PLATE II.)

AT the meeting of the British Association, held at York a few months ago, an opportunity was afforded of welcoming back to Yorkshire many prominent workers in the field of natural science. To few, however, was a more sincere welcome accorded than that given to the veteran Yorkshire botanist, Mr. John Gilbert Baker, who, notwithstanding his years, took a prominent part in the work of Section K, and was as enthusiastic and active as many there who were about half-a-century his junior, and surely it must be admitted that few have done so much for the furtherance of the study of plants of their county as Mr. Baker has done for Yorkshire. Without the many years of untiring energy in studying the flora of the broad-acred shire which he has given, our knowledge of the plants of this beautiful county would be much more meagre. As it is, we can take pride in the fact that the plants of our county are as well known as are those of any similar area in the British Isles, and this has been made possible by the industry, example, and encouragement of John Gilbert Baker.

Whilst listening to the paper and debates in Section K, in August last, Mr. Baker must have been forcibly impressed with the difference in the state of our knowledge of botany at the present time compared with the year 1847, when, as a scholar at the Friends' School, at Bootham, in the same city, the subject of our sketch had developed sufficient interest in botany to be appointed Curator of the herbarium of that school. And from that time until to-day he has had his hobby well before him, and has kept abreast with the study of plant life as it has progressed with the years. This, of course, so far as was humanly possible, as, particularly during the last decade, such rapid strides have been made in such a multitude of directions that it is not now possible for any single individual to be familiar with the details of the various branches of even botanical science.

Mr. Baker was born at Guisborough, in Cleveland, on January 13th, 1834, and naturally has always an affectionate regard for the plants of his native place. At the meeting of the Yorkshire Naturalists' Union held at Guisborough last year, he kindly contributed an account of the botanical attractions of the area, which proved very useful to those taking part in the excursion.

Soon after he was born, his parents removed to Thirsk, and we find that in 1843 he was attending the school belonging to the Society of Friends at Ackworth, a school which has trained so many first-rate naturalists. By 1846 he was carefully collecting and studying the plants found growing in the vicinity of his school. In 1850 he had made his first contribution to botanical literature, in a paper entitled 'On the Occurrence of *Carex Persoonii* in Yorkshire,' printed in the 'Phytologist.' Four years later he was so thoroughly familiar with the flora of his county that he issued a valuable Supplement to Baines' 'Flora of Yorkshire.' In the following year he read a paper, at the Glasgow meeting of the British Association, on the classification of British plants according to their geological relations. This paper was one of the first, if not the first, on this important subject.

For seven years, commencing 1859, he acted as distributor for the Botanical Exchange Club, and wrote its reports. It is interesting to remember that this Club is still in existence, and is carried on on almost the same lines as it was when Mr. Baker was its secretary.

By 1863 he had published his 'North Yorkshire : Studies of its botany, geology, climate, and physical geography.' This volume, which was dedicated to Hewett Cottrell Watson, was printed at Thirsk. It contained 366 pages and four maps. Unfortunately, a fire at the author's residence, which occured in 1864, soon after the publication of the work, consumed the entire stock, thus making this a scarce volume. In 1888 the Yorkshire Naturalists' Union commenced to reprint this work in its Transactions, and early last year the final part of this second edition was completed, and issued to the public. This second edition, which has been brought down to date, was reviewed in these columns for June last. It contains 688 pages, and maps, and a carefully compiled list of the mosses, prepared by Mr. M. B. Slater of Malton, a life-long friend of Mr. Baker's.

Not only did the fire referred to destroy the stock of the first edition of 'North Yorkshire,' but it also burnt Mr. Baker's extensive library and herbarium. By the efforts of the members of the Botanical Exchange Club, and other botanists, however, the loss was to some extent repaired.

In the same year the 'Naturalist'—then under the editorship

of the late C. P. Hobkirk—was favoured with its first contribution from Mr. Baker's pen in a 'Review of British Roses.' A 'Monograph of British Mints' was published by him in 'Seemann's Journal of Botany' in 1865, and in the following year Mr. Baker's worth was recognised in London, and he was appointed first Assistant in the Herbarium of the Royal Gardens at Kew. Prof. D. Oliver was at that time the Keeper, and Sir J. D. Hooker was the Director.

Conjointly with Sir W. J. Hooker, who died in 1865, Mr. Baker published, between 1865 and 1867, 'Synopsis Filicum : a synopsis of all known Ferns, with figures of the Genera and Sub-Genera.' A second edition was called for in 1875, and in 1892 Mr. Baker brought the list of species to date in the 'Annals of Botany.'

With Dr. G. R. Tate, he, in 1866, published the 'New Flora of Northumberland and Durham.' In the following year the Linnean Society published his 'Geographical Distribution of Ferns.' Between 1868 and the present time he has frequently contributed to Oliver and Dyer's 'Flora of Tropical Africa,' and between 1867 and 1899 he wrote various popular monographs of Narcissus, Lilium, Iris, Crocus, Agave, Yucca, Aquilegia, Hellebore, etc., to the 'Gardeners' Chronicle.' Between 1868 and 1872 he contributed descriptions of new and rare plants from the garden of Mr. W. Wilson Saunders, of Reigate, to Vols. I., III., IV., and V. of 'Refugium Botanicum.' During the years 1870-1880 Volumes XI. to XVIII. of the Journal of the Linnean Society contained Baker's Monograph of the Liliaceæ. And so on, year after year, have appeared most valuable and extensive monographs and papers dealing with the plants of England and abroad.* His latest piece of work has been the preparation of an account of the Botany of the County for the Victoria History. At the present moment he has in preparation, in conjunction with Miss Ellen Wilmott, 'A Book of Roses,' the coloured plates for which have been drawn by Mr. Alfred Parsons, A.R.A.

A large number of genera and species of plants has been named in honour of Mr. Baker. Amongst them are the following :—Genera : *Bakeria*, Seemann (Araliaceæ), sunk by Bentham and Hooker ; *Bakeria*, André (Bromeliaceæ) ; and

---

* An idea of the number of these may be gathered from the fact that 83 papers on Ferns are enumerated in Christensen's 'Index Filicum,' and 43 on various botanical subjects are listed in the Royal Society's Catalogue for 1864-1873.

Bakererella, · Van Tieghem (Loranthaceæ).   Species : *Rosa Bakeri* Dereglise ; *Rubus Bakeri,* F. A. Lees ; *Galium Bakeri* Lyme ; *Anthurium Bakeri,* Hook. fil ; *Allium Bakeri,* Regel ; *Eucharis Bakeri,* N. E. Brown ; *Crinum Bakeri,* Schumann ; *Iris Bakeri,* Foster ; *Dracæna Bakeri,* Scott-Elliott ; *Rhodolæna Bakeri,* Baillor, etc.

A valuable part of Mr. Baker's work has been his carefully prepared biographical notices of various botanists.   In 1902 the Tyneside · Field Club printed his paper on the ' Early Botanists of Northumberland and Durham,' and to the Yorkshire Naturalists Union, in 1885, he gave an address on ' Fathers of Yorkshire Botany,' which was printed in the Union's Botanical Transactions.

Notwithstanding the enormous amount of his writings, and his practical botanical work, Mr. Baker has done much by lecturing, etc., to train other botanists, not a few of whom, occupying prominent positions to-day, express their indebtedness to him for his painstaking addresses.

In 1869 he was appointed lecturer on botany to the London Hospital ;   for thirty years (between 1874 and 1904) he was lecturer at Kew Gardens ; from 1882 to 1896 he was lecturer on Botany to the Society of Apothecaries at Chelsea Gardens.   In 1866 he was elected a Fellow of the Linnean Society.   The Victoria Medal of the Royal Horticultural Society was awarded to him in 1897, and the Gold Medal of the Linnean Society in 1899.   These are only some of the many well-deserved honours that have been bestowed upon him by the various learned societies in Britain and abroad.   As might be expected, he was amongst the first of the Presidents of the Yorkshire Naturalists' Union, occupying the presidential chair in 1884-5, during which he gave addresses on ' Recent Progress in English Botany ' and ' Fathers of Yorkshire Botany.'—T. S.

———————

Arrangements are being made for a Course of University Extension Lectures which Professor L. C. Miall has consented to give in the last year of the tenure of his Professorship.   The subject of the Course will be ' The Early Naturalists, their lives and works.'   Final arrangements are not quite complete, but the course will probably begin on January 31st.   The lectures will deal with Natural History before John Ray ; Early workers with the microscope ; The Growth of System from Linnæus onwards ; Life Histories of Insects ; Theories of the Earth and the origin of Species ; Summary of progress from Ray to Cuvier, etc.   Further particulars may be had on application to Dr. W. G. Smith, the University, Leeds.   A proposal has been made by some of Professor Miall's friends inside and outside the University that, on the occasion of his retirement in June next from his Chair, his services should be commemorated by the painting of his portrait· for the University, and a committee has been formed to carry this out.

NOTE ON A CURIOUS FACULTY IN SPIDERS.

Rev. O. PICKARD-CAMBRIDGE, M.A., F.R.S.
Wareham.

In the 'Naturalist' for November, 1906, No. 598, p. 401, Mr. W. W. Strickland (Singapore) gives an account of two spiders of the family *Salticidæ*, in whose fore-central eyes he noticed a change from one colour to another, 'evidently under the control of the spider's will.' Being unable to account for this, and the Curator of the Singapore Museum being equally ignorant of such a faculty in spiders, Mr. Strickland tells us that he wrote to Mr. R. I. Pocock, giving an account of this wonderful faculty, and he (Mr. Pocock) replied, saying that the faculty 'was quite unknown to naturalists.' Will you allow me to point out that such a faculty is not quite unknown to naturalists. The late Mr. John Blackwall, whose researches on Araneology were extensive, and are well known, writes (Annals and Mag. Nat. Hist., Pt. 1, Vol. XVIII., No. 120, p. 299) where he is speaking of one of the *Thomisidæ* (*Thomisus*—now *Oxyptila*), *pallidus*, 'this species, with *Thomisus cristatus*, *Thomisus bifasciatus*, and some others, has the power of changing the colour of the anterior intermediate pair of eyes from dark red-brown to pale golden-yellow by a very perceptible internal motion. No such motion appears to occur in the other eyes, which are always black.' Mr. Blackwall has also, I believe, noticed this faculty in some other publication, to which I am unable at this moment to give the reference. I may add, however, that I have on many occasions noticed similar changes, but always in my experience in some spider of the family *Salticidæ*. The only doubt I had was about the 'faculty,' whether the phenomenon was produced at the will of the spider, or whether it was simply caused by a movement, very slight perhaps, and almost imperceptible at times, of the spiders caput, occasioning, of course, the light to fall upon the convex transparent cornea of the eye at a different angle, and so reflecting a different light within the eye. The internal variation in colour, however caused, is very noticeable in some at any rate, perhaps in all of the *Salticidæ*; I have observed it especially in one, a tolerably common species, *Hasarius falcatus* Clerck (*Salcicus coronatus* Blackw). In an adult male example of this spider, examined not very long ago (which had been in spirit for over nine months), the large fore-central pair of eyes was as transparent and as brilliant as if the spider were still alive; and certainly in this dead spider there was a change of internal colour visible when the caput was slightly moved one way or another.

In reference to the supposed internal movement of the eyes *at the will of the spider,* I may perhaps refer to a published observation of my own at p. 538 of 'Spiders of Dorset,' 1881. This is to the same effect as the remarks I have made above, but I had for the moment forgotten it.

NOTE ON A LARGE BASKING SHARK AT REDCAR.
(PLATE III.)

T. SHEPPARD, F.G.S.
Hull.

EARLY in August last an enormous shark, measuring 23 ft. 10 ins. in length, and '$11\frac{1}{4}$ ft. across the tail,' became entangled in the salmon nets at Redcar, and for a time was a great attraction to the visitors. It was variously described as a grampus, a blue shark, a basking shark, etc. Mr. T. H. Nelson, of Redcar, to whom we are indebted for the accompanying photograph (Plate III.), was absent whilst the shark was on view, but afterwards, in order to settle the matter, kindly obtained a portion of a comb-like body from the gills, and a piece of the skin. These were submitted to Sir William Turner, F.R.S., of the Edinburgh University, who has been good enough to examine them, and reports that they belong to the Basking Shark (*Selache maxima*).

The comb-like bronchial appendages, which were so interesting a feature in the Redcar specimen, are exceedingly curious, and formed the subject of a paper[*] by Sir William Turner, which was printed in the *Journal of Anatomy and Physiology*, Vol. 14, 1880, pp. 273-286.

This comb-like structure was at first thought to be a variety of whalebone, to which material it is to some extent similar, and indeed the fish was referred to as a whale by the early writers. Sir William, however, proved that the type of the structure resembles the dentine of a tooth, and is not of the same nature as whalebone. Its purpose was probably as a sort of a filter, and answered the same purpose as the baleen plates in the mouth of a whale, consequently "they provide us with an excellent example of objects which, though different in structure and mode of origin, yet fulfil corresponding physiological properties."

[*] The Structure of the Comb-like Bronchial Appendages and of the Teeth of the Basking Shark (*Selache maxima*).

Basking Shark at Redcar.

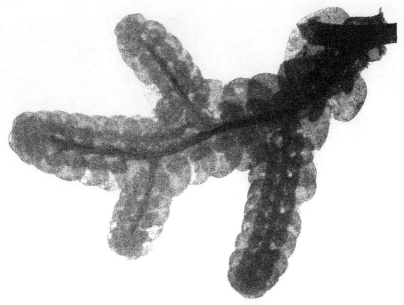

Frullania tamarisci.

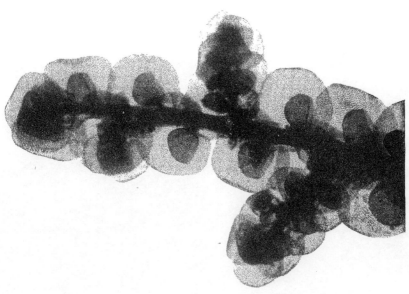

Frullania dilatata.

NOTES ON YORKSHIRE BRYOPHYTES.*
IV. *Frullania* and *Jubula*.

F. CAVERS, D.Sc., F.L.S.
Professor of Biology, Hartley University College, Southampton.

THE genus *Frullania*, one of the most sharply defined and easily recognised among leafy liverworts, is represented by over 300 species, the great majority of which are tropical. Only six species occur in Europe ; five of these are found in Britain, and four in Yorkshire (*F. dilatata, F. tamarisci, F. microphylla, F. fragilifolia*). The two first named are the commonest, and they are perhaps the most elegant of our native liverworts (Plate V.). Most of the species of *Frullania* grow as epiphytes on bark of trees, but some grow on rocks ; for example, *F. dilatata* is nearly always found on trees, whereas *F. tamarisci* appears to grow most frequently on rocks, though also found sometimes on trees. Our native species often form fairly large matted layers, the overlapping branches being closely pressed to the substratum, but in some tropical species the plants hang from the branches of trees in 'huge masses, sometimes half a yard long, and too bulky to be grasped in the arms.' †

The genus *Jubula* has a curious geographic range. The type-species, *J. hutchinsiæ*, was discovered in Ireland, and besides occurring in several scattered Irish localities, plants more or less varying from the type have since been found along the western side of England, Wales, and Scotland, and a few years ago in West Yorkshire. *Jubula* apparently occurs nowhere else in the temperate regions, but it has a wide distribution in the tropics, especially in Central and South America. Further reference will be made to this genus, which is by some writers merged in *Frullania*, though it stands well apart from the latter in many respects. The greater part of this paper, however, deals with *Frullania*, especially *F. dilatata* and *F. tamarisci*, abundant living material of which was available for investigation.

In *Frullania*, as in leafy liverworts in general, there is a main axis bearing lateral branches in two opposite rows, and both main axis and branches consist of a cylindrical stem from which arise two rows of sideleaves and a single row of under-

---

* For previous articles see the ' Naturalist' for September and November, 1903, and July and August, 1904.

† Spruce, *Hepaticæ amazonicæ et andinæ*, p. 38.

leaves.   Here and there the plant is attached to the substratum
by a tuft of rhizoids ('root-hairs') springing from the base of an
underleaf.   The leaves are spirally arranged, so that one underleaf corresponds to two sideleaves.   Each sideleaf is attached
to the stem by a narrow insertion, which is exactly transverse
in the young leaf, but later becomes shifted so as to run slightly
forwards above (the line of insertion changes from I to \, looking
at the stem from the side, with the growing forward end on the
left), and the leaf curves so that its fore edge covers the hinder
edge of the next leaf in front (the 'incubous' arrangement).   The
underleaves, however, keep their original transverse insertion.

Each leaf is divided into two lobes.   In the underleaves,
this division is not very deep ; the middle of the free part of the
leaf is more or less deeply notched, the two teeth being of about
the same size and shape.   In the sideleaves, the two lobes are
sharply separated to the base ; the relative size of the lobes
varies in different species and individual plants, even on the
same plant, but the upper lobe is always the larger.   While
the upper lobe is a flat or curved plate, the lower lobe has the
form of a pitcher which serves for storing water.   This pitcher
('lobule' or 'auricle') is open behind and is joined to the upper
lobe by a narrow stalk inserted near the mouth of the pitcher,
and on this stalk there is usually a short outgrowth ('stylus')
consisting of a single row of cells or a triangular plate (Fig. 1).
The pitchers, which sometimes contain small organisms (*e.g.*
Nostoc and other algæ, rotifers, insect-larvæ), differ considerably
in shape in different species, besides varying to some extent in
the same species or on the same plant.   In *F. dilatata* they are
usually helmet-shaped, with a wide oblique opening; in the other
British species they are more cylindrical with a narrow transverse opening.   In many of the tropical species (Fig. 1) the
pitchers have curious shapes; the mouth is often drawn out like
a scoop and is sometimes toothed, the closed end may be covered
with small outgrowths, and in one case (*F. replicata*) the pitcher
is reversed, having the opening facing forwards.

In order to get a clear idea of the organisation of *Frullania,*
it is necessary to study the early stages of development, by
means of sections through the tip of the main stem or of a
branch.   The growing point has an apical cell, shaped like a
pyramid with a slightly curved base, projecting forwards, and
three flat sides, two of which meet in the middle line above,
while the third is parallel with the lower surface of the stem
(Fig. 2).   From the three flat sides are cut off three sets of seg-

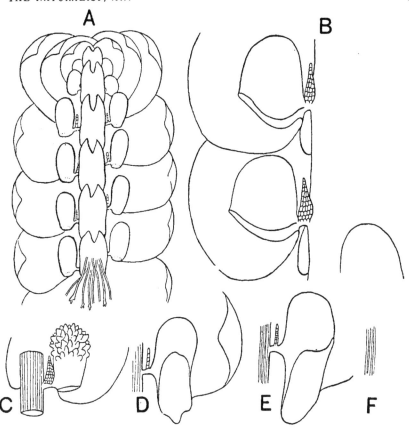

Fig. 1.—Various species of *Frullania*, seen from lower surface, magnified.
A, *F. tamarisci*; B, *F. dilatata* (the underleaves removed); C, *F. repandistipula*;
D, *F. ringens*; E, *F. arecae*; F, *F. ecklonii*.

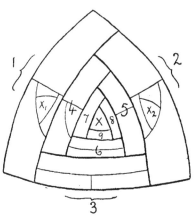

Fig. 2.—Diagram of the Apical Growing-point, seen from the front.

X, is the apical cell of the main axis; the last nine segments cut from it are numbered; X1, X2, are the apical cells of branches.

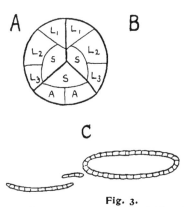

Fig. 3.

A. Diagram to show the early divisions in the seg (see Fig. 2); S, S, S, are the cells which contribute stem; L1, L2, L3, form respectively the lobe, lobul stylus of a sideleaf; **A, A** the two lobes of an und B. Leaf of *F. dilatata* grown in moist surroundings; the has remained flat instead of forming a pitcher. **C.** section of *F. dilatata*, showing stem, lobe, pitcher, and underleaf.

ments in spiral succession, the apical cell growing in size after each segment is cut off by a wall parallel with one of the sides. In each lateral (more strictly dorso-lateral) segment there first appears a wall dividing the segment into an upper and a lower cell, the latter being the larger; then the lower cell divides by a curved wall cutting off an inner cell which contributes to the formation of the stem, while the two outer cells of the segment give rise to a sideleaf. In each of the ventral segments, the first division separates an inner cell, which contributes to the stem, from an outer cell which produces an underleaf, and which divides by a vertical wall. In each case, therefore, the young leaf begins its growth with two cells placed side by side around the stem. At first these two cells grow out independently, their independent growth lasting for a short or long time in different cases, and it is to differences in this respect and in the mode of growth of the two primary leaf-lobes that the immense variety of leaf-forms in the foliose liverworts is due. The later growth of the leaf takes place at the base, near the stem, the oldest cells being therefore at the apex and margin. In the underleaf the two lobes become triangular, and after a time grow together at the base, and grow at about an equal rate, but in the side leaf the upper lobe soon outstrips the lower and becomes much larger, bending over the growing point; all the leaf-lobes bear a club-shaped gland-hair at the tip, and sometimes also at the margin and the base, secreting mucilage which keeps the growing point moist. The lower cell of the lateral leaf (two-celled stage) usually divides at first into two, and the cell nearest the stem forms the 'stylus' which is also tipped by a mucilage hair but undergoes little development. The rest of this lower half-segment then grows out to form a rounded or oval plate, and after a time its growth becomes restricted to the middle, so that it gradually becomes hollow and eventually forms a pitcher.

When *Frullania* plants are grown in air saturated by moisture, *e.g.*, by keeping pieces of bark with *F. dilatata*, or rock with *F. tamarisci*, in a covered glass dish and watering liberally, the new branches are found to bear, instead of pitchers, flat rounded or triangular lobules (Fig. 3). In this connexion it is interesting to compare with *Frullania* the closely allied genus *Jubula*, in which the plants are dark green in colour and more delicate in texture than in *Frullania* (Fig. 4). Lett[*] gives the habitat

---

[*] 'Hepatics of the British Islands,' 1902, p. 54.

J. hutchinsinæ as 'rocks and stones over which water trickles, and very moist places in shaded situations near waterfalls,' and Spruce gives a similar habitat for the South American plants referred to this species. In the numerous British and South American specimens I have examined, the lobules though varying considerably in form, even on the same branch, are flat or slightly concave, strongly recalling the young lobules of *F. dilatata* and *F. tamarisci* as well as the mature lobules of *Frullanias* cultivated in very damp surroundings. In the Hawaiian species, *J. piligera*, described and figured by Evans,* the lobules are mostly of the pitcher type, and this species grows 'on the ground and on trunks of trees in damp places,' resembling in habitat the pitcher-bearing *Frullanias* and not the thoroughly moisture-loving *J. hutchinsinæ.*

Water-storing arrangements of various kinds exist in several genera outside of *Frullania* and its allies, but well developed and stalked cylindrical pitchers of the *Frullania* type are rarely met with in other genera. One of the few examples is the genus *Polyotus* (*Lepidolaena*) which is restricted to the temperate and cold regions of the Southern Hemisphere. In this genus (Fig. 5) the pitchers are developed more freely than in *Frullania;* each sideleaf usually has a pair of pitchers, and they are sometimes borne on the underleaves as well.

In *Frullania* the stem consists of uniform cells, and is somewhat flattened above and below; the number of cells rarely exceeds ten in the median horizontal plane (Fig. 3). The leaves are only one cell in thickness, except at the base, where they are often two-layered close to the junction with the stem. The cells are about uniform in size and shape, but in *F. tamarisci* and *F. microphylla* there is usually a line of longer and wider cells along the middle of the upper lobe, which doubtless serves the function of a mid-rib, and in *F. fragilifolia* similar cells occur singly or in groups. Each cell contains several chloroplasts lying in the layer of protoplasm lining the cell wall, and numerous oil-bodies occur in the central part of the cell. The originally thin and colourless cell walls usually become strongly thickened, especially at the angles, and deeply stained red, brown, or purple. The outer walls usually project somewhat, especially on the upper side of the leaf; these projections are usually well marked in *F. tamarisci*, and doubtless give this species its characteristic glossy appearance. The lateral

---

* The Hawaiian *Hepaticæ* of the tribe *Jubuloideæ.* 'Trans. Connecticut Acad.', 1900, p. 407.

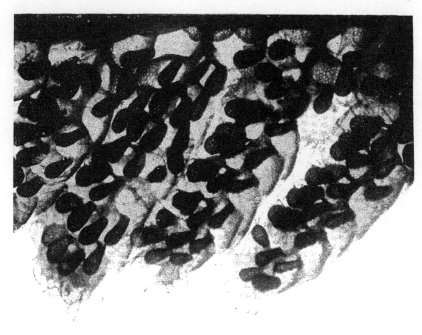

Fig. 5.—*Polyotus magellanicus*, an exotic liverwort (see text). Part of a plant showing lower surface of four of the ultimate branches, with numerous water-sacs.

Fig. 4.—*Jubula hutchinsiæ*. Part of plant seen from lower surface.

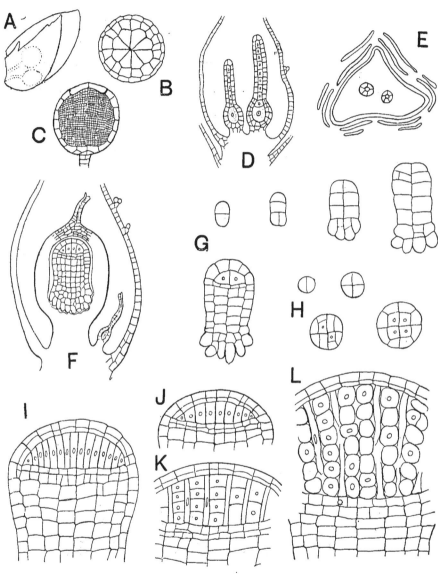

Fig. 6.—*Frullania dilatata*.

A, a perigonial leaf (male bract), from ventral surface, showing two antheridia ; B. C, antheridium in apical view and in longitudinal section ; 'D, longitudinal section of, perianth with two archegonia ; E, transverse section of same ; F, longitudinal section of perianth showing embryo in fertilised archegonium, with unfertilised archegonium on the right ; G, a series of longitudinal, and H, a series of transverse, sections of embryos of different ages ; I, J, K, L, longitudinal sections of older sporogonia, to show development of spores and elaters.

walls of the cells bear very thin places, or pits, through which dissolved substances can pass more readily. The degree of wall-thickening and of coloration depends largely on the habitat of the plant, especially as regards light, the thickest and most deeply coloured cell walls being found in plants growing in the most exposed places. From the results of cultures, I have found that in *Frullania* and various other liverworts the thickness of the cell walls is largly influenced by the amount of moisture in the surroundings, while their coloration depends solely on the intensity of the light. Cutin is not developed on any part of the plant (except on the outer surface of the capsule and in the outer coat of the spores), hence water can be absorbed at all parts, and is stored during drought, not only in the pitchers and in the cavities of the leaf cells, but also in the thick cell walls. Plants kept constantly moist not only dispense with pitchers, but also with thickened cell walls. The coloration of the walls may act as a screen against too strong light, or may be a protection against cold in addition. Experiments have shown[*] that in plants of *F. tamarisci* with deep coloration the gaseous exchanges concerned in assimilation and respiration are less active than in green plants of the same species.

Each branch of the stem arises just behind one of the side leaves. According to Leitgeb, [†] the whole of the lower half of a segment is used up to form the growing point of the branch (Fig. 2), so that the leaf behind which the branch arises has no lobule. This is usually the case in *F. dilatata* and *F. tamarisci*, which Leitgeb investigated, but from sections and cleared apices of several other species of *Frullania*, I am inclined to believe that in some cases, at any rate, the branch represents only the posterior part of the half segment, and that what appears to be the first underleaf of the branch is really the lobule and stylus derived from the rest of the half segment; this would agree with the interpretation of the branching in *Frullania* suggested by Spruce and by Evans.

All the British species of *Frullania* are dioecious, though *F. dilatata* is said to have the male and female branches sometimes on the same plant. In this species, as was noted by Hofmeister, [‡] the male plants usually occur at a higher level on the substratum than the female plants, to which the antherozoids are washed down by rain or dew, thus reaching the

* Jonsson, *Comptes rendus*, 1894.
† *Untersuchungen über die Lebermoose*, Heft 2, pp. 22, 25.
‡ *Vergleichende Untersuchungen*, p. 37.

archegonia. The same is the case with *F. tamarisci*, and in both species the plants are always dioecious, so far as I have seen, and the males are produced very sparingly towards the upper edges of the patch, and are much thinner and smaller-leaved than the females. The antheridia are developed on special short branches, with the under leaves rudimentary or absent except at the base of the branch, and the side leaves ('male bracts' or 'perigonial leaves') consist of two nearly equal lobes, which are joined below and curve towards each other to form a pocket; the lobule is not developed as a pitcher. In the axil of each of these leaves stand two anthe-ridia, as a rule, but the number varies from one to four (Fig. 6 A). The antheridium is somewhat egg-shaped, and is carried on a slender stalk. The cells forming the apical part of the antheridium wall are elongated (Fig. 6, B C), and when the ripe antheridium absorbs water these long cells swell up, sepa-rate from each other, and spring outwards (or perhaps are forced out by the swollen antherozoid mother cells) and leave a wide opening through which the antherozoids escape.

The archegonia are borne in terminal groups on the main axis and on short branches; each group contains, as a rule, only two archegonia, but three are often seen, and less frequently four. The first archegonium arises from the apical cell, the second from its youngest segment. The three seg-ments below grow out together and remain joined, forming a tube (perianth) which is at first narrow, but later widens at the base (where the growth takes place, as in ordinary leaves), and eventually forms a wide sac with a narrow tubular pro-longation (the earliest formed part) at the top. The perianth consists of a single layer of cells, except at the very base, which is two-layered; in *F. dilatata* the outer surface bears numerous small projections (Fig. 6, D, E, F; Fig. 8). The leaves just below the perianth begin to grow in the usual way, but the two lobes become equally developed, or nearly so, and are usually triangular and more or less sharply pointed; these 'involucral leaves' or 'female bracts' bend over the developing archegonia and help in protecting and keeping them moist.

(To be continued.)

Mr. W. Jerome Harrison favours us with a copy of his paper on 'The Desirability of Promoting County Photographic Surveys,' read at the York meeting of the British Association. (See 'Naturalist,' September, 1906, p. 290.) With this is printed the remarks made in the discussion following the paper.

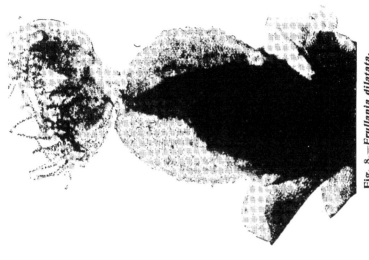

Fig. 8.—*Frullania dilatata.*

End of a female branch (dorsal view) with capsule, which has exploded and shed the spores, after having (by the elongation of the seta), broken out of the calyptra and the perianth. The calyptra, bearing the withered archegonium-neck, can be seen inside the perianth; the tubercles on the outer surface of the perianth are plainly shown on either side.

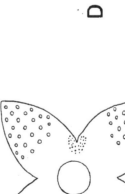

Fig. 7.—*Frullania dilatata.*

A, longitudinal section of well-grown capsule, with spore-mother-cells dividing; B, cross-section of seta; C, part of wall of capsule (upper end of one of the four valves) seen from outer surface; D, part of inner surface of capsule-wall, with an elater; E, diagram (from Jack, 'Bot. Zeit.,' 1877) showing the arrangement of the elaters on the four valves of the capsule-wall, as seen from above.

LIFE ZONES IN BRITISH CARBONIFEROUS ROCKS.*

Part II.—The Fossils of the Millstone Grits and Pendleside Series.

WHEELTON HIND, M.D., B.S, F.R.C.S, F.G.S.

THERE is much to be done yet to work out the life-zones of the Millstone Grits and Pendleside series. A certain amount of detail has already accumulated, and I think it may be well to put on record what is known on the subject.

It must be remembered that the Pendleside series and Millstone grits are a very local deposit, the maximum thickness of each coinciding. I have pointed out on many occasions that the Pendleside series thins out north and south. I have not met with the deposit, in the Pennine area, north of Settle. It is absent in Coalbrook Dale and Cannock Chase, where the Carboniferous Limestone is succeeded by Coal Measures, without unconformity, but the series is represented at Chokier, Mons, and Clavier in Belgium, and in the homotaxial equivalent of the Culm of Magdeburg and other German localities.

In the West I have estimated that it is represented by some 60 feet of shales and concretionary limestones in Cos. Clare and Limerick. A comparison of the faunas demonstrates that the Lower Culm of North Devon is the homotaxial equivalent of the Pendleside series of the Midlands.

So also with regard to the Millstone Grit series; the maximum thickness of this series occupies an area which may be roughly said to extend from North Derbyshire to Bolland and Craven. In North Staffordshire the Grits thin out very rapidly, so that they are not represented in the Carboniferous sequence further south. Northwards, beds of grits succeed the Yoredale phase of the Carboniferous Limestone as far as Northumberland, but, as grits, are absent in the Scotch sequence where they are represented by about 687 feet of sandstones and shales, which intervene between the Castle Carey Limestone and the Slate band ironstone, the roof of which is characterised by the presence of *Carbonicola robusta*.

In the coalfields of the Midlands we find the marine beds of the Ganister series, with *Gastrioceras listeri*, *G. carbonarium*, *Dimorphoceras gilbertsoni*, *Pterinopecten papyraceus*, *Posidoniella lævis* and *Orthoceras*, below the maximum of *C. robusta*, and at present only a fragment of *P. papyraceus*, and some two or three specimens of *Posidoniella lævis* have been obtained in the

---

* For previous paper see 'Naturalist' for August, 1906.

series assigned to the Millstone Grit in Scotland, though some of the beds contain a rich marine fauna, largely new to the Eastern Hemisphere.

Mr. Kidston has shown that the change from a Lower Carboniferous flora to the Upper occurs in the Millstone Grit series of Scotland, about the horizon of the Roslin sandstone.

In the South, the sandstones of the Middle Culm most probably represent the Millstone Grits, but it is doubtful if they are presented to any extent in the South Wales or Bristol coal refields.   In Belgium, we know that the Millstone Grits are absent.   In the West of Ireland, a fairly well developed series of flagstones, grits, and inter-bedded shales with marine bands evidently represent the Millstone Grit.

Although in the localities mentioned above, the Carboniferous sequence differs enormously, as yet no apparent unconformity has been proved anywhere between the Carboniferous Limestone and the Coal Measures.   Whatever beds overlie the Carboniferous Limestone appear to be conformable to it.

Within the last two years an important fact has been made out for the Carboniferous areas of North Wales, the Pennine system, and the East of Ireland, that is, the upper beds contain everywhere a definite coral fauna, characterised by the presence of *Cyathaxonia*, *Amplexizaphrentis*, and *Cladochonus* ; thus a definite top is obtained for the Carboniferous Limestone series, and in other words a definite base for the succeeding Pendleside series.

It is at the top of these *Cyathaxonia* beds that the great change takes place in the Molluscan fauna, and in the Midlands in the Fish fauna.   The Corals, Echinodermata, and Polyzoa became almost entirely annihilated, and a new fauna of mud-loving animals took their place, but land conditions remained constant, for the great floral change only took place at a much later period, as I have previously stated.   The great Carboniferous faunal break must therefore be placed at the top of the *Cyathaxonia* beds, and it is here that the line dividing Upper and Lower Carboniferous must be drawn.

The *Cyathaxonia* beds at Warsoe-end House, near Pendle Hill ; Congleton Edge quarry, Cheshire ; Torrs' quarry near Bradbourne, Derbyshire, contain *Prolecanites compressus*, a fossil which characterises the junction of Upper and Lower Carboniferous rocks, and passes up into the lower part of the Pendleside series, but does not range far up.

This fossil is plentiful in the Coddon Hill beds of the Culm, which I consider to be at the base of the Culm series. It is associated there with Trilobites, and rare casts of corals, and an important lamellibranch *Chœnocardiola footii*, that is found only in the base of Pendleside series in the Midlands and in Ireland.

P. compressus occurs in the Carboniferous Limestones of Little Island, Co. Cork; Scarlett, Isle of Man; and Kendal, Westmorland; associated always with an upper Dibunophyllum fauna.

In the Pendleside series I have obtained specimens from ? Telia, N. Wales, the shales of the Hodder, at Hodder Place, Pendle Hill, Lancashire; Morredge, N. Staffordshire, so that in *Prolecanites compressus* we have the index of the junction of upper and lower Carboniferous rocks and *Stroboceras sulcatum* seems to have the same range.

There is not always a cessation of the deposit of limestone associated with the faunal change, but the limestones associated with the incoming Pendleside fauna are different in texture, cleavage, colour, and ring.

The new fauna is characterised by the immediate appearance of *Pterinopecten papyraceus*, *Posidonomya membranacea*, *P. becheri*, *Posidoniella lævis* and *Glyphioceras reticulatum*, *Nomismoceras rotiforme*, to mention certain members of a fauna which is given in full lower down.

This faunal succession has been demonstrated at the following places :—In North Staffordshire ; at Pepper Mill, Wetton, limestones with *Cyathaxonia* and *Amplexizaphrentis* are succeeded by Shales and Limestones with *Pterinopecten papyraceus*, *Posidonomya becheri*, *Posidoniella lævis* and *Nomismoceras rotiforme*. At Mixon Hey *P. papyraceus*, *Posidonomya becheri* *Posidoniella lævis*. A station in the River Hamps, near Onecote, yields *Posidonomya membranacea* and *Prolecanites compressus* at Tissington Station, Derbyshire, *Pterinopecten papyraceus* and *Posidonomya becheri* succeed the uppermost beds of the Carboniferous Limestone.

In the river Hodder the same fossils are found as at Wetton, and in addition *Prolecanites compressus*, *Phillipsia polleni*. The series exposed in the Hodder is a most interesting one, and I hope to publish a paper on the subject at an early date. In Yorkshire the *P. becheri* beds have been found succeeding the *Cyathaxonia* beds at the Cracoe fells.

The same fauna is found at the following localities :—in the River Wharfe a little west of Linton Church, Linton ; at Lothersdale; Mill Beck, near Hetton ; Dinckley Hall, R. Ribble;

Ramsclough, near Thornley Hall; West Bradford, Bolland; Holden; Agden Clough; and Stream near Browsholme Hall, Bolland; Sulber Lathe, Flasby; Newton Gill, 1 mile E. of Long Preston.

The Black Limestones of the Isle of Man contain, as had been known for a long while, *Posidonomya becheri* in abundance, *Pterinopecten papyraceus, Nomismoceras rotiforme, Glyphioceras reticulatum, Orthoceras morrisianum, O. sulcatum,* and many plants.

In North Wales at Teilia, Prestatyn and Holloway, the junction of *Cyathaxonia* beds with thin black Limestones containing *Pterinopecten papyraceus, Posidonomya becheri* and *P. membranacea* and other members of the fauna and flora are well seen.

This Life Zone is traced West into Ireland, and is well seen on the Coast near Lough Shinney, Co. Dublin, and in Co. Clare, where I have demonstrated the presence of *Posidonomya membranacea, Pterinopecten papyraceus, P. becheri,* in beds which succeed the Carboniferous Limestone.

Here, then, is a wide and extensive Zone which may well be called the Zone of *Posidonomya becheri,* which may be considered to include the first 200-300 feet of the Pendleside Limestones.

I am purposely vague as to the extent, because the whole series, as I stated above, thins out rapidly N. and S., and at Pendlehill I believe this Zone to be thicker than elsewhere.

In certain localities I fear that the other Zones in the series are not so perfectly demonstrable, nevertheless they exist, and if I say what I know about them, others may be able to fill in details.

Above the *P. becheri* Zone we find a horizon in which *Glyphioceras spirale* occurs abundantly. The maximum of this fossil occurs in Cheshire, at Congleton Edge, about 500 below the 3rd grit and about the same distance above the top of the *Cyathaxonia* beds. This shell is very common in the Lower Culm of North and South Devon, and has occurred at Foynes' Island, Co. Limerick, Lough Shinney, Co. Dublin, and Pendle Hill.

At Congleton Edge occurs a most interesting section, which has been described by me so frequently. The quarry is opened to work some strong Quartzose Sandstones containing plant remains and thin streaks of Coal, for the purpose of forming the floors of iron furnaces.

The upper band of this rock is succeeded by two feet of greyish yellow marl, with calcareous nodules, rapidly thinning out towards South. This is succeeded by a foot of black stratified hardened shale, crammed full of *G. spirale*. In the same bed are a few specimens of *G. diadema* and *Orthoceras steinhaueri*. Above are 14-15 feet of Shales with calcareous nodules, which contain the following fauna ; succeeded by 5 feet of a shell bed composed of closely-packed valves of *Schizophoria resupinata*, which shows a curious character in its radiating ribs, which probably may be relied upon to distinguish the form at this horizon, viz., the lines every now and then thicken and come to an abrupt end, as if there had been an attempt to commence to form spines and this had failed. I have seen the same character in specimens from an Orthis bed in the Millstone Grit elsewhere.

Corals.
Amplexizaphentis (rare).

Brachiopoda.
Seminula ambigua.
Chonetes cf. *laguessiana.*
Dialasma hastata.
Lingula mytiloides.
„ *scotica.*
Productus aff. *cora.*
„ *longispinus.*
„ *scabriculus.*
„ *semireticulatus.*
Schizophoria resupinata.
„ (*Rhipidomella*) *michelini.*
Spirifer glaber.
„ *trigonalis.*
„ *bisulcatus.*
Derbya sp.

Lamellibranchiata.
Aviculopecten gentilis.
Actinopteria persulcata.
Ctenodonta laevirostris.
Allorisma sulcata.
Edmondia sulcata.
„ *rudis.*
„ *maccoyi.*
Leiopteria squamosa.
Modiola transversa.
Myalina peralata.
Mytilomorpha rhombea.
Nucula gibbosa.
„ *aequalis.*
Nuculana attenuata.
Parallelodon obtusus.
„ *bistriata.*
Posidoniella lævis.

Posidoniella semisulcata.
Pteronites angustatus.
Palaeosolen parallela.
Protoschizodus orbicularis.
Sanguinolites, v. *scriptus.*
Scaldia benedeniana.
Sedgwickia ovata.
Schizodus axiniformis.

Gasteropoda.
Loxonema sp .
Macrocheilina sp.
Raphistoma junior and others not identified.
Euphemus urei.
„ sp.

Pteropoda.
Conularia quadrisulcata.

Cephalopoda.
Temnocheilus coronatus.
Ephippioceras bilobatum.
Stroboceras sulcatum.
Coelonautilus, cf. *cariniferus.*
Glyphioceras diadema.
„ *spirale*
Orthoceras steinhaueri.
„ *teres.*

Crustacea.
Ceratiocaris oretonensis.
Dithyrocaris testudinea.

Bryozoa,
Millepora interporosa.

Plantæ.
Smooth stems.
Stigmaria
Trigonocarpon.

This is a large and curious fauna to find high up in the Pendleside series, but although elsewhere in the Midlands this bed has not been positively identified, other rich faunas occur in the grits at higher horizons at Pule Hill and in the neighbourhood of Pateley Bridge.

Owing to the fact that the Millstone Grits are thinning out along Congleton Edge, the exact horizon of the shell bed here is equivocal. 12 miles N.W. the grits are in full force, and consist of 5 beds :—Rough Rock, 2nd Grit, 3rd or Roches Grit, Kinderscout Grit, and Farey's or the Pendle Grit; at Congleton Edge only two grits are present, and these are thin. They are presumed to represent the 1st and 3rd Grits. It may be that the Crow stones immediately below the fossil bed represent Farey's Grit, but the presence of *G. spirale* and *G. diadema* points to a Pendleside facies rather than Millstone Grit.

Glyphioceras diadema has a broad horizontal distribution. It occurs in profusion in the North of County Clare, some 2-4 feet above the top of the Carboniferous Limestone. I have suggested that here the *Posidonomya becheri* beds are absent owing to an overlap of higher beds. (Proc. Roy.-Irish Acad., Vol. XXV., section B., pp. 93-116.)

At Chokier, Belgium, the same fauna occurs, and it is interesting to note that in each of these localities, ordinary specimens of *G. diadema* are accompanied by a variety with very coarse ribs and a wide umbilicus. (*Op. Supra. cit.*, pl. vi., figs. 10-11.)

The Zone of *Glyphioceras reticulatum* (maximum) requires working out in more detail. In the *Posidonomya becheri* beds it appears first as a large shell with coarse reticulate marking. Higher up, and probably below the maximum of *G. spirale*, it occurs in profusion in some localities, persists through the grits, appearing in force in the shale below the 3rd grit at Wadsworth Moor and Eccup, near Leeds. In the Pendleside series it is at its maximum in the black shales and nodular limestones of the Pendleside series, of Horsebridge Clough and High Green Wood, near Hebden Bridge.

The fauna here contains the following :—

Brachiopoda.
Productus (scabriculate form).
Lamellibranchiata.
Pterinopecten papyraceus.
Posidoniella lœvis.
    ,,   *kirkmani.*

Posidoniella minor.
    ,,   *variabilis.*
Schizodus antiquus.
Sanguinolites cf. *sulcatus.*
Nucula æqualis.
Leiopteria longirostris.

Gasteropoda.

Macrocheilina elegans.
,,   *gibsoni.*
,,   *reticulata.*

Cephalopoda.

Glyphioceras phillipsi.
,,   *vesica.*
,,   *implicatum.*
,,   *reticulatum.*
Glyphioceras davisi.
,,   *diadema.*
,,   *calyx.*
,,   *platylobum.*
Dimorphoceras discrepans.
,,   *gilbertsoni.*
,,   *loonyi.*

Gastrioceras listeri fide Spencer.
Nomismoceras spirorbis.
Orthoceras morrisianum.
,,   *steinhaueri.*
,,   *konickianum.*
,,   *aciculare.*
,,   *brownii.*
Cælonautilus quadratus.
Solenocheilus cyclostomus.
Temnocheilus carbonarius.
Pleuronautilus pulcher.

Pisces.

Cladodus mirabilis.
Orodus elongatus.
Acrolepis hopkinsi.
Elonichthys aitkeni.

I know no other locality so rich species as the Hebden Bridge localities, but bullions rich in *G. reticulatum* occur in the Dane Valley and below Mórredge, near Leek. In both localities I estimate the bed to be about 250 feet below the Kinderscout Grit.

Glyphioceras bilingue is a much rarer shell. It appears in the Pendleside series, for the first time, only some little way above the *Posidonomya becheri* zone.

In the North Staffordshire district I have obtained it from Shales in contact with the red rock fault, River Dane East of the Railway Viaduct, and in the River Dove at the foot of Park Hill, where it occurs with *Chaenocardiola footii.* I also obtained it in the banks of the stream at Wildmoor Bank Hollow, off the Macclesfield and Buxton Road. Here its position is apparently below the Kinderscout Grit.

(To be continued.)

————◆◆————

There are apparently different ways of studying nature. Judging from a report in the *Hull Daily Mail*, 'fifty of the senior scholars attending the National School,' at an East Riding village, 'had their final, and most enjoyable, ramble of the season.' After noting the flowers, birds and insects, 'a move was made to St. Augustine's Stone . . . *and each scholar possessed himself of a small specimen of the rock* . . . prayers brought the ramble to a close'—prayers for the preservation of objects of natural beauty, we presume!

At a recent meeting of the Lancashire and Cheshire Entomological Society, Mr. W. Mansbridge read a paper entitled 'Notes on a Melanic Race of *Agrotis ashworthii*,' and exhibited a long series of moths bred in 1905 in illustration of his remarks. Mr. Mansbridge reviewed the evidence for and against the view that *A. ashworthii* and *A. candelarum* are the same species, and suggested the name *substriata* to distinguish the new form. The opinion of the meeting was to the effect that more evidence of identity was required, especially as regards early stages and structural details of *candelarum.*

1907 January 1.

THE CHEMISTRY OF SOME COMMON PLANTS.

P. Q. KEEGAN, LL.D.,
Patterdale, Westmorland.

ROCK LICHEN (*Parmelia saxatilis*).—This organism is very commonly observed on roadside unmortared walls, and sometimes on withered time-worn hawthorn barks, etc., and is a fairly good representative of the class to which it belongs. It is divided by systematists into two varieties : one, *retiruga*, containing, according to Hesse, atranorin, protocetraric acid, and saxatic acid ; the other, *omphalodes*, containing atranorin, saxatic acid, and an acid like protocetraric acid. It may be advisable to observe here that the chemistry of lichens is very difficult and greatly confused for two reasons, viz., the difficulty of purifying the constituents extracted by solvents, and the fact that these constituents are liable to vary in composition and in relative amount at any particular period. Therefore, I shall merely present the results of my own rough analysis of this species, prosecuted in the ordinary way. The dried substance treated with boiling benzene yielded 0.75 per cent. of a brownish-yellow extract, which with sulphuric acid gave green, red and brown colours, and has a white opaque matter insoluble in cold benzene like stearine. The treatment with boiling alcohol (after benzene) afforded a crystalline deposit on cooling, and also on adding water, and the liquid gave with perchloride of iron a violet colour, with solution of bleaching lime a transient violet, with ammonia water it dried up to a red-brown mass, with lime added and the filtrate acidified by HCl a bright red precipitate was presently deposited ; the liquid contained no free phloroglucin or sugar. The hot water extract gave reactions similar to the foregoing, it had neither sugar nor albumenoid. Dilute caustic soda further withdrew a small quantity of mucilage coloured red-brown, but still no albumenoid. Dilute HCl did not extract any starch or lichenin. The residue (crude 'fibre') amounted to 63.7 per cent. of the original. The ash amounted to 5.4 per cent., and yielded 13.3 per cent. soluble salts, 55 silica, 2.2 lime, 17.5 oxide of iron, 4.1 P^2O^5 and 1.1 SO^3 ; there was a little manganese, but no carbonates. It would appear that in the above analysis the benzene extracted atranorin, and the alcohol a mixture of protocetraric and saxatic acids. The well-known dyeing property of the lichen is mainly due to the former acid, which seems to be a derivative of betaorcin $C^8H^{10}O^2$ = dimethyl resorcine, and to result from the hydrolysis of the lichen proteids.

Like the indigo of certain plants, it is thrown out or excreted in the form of minute granules on the exterior of the hyphae in the cortical portion of the upper surface. It would appear that when a decrease of the albumenoids takes place the quantity of this lichen-acid increases, so that it may be regarded as a sort of waste-product of the living plant, and not a true product of deassimilation.

PARSLEY FERN (*Allosorus crispus*).—This plant is distinctly local in habitat ; it nestles under huge boulders or largish stones on the mountain side, or under walls in shady lanes. Generally it affects rough and stony ground appurtenant to wild and well watered areas. The exterior aspect of its fillets differs from most of its class, resembling more those of certain dicotyledons ; but its chemistry is pretty similar to that of its ally and associate, the common bracken. The dried overground parts on 19th July yielded 3.25 per cent. of wax with only a little carotin or glyceride ; there was much tannoid reacting like quercitrin, also a resin and a little tannin (insoluble in strong alcohol) which precipitated gelatine and bromine water, much sugar, proteid and starch, some mucilage non-coagulable by acids, but no oxalate of calcium ; the ash amounted to about 5 per cent., and contained 60.1 soluble salts, 14.2 silica, 5.2 lime, 5.9 magnesia, 8 P^2O^5, 3.5 SO^3, and 4.2 chlorine, with much manganese and very much soluble carbonates, thus attesting the eminent richness of the plant in organic acids. The remarkable feature, however, is the large quantity of soluble salts conjointly with the considerable amount of silica. We conclude that the carbohydrates engaged in the fruiting process have undergone a very active though incomplete oxidation, while the silica depositing incidental to a failing life-energy has already become manifest. Later on in November, the proportion of soluble salts diminished to below one half, and the silica percentage rises to about 31. There is thus a considerable similarity in the life-cycle and the physiological processes to those of the more splendidly developed bracken ; but this fern evidently possesses more vitality and enjoys a longer life.

———◆———

ENTOMOLOGY.

Chærocampa celerio at **Wakefield.**—I have recently added to my collection a specimen of *Chærocampa celerio*, which was taken on the outside of a shop window in Wakefield on October 24th last by Mr. H. Lumb.—GEO. T. PORRITT, Huddersfield. December 10th, 1906.

YORKSHIRE NATURALISTS AT YORK.

THE forty-fifth annual meeting of the Yorkshire Naturalists' Union was held at York on Saturday, December 15th. The meeting was of peculiar interest, from the fact that it was the two hundredth meeting of the Union.

In the morning those who arrived early were conducted round the premises of the British Botanical Association at Acomb, where Dr. Burtt, the director, kindly exhibited the various botanical preparations made under his supervision.

The sections met in the Museum at 3 p.m., and at 3-30 an exceptionally well attended gathering of the general committee was held in the Lecture Theatre, under the presidency of Mr. W. Eagle Clarke. The report on the year's work was carefully considered, and the opinion was generally expressed that the work of the Union during the year was quite equal to that of any previous year. In some sections work of exceptional importance had been accomplished, details of which had appeared in the Union's official organ, the ' Naturalist.' For the forthcoming year it was announced that some of the committees had arranged to undertake special investigations.

It was decided to hold meetings for 1907 as under :—

For York, N.E.—Robin Hood's Bay (Whit week-end, May 18th to 20th).

,, S.E.—South Cave (Saturday, June 22nd).

,, S.W.—Thorne Waste (Thursday, July 11th).

,, Mid.W.—Kettlewell for Arncliffe (August Bank Holiday week-end).

,, N.W.—Horton-in-Ribblesdale (Sept. 7th).

Fungus Foray—Grassington for Grass Woods and Bolton Woods (September 21st to 26th).

An invitation from the Halifax Scientific Society for the Union to hold its next annual meeting at Halifax was accepted, the place being particularly appropriate, seeing that the Union's new president, Mr. C. Crossland, F.L.S., is a Halifax man.

The hon. treasurer, Mr. J. H. Howarth, J.P., had pleasure in reporting that, probably for the first time for many years, the expenses of the Union for the year had been slightly less than the receipts, and that when the arrears of subscriptions had been paid the financial position of the Union would be most satisfactory. This success was partly due to the new arrangement with the ' Naturalist,' which had worked so well for all concerned.

After tea the annual general meeting was held in the Museum. At this, reference was made by several members to the recent press reports in which it appeared that allegations had been made at a meeting of the Hull City Council against the suitable nature of the exhibits at the Hull Museum. On the proposition of Mr. G. T. Porritt, of Huddersfield, seconded by Mr. H. H. Corbett, of Doncaster, the following resolution, to be sent to the Hull City Council, was unanimously passed :—
'That the members of the Yorkshire Naturalists' Union, assembled in annual meeting at York, desire to congratulate the Corporation of the City of Kingston upon Hull, upon the excellent condition of their Municipal Museum, which now occupies a foremost position among English provincial museums, and upon the immense advance it has made since the appointment of the present curator, whose energy, tact and scientific judgment are fully manifested throughout the galleries of the institution. The members of the Union are well acquainted with the Museum, and an annual meeting of the Union has been held there. The Union, therefore, feels justified in expressing an opinion which it believes may not be unwelcome.'

Mr. W. Eagle Clarke, F.R.S.E., of the Royal Scottish Museum, Edinburgh, who was warmly received, then delivered his presidential address, entitled 'Bird Life in the Antarctic.' The president referred to the fact that the ornithological collections acquired by the Scottish and National Antarctic Expeditions had been submitted to him for examination and description. Amongst the numerous specimens brought home by these two expeditions were several which, from the fact that they were previously unknown, rendered them of the utmost value to zoological science. In connection with the nesting habits and life history of several hitherto little-known birds, much interesting information was imparted by Mr. Clarke (see 'Naturalist' for 'Naturalist' for July, 1906, p. 201).

Of particular interest was a fine series of charming lantern slides from photographs taken in the Antarctic on the expeditions referred to. These showed many quaint phases of bird life, and were much appreciated.

A vote of thanks was accorded to Mr. Eagle Clarke for his address, on the proposition of the Right Hon. the Lord Mayor of York, and a similar compliment was accorded to Dr. Gramshaw, the president of the York and District Field Naturalists' Society, for taking the chair during the delivery of the presidential address. Votes of thanks were also accorded to the

York Philosophical Society for the use of the Museum and Lecture Theatre, and to the York Naturalists' Society for their entertainment and efforts to make the meeting the success it was.

After the address a pleasant evening was spent in the Museum, where the York Society had arranged some special exhibits, a concert, and also provided refreshments.—T. S.

ON THE SMALL QUANTITY OF AIR NECESSARY TO SUSTAIN LIFE IN A BAT.

H. B. BOOTH, M.B.O.U.
Shipley.

On September 23rd I received a securely fastened brown paper parcel from a friend in Shropshire. Inside were two Pipistrelle Bats, each closely wrapped several times round with tissue paper, and just filled a 2-oz. tobacco tin. Around the tin was a short note saying that one of the bats was dead, and, although the other one was just alive when put in, it would not reach me so. The time occupied in the post was 18 hours. To my great surprise, on taking off the numerous wrappings, I found one of the bats was still warm, and showed signs of life. My first impulse was to immediately put it out of its misery, but as it gave some signs of moving, I put it aside in a larger box. Several hours after it was crawling slowly about, and squeaked, and snapped at my fingers when touched. I now spent fully half an hour catching house-flies, and it took about a score from the point of a pin, and became quite lively. It lived for a fortnight longer, and during most of that time it was on view to the public in the Museum Room of the Cartwright Hall, and may possibly have succumbed eventually to an excessive house-fly diet. Both bats had been in a cigar box parcel about 24 hours before being sent on to me.

Although I much regret if any cruelty has been thoughtlessly imposed on the above individual, yet I think the facts of the case worth recording, as it appears wonderful to me what a small quantity of air is really necessary in order to keep a bat alive. No doubt this will be partly due to the out-of-the-way crannies in which many bats spend most of their lives, and where, in many cases, there is no circulation of air, and partly because they are hibernating animals. The bat which survived was mature, and the one received dead was an immature, although a full-grown, specimen.

FIELD NOTES.

MAMMALS.

Otter near Barnsley.—An otter was shot a fortnight ago in the Dearne Valley about 1½ miles *north* of Barnsley. Some years they are not uncommon *south* of Barnsley, in the neighbourhood of Darfield, but I do not remember hearing of them so far up stream before.—E. G. BAYFORD.

—: o :—

BIRDS.

Little Auk and Albino Swallows at Skipton.—A mature specimen of the Little Auk, in splendid plumage, was killed at Skipton on November 8th, 1906, immediately after a two days' north-west gale. Two Albino Swallows were bred at Embsay, near Skipton, this summer, and got away safely. It is hoped they may return next year.—W. WILSON, Skipton, December 8th, 1906.

Purple Sandpiper in Upper Airedale.—On December 2nd I received a bird for identification from Malham, which proved to be a Purple Sandpiper, and new to the local list of the Bradford Natural History and Microscopical Society. For several days before there had been strong westerly winds, almost amounting to a gale at times, and no doubt this bird, which was rather battered, had been blown out of its course from the sea shore.—H. B. BOOTH, Shipley.

Albino Blackbird in Lincolnshire.—In the summer of 1904 I saw a male blackbird, with a perfectly white head, a little way out of Great Ponton. In 1905 I saw the same bird; and again this year, I notice not only the white-headed bird, but also one with much white about its neck and nearly a white tail; and another with many white feathers in its wings, possibly descendants of the first-mentioned bird. I and others have seen them several times lately, always within a range of about four fields.—H. PRESTON, Grantham, December 2nd, 1906.

—: o :—

FISHES.

Short=finned Tunny landed at Grimsby.—During the third week in December there was quite a mild sensation in Bradford, caused by a large fish which was on view at the game shop of Mr. W. L. Blakeley, at the bottom of Horton Road. It was supposed to be a Salmon hybrid, and it was said that none of the fisherfolk at Grimsby had ever seen anything like it before. Mr. F. King, of Grimsby, who sent this fish to Bradford, states that it was caught five miles to the north-east of the Dogger Bank on December 10th, 1906. The local evening papers of the 17th December stated that it had the head of a Salmon, the body of a Porpoise, and the tail of a Shark (! !), so that it attracted considerable attention. I found it to be one of the larger species of the *Scombridæ* (Mackerels), and with the aid of ' Our Country's Fishes,' was able to identify it with certainty as the Short-finned Tunny (*Orcynus thynnus*).

Although the Tunnies are common further south, and are a source of much economic value as food in the south-west of Europe and on the shores of the Mediterranean, yet they are of sufficient rarity in the British seas as to merit the following description of the present specimen :—

Colour and Scales.—Back, dark blue, shading to netted grey on the sides, and to silvery white below. The scales above the pectoral fins larger and forming a corslet, from which smaller scales extended towards the tail. Otherwise almost scaleless.

Measurements, etc.—The total length from the snout to the centre, or root, of the tail (measured along the back) was 3 feet 9½ inches. The tail, which was keeled at the sides, was large and deeply forked ; the span from tip to tip of the lobes being 13½ inches. The first dorsal fin contained 14 spines (the last one very small), and almost joined the second dorsal fin, which, although damaged, still showed its triangular shape. There were ten finlets behind, one being very small. The pectoral fins (which fitted into grooves), did not reach to within about 4 inches of the parallel of the second dorsal fin, and measured 9 and 7 inches respectively along the upper and lower edges, and were edged with white on the inner lower margins. Ventral fins, almost joining, not more than an eighth of an inch apart. Eight finlets underneath, between the anal fin and tail. The greatest girth would be about 30 inches ; it was very solid and fleshy and I was assured by the owner that this fish scaled nearly 100 lbs. Unfortunately its flesh was wasted, as it was exhibited and unidentified until it was too stale for food.—H. B. BOOTH, Bradford.

—:o :—

ENTOMOLOGY.

Sphinx convolvuli at Bradford.—Two specimens of *S. convolvuli* were taken at Saltaire during August, on the 8th and 11th respectively. Both were taken to Mr. S. Hainsworth, one of which he exhibited at a meeting of the Bradford Natural

History and Microscopical Society.—J. W. CARTER, Bradford, November 21st, 1906.

Acherontia atropos at Bradford.—The capture of two specimens of *A. atropos*, in August, has been reported to me. Both were taken at Shipley, one of which is recorded by Mr. Hainsworth, the other was brought to Mr. F. Rhodes at the Cartwright Hall by Mr. Pitts.—J. W. CARTER, Bradford, November 21st, 1906.

Leucophæa surinamensis, Linn., etc., at Bradford.—A few months ago a specimen of a Cockroach, which had been taken in the Bradford market, was given to me, and was put on one side for further examination.

A few days ago I determined it to be *Leucophæa surinamensis*, Linn., and sent the specimen to Mr. Porritt, who kindly confirms my determination, and states it is the first recorded example for Yorkshire.

Since my list of Cockroaches for this district was published ('Nat.' Jan., 1897, p. 26), four species have been added, viz., *Periplaneta australasiæ*, Fabr. ; *Panchlora exoleta*, Klug., which is frequently taken in the market ; *Rhyparobia maderæ*, Fab. ; and *Leucophæa surinamensis*, Linn., not a bad list of these generally detested insects.—J. W. CARTER, Bradford, November 17th, 1906.

—: o :—

MOLLUSCA.

Succinea oblonga Drap. in Westmorland.—In August last, whilst staying at Grange-over-Sands, I spent some time in collecting Mollusca in the surrounding district, and at Meathop, not far from Grange, I obtained a number of *Succineæ*, which I have submitted to Mr. Charles Oldham and others, who pronounce them to be, what I had myself suspected, *Succinea oblonga* Drap. The shells approach the form now generally recognised as var. *arenaria* Bouchard, and agree closely with specimens from Brunton Burrows, Devon, and from Irish localities. This find constitutes an important record for the county.—H. BEESTON, Havant.

—: o :—

GEOLOGY.

Exposure of New Red Sandstone at Middlesbrough.— The extensive enlargement of the Middlesbrough docks, still in progress, has in the excavation in the north-west part of the workings, exposed upper New Red Sandstone to a depth of

35 feet (about 12 feet below made ground), containing, irregularly stratified in the loose marl, a large quantity of gypsum, varying in tint from an almost pure white to a very deep red. In the marl, numerous large rhomboidal crystals of selenite were obtained, being found in separate pockets. Although there are disused gypsum pits in the neighbourhood, I have been unable to find any record of the presence of such crystals. I collected some specimens, which are preserved in the Dorman Memorial Museum, Middlesbrough.—W. Y. VEITCH, Middlesbrough.

REVIEWS AND BOOK NOTICES.

The Victoria History of the Counties of England. Sussex—Entomology. A. Cònstable & Co.

The Entomological portion of another of the Counties of the Victoria History has been issued. Sussex, entomologically, is one of the best investigated of our Counties, and as the compilation of the lists of species in the various orders has been done by thoroughly competent specialists, we are not surprised to find this portion of Vol. I. as satisfactory as it is voluminous. The whole is edited by Mr. Herbert Goss, F.L.S., and he and Mr. W. H. B. Fletcher, M.A., are chiefly responsible for the Lepidoptera, though it is easy to see that the veteran Sussex entomologist, the Rev. E. N. Bloomfield, M.A., has had a considerable share in the work on most of, if not all the orders. Besides these, the services of other equally well-known authorities, have been enlisted. The Rev. Canon Fowler, M.A., F.L.S., for the Coleoptera; Messrs. Edward Saunders, F.L.S., and Mr. Claude Morley, for the Hymenoptera; Mr. W. J. Lucas, F.E.S., for the Neuroptera and Trichoptera; Mr. Malcolm Barr, F.L.S., for the Orthoptera; Mr. J. H. A. Jenner, for the Diptera; and Messrs. E. A. Butler and A. C. Vine for the Aphides, etc. Every order seems to have been carefully, and so far as the species are known, exhaustively done, and an improvement on some of the earlier County lists, is that in all the orders, precise localities for the species are given. It would be invidious to select any one list as better than another, though naturally, from the fact that one of our hardest working, and best authorities on the 'Micros,' has spent so large a portion of his life in the County, the Lepidoptera take up a far greater space than any other. The Coleoptera come next, then the Hymenoptera, Diptera. Hemiptera, Neuroptera and Trichoptera, and Orthoptera, respectively. It may be worth while stating here, for future reference, that the only Sussex specimen of *Hadena peregrina* (of which there are probably only some half-dozen British caught examples known), recorded in this Sussex list, has, since the dispersal of the late Dr. P. B. Mason's collection, in which it stood, found a resting place in my own cabinet.

The value of these county histories to entomological science cannot be estimated, for although there are a few most admirable county lists of Lepidoptera, and more would undoubtedly have been forthcoming; this splendid undertaking has stimulated the publication of them at an infinitely more rapid rate than would otherwise have been the case. Besides this, we now get the advantage of lists of species in all the orders in one volume, and can ascertain at a glance what has been done entomologically in each county; and every specialist knows where to refer to for the information he requires. Every naturalist should possess the volume relating to his own county, and every Municipal Library ought to contain the whole series.—G. T. P.

FEBRUARY 1907.

No. 601
(No. 379 of current series).

THE NATURALIST.

A MONTHLY ILLUSTRATED JOURNAL OF
NATURAL HISTORY FOR THE NORTH OF ENGLAND.

EDITED BY

T. SHEPPARD, F.G.S.,

THE MUSEUM, HULL;

AND

T. W. WOODHEAD, Ph.D., F.L.S.,

TECHNICAL COLLEGE, HUDDERSFIELD;

WITH THE ASSISTANCE AS REFEREES IN SPECIAL DEPARTMENTS OF

J. GILBERT BAKER, F.R.S. F.L.S., GEO. T. PORRITT, F.L.S., F.E.S.,
Prof. P. F. KENDALL, M.Sc., F.G.S., JOHN W. TAYLOR,
T. H. NELSON, M.B.O.U., WILLIAM WEST, F.L.S.

Contents :—

LONDON :

A. BROWN & SONS, LIMITED, 5, FARRINGDON AVENUE, FEB 12 17
And at HULL AND YORK.

Printers and Publishers to the Y.N.U.

PRICE 6d. NET. BY POST 7d. NET.

(Of a few of these there are several copies.)

1. **Inaugural Address,** Delivered by the President, Rev. W. FOWLER, M.A. in 1877. **6d.**
2. **On the Present State of our knowledge of the Geography of Britis.** Plants (Presidential Address). J. GILBERT BAKER, F.R.S. **6d.**
3. **The Fathers of Yorkshire Botany** (Presidential Address). J. GILBER BAKER, F.R.S. **9d.**
4. **Botany of the Cumberland Borderland Marshes.** J. G. BAKER, F.R.S. **6d**
5. **The Study of Mosses** (Presidential Address). Dr. R. BRATHWAITE F.L.S. **6d.**
6. **Mosses of the Mersey Province.** J. A. WHELDON. **6d.**
7. **Strasburger's Investigation on the Process of Fertilisation in Phanero gams.** THOMAS HICK, B.A., B.Sc. **6d.**
8. **Additions to the Algæ of West Yorkshire.** W. WEST, F.L.S. **6d.**
9. **Fossil Climates.** A. C. SEWARD, M.A., F.R.S. **6d.**
10. **Henry Thomas Soppitt** (Obituary Notice). C. CROSSLAND, F.L.S. **6d**
11. **The Late Lord Bishop of Wakefield** (Obituary Notice). WILLIA WHITWELL, F.L.S. **6d.**
12. **The Flora of Wensleydale.** JOHN PERCIVAL, B.A. **6d.**
13. **Report on Yorkshire Botany for 1880.** F. ARNOLD LEES. **6d.**
14. **Vertebrates of the Wertern Ainsty (Yorkshire).** EDGAR R. WAITE F.L.S. **9d.**
15. **Lincolnshire.** JOHN CORDEAUX, M.B.O.U. **6d.**
16. **Heligoland.** JOHN CORDEAUX, M.B.O.U. **6d.**
17. **Bird-Notes from Heligoland for the Year 1886.** HEINRICH GÄTKE C.M.Z.S. **1s.**
18. **Coleoptera of the Liverpool District.** Part IV., Brachelytra. JOH W. ELLIS, L.R.C.P. **6d.**
19. **Coleoptera of the Liverpool District.** Parts V. and VI., Clavicorni and Lamellicornia. JOHN W. ELLIS, L.R.C.P. **6d.**
20. **The Hydradephaga of Lancashire and Cheshire.** W. E. SHARP. **6d.**
21. **The Lepidopterous Fauna of Lancashire and Cheshire.** Part I. Rhophalocera. JOHN W. ELLIS, L.R.C.P. **6d.**
22. **The Lepidopterous Fauna of Lancashire and Cheshire.** Part II. Sphinges and Bombyces. JOHN W. ELLIS, L.R.C.P. **6d.**
23. **Variation in European Lepidoptera.** W. F. DE VISMES KANE, M.A. M.R.I.A. **6d.**
24. **Yorkshire Lepidoptera in 1891.** A. E. HALL, F.E.S. **6d.**
25. **Yorkshire Hymenoptera** (Third List of Species). S. D. BAIRSTOW F.L.S., W. DENISON ROEBUCK, and THOMAS WILSON. **6d.**
26. **List of Land and Freshwater Mollusca of Lancashire.** ROBER STANDEN. **9d.**
27. **Yorkshire Naturalists at Gormire Lake and Thirkleby Park.** **6d.**

WANTED.

Quarterly Journal Geological Society, Vols. 1-22.
Phillip's Life of William Smith.
British Association Reports for 1839 and 1840.
Proceedings of Yorkshire Geological and Polytechnic Society, Vol. I.
Barnsley Naturalist Society's Quarterly Reports, Set.

GOOD PRICES GIVEN.

Apply—Hon. Sec., Y.N.U., Museum, Hull.

NOTES AND COMMENTS.

THE LINCOLN BORING.

At a recent meeting of the Geological Society of London, Prof. E. Hull read a paper on the boring for water at Lincoln, of which an account has already appeared in these columns.* In the discussion which followed, Mr. Percy Griffith described the boring as the deepest and longest extant, and the difficulties met with as of an altogether unusual character. The loss of the boring-tool, for instance, involved a delay of sixteen months for its recovery. The corporation of Lincoln then authorised the continuation of a 9-foot shaft to a depth of 1500 feet from the surface at their own expense. A pilot boring, 3 inches in diameter, was then driven from the bottom of the well, and at a depth of 1561 feet, water was met with which resisted all efforts to prevent it from rising into the well. The concrete well-bottom was, however, put in successfully, having a guide pipe fixed in it to allow of the boring being continued through it. The boring was then resumed of a diameter of 32 inches, and, on the depth of 1561 feet being reached, the water rushed into the well, and in thirty-six hours overflowed at the surface. As the boring was continued to a lower level, the surface-flow increased until it reached a maximum of 180,000 gallons per day (24 hours). An enormous quantity of fine sand was blown into the well by the first rush of water, and some time was lost in removing this before boring could be resumed. The 32-inch boring was carried to a depth of 2015 feet from the surface; and steel tubes, of 30 inches internal diameter, were lowered into the boring to a depth of 1600 feet from the surface, or about 40 feet into the sandstone. The work was commenced in August 1901, and had therefore been five years and four months in progress. The cost to date was about £20,000. Pumping was now going on, and so far had proved the yield of the well as being about 750,000 gallons per 24 hours, at a depth of 200 feet.

EDUCATIONAL MUSEUMS

In No. 8 of the *Museum Gazette*, which is published at the Hazelmere Educational Museum, an editorial note appears, without warning and apparently without incentive, as under :—
'We differ *toto cælo* from those who hold that museums in

---

* 'The Artesian Boring for the Supply of the City of Lincoln from the New Red Sandstone,' by Prof. E. Hull. 'Nat.,' Sept. 1906, pp. 338-339.

villages or small towns ought to restrict themselves to local objects. By all means have in such museums the best collection possible in illustration of what can be got in the district, but let it be a department, not the whole affair. What are called local museums are difficult to make anything like complete, and the endeavour to make them so may stimulate some of the worst vices of the mere collector. However successful, they remain meagre and unattractive. For educational purposes they cannot approach one of general scope. The curators of such museums are like a [*sic.*] wing-maimed bird, or a [*sic.*] pugilist in shackles. As a rule, we fear that merely local museums rarely attain the dignity of a curator, but languish for a few years under an honorary secretary, who once was zealous, and finally hand over to some more liberally constituted institution the remains of the Herbarium, and a few stuffed birds. It is better to do things well whilst we are about them.'

AND LOCAL MUSEUMS.

Just so ! The last sentence quoted is just the point. But is it possible 'to do things well' in a village or small town, in any other way than by making the collection strictly local ? Nothing is so conducive to making such museums 'receptacles for rubbish' (respecting which we have heard so much of late), and of little value educationally, or in any other way ; and a curator of a small local museum had far better be like a 'pugilist in shackles' than a pugilist let loose, in which latter case the harm he might do might be serious ! To some extent it depends on what the *Museums Gazette* calls 'a village or small town,' but in the north we should put Chesters (Northumberland) and Pickering and Driffield in Yorkshire under that head. In each of these places (and others might be cited) is a local museum of very great value—historically, educationally, and in other ways too. We might mention larger places, such as Perth, York, and Chester, where there are strictly local museums of the greatest value and importance, museums which are well known in the north at any rate, if not at Hazelmere, and each of these museums is particularly valuable because of the local character of its exhibits, and each also 'attains the dignity of a curator' !

THE SELBY MUSEUM.

We have in Yorkshire, at Selby, an 'educational museum' well known to the editor of the *Museum Gazette*, seeing that

Selby is indebted to its son, Dr. Jonathan Hutchinson, for it. We presume therefore we may take this as a type of an educational museum. We were sorry to see from a report of an address, recently delivered at Selby by Dr. Hutchinson, that this museum is not appreciated as it ought to be. We are grateful to Dr. Hutchinson for what he has done for Selby, and should certainly be the last to in any way deprecate the good work he has accomplished there. But might not the museum be more interesting and more educationally valuable, if it were strengthened in its local collections? We are not quite sure whether the curator in this educational museum is like a pugilist who is not in shackles, or like a bird which is not wing-maimed, or whether the Selby institution 'attains the dignity of a curator' at all. But we feel sure that the Selby Museum would be more valuable, educationally, if within its walls we could see a representative collection of the antiquities, plants, shells, birds, insects, etc., of the interesting district in which the museum is situated. It is all very well for a museum in a small town to be 'frankly fragmentary—here a little and there a little' as the *Museum Gazette* prides itself in being, but as to the best kind of a museum, educationally, in a small town, there are more opinions than one. We hold a brief for the *local* museum, and whilst no museum can or should ever be 'complete,' a local museum can surely be more so than a general museum.

LIASSIC DENTALIIDÆ.

In a recent issue of the 'Quarterly Journal of the Geological Society,' Mr. Lindsall Richardson has an important paper on

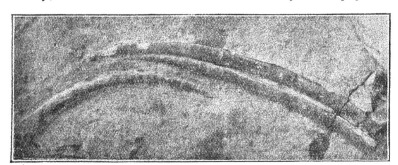

Dentalium giganteum, **Phillips.** (Natural Size).

the above subject, in which the various Liassic species of Dentalium are figured and described. Particular reference is made to *Dentalium giganteum*, Phillips, myriads of which 'may be seen covering the upper surfaces of some of the sandstones situated near to the base of the Zone of *Ammonites margaritatus* at Hawsker, Staithes, Rockcliff, Hummersea, Huntcliffe, and Coatham Scars, on the coast, and inland at Hutton, near Guisborough, and in Danby Dale.' The specimens are often three inches long. The illustration herewith, reproduced by the permission of the Geological Society, is from the Moore collection at Bath, and is from the Middle Lias, Cleveland.

COAST EROSION AGAIN.

The interest which has been aroused in reference to coast erosion is resulting in some valuable information being placed on record. From a recent impression of the *Hull News*, we find a letter in which the writer 'Having carefully followed the evidence given by a number of witnesses before the Royal Commission on coast erosion, as to the cause *or otherwise* [!] of the wasting away of the Holderness Boulder Clay Cliffs,' informs his readers that ' My opinion as to the wastage of these clay cliffs is based on close observation. I include boulder clay as a rock. It is by its mechanical effects that the sea accomplishes most of its erosion. The mere weight with which the ocean waves fall upon exposed [*sic*] coasts breaks off fragments of rock from the Cliffs.' This is just the sort of information the Royal Commission wanted !

MAMMALS.

Otters near Barnsley.—Every season during this last few years, otters have bred in Bretton Park, near Barnsley. They have come up the river Dearne as far as Clayton West and Skelmanthorpe, and have been seen near Darton.—FRED LAWTON, Skelmanthorpe.

Otters near Barnsley.—The otter is not so uncommon in the upper waters of the Dearne, to the north of Barnsley, as Mr. Bayford's paragraph (' Naturalist,' January, p. 29) might lead us to suppose. Two otters were seen about two years ago by a friend of mine in Bretton Park, and one was seen last summer in Cannon Hall Park. In November, two old ones and two young ones were seen at very close quarters by several in the Mill Dam, to the West of Cannon Hall Park ; and the head keeper has told me, that, when the recent snow was on the ground, he could distinctly trace three otters into some shrubs near the Dam. An otter-hunting friend told me in the summer of 1905, that he saw the footprints of an otter in the Park.—C. T. PRATT, Cawthorne Vicarage.

THE AMERICAN GREY SQUIRREL IN YORKSHIRE.

W. H. ST. QUINTIN, J.P.,
Scampston, E. Yorks.

AT the end of June 1906 I was given about three dozen of the American Grey Squirrel, bred at large in one of the home counties.. I ought to have announced the fact at the time, but I did not realise how widely such small animals might stray from the place where they were enlarged. Now, however, I find that they have established themselves at various places within a radius of four miles from Scampston, and I hear of one being killed near Pickering, which is over six miles away, 'as the crow flies,' with the River Derwent to cross! I therefore now think it time to appeal to readers of the 'Naturalist' to do what they can to protect these beautiful little animals, which, I am assured, have been found quite innocent of any damage to woods and plantations in other parts of England, where they have been at liberty for some years. In my own case I the more gladly accepted these little 'aliens,' as some disease closely resembling mange had reduced our native Red Squirrels almost to vanishing point. These Grey Squirrels seem to find a great deal of food on the ground, even in summer time. As far as I could tell, last June and July they were largely eating grass. But they readily avail themselves of any hospitality offered, and mine regularly visit a food-box fixed to a prominent tree in sight of the windows, in which we place maize, hemp-seed, and nuts. Recent visitors to the London Zoological Gardens will have been charmed by the numbers and familiarity of the Grey Squirrels, which soon made their escape from the enclosure intended for them, but fortunately had the good sense to remain within the grounds, to the great entertainment of the children with their paper bags of nuts and buns.

We regret to record the death of Mr. John Ward, of Longton, Staffs., a keen geologist, whose work we have more than once referred to in these columns.

Mr. Clement Reid's notes on 'Coast Erosion,' read at the York Meeting of the British Association (See 'Naturalist,' Sept., 1906, pp. 327-9), as well as Mr. E. R. Matthew's remarks thereon, appear in the November *Geographical Journal.*

The Rev. O. Pickard-Cambridge favours us with a copy of his paper 'On some New and Rare British Arachnida,' which has recently appeared in the 'Proceedings of the Dorset Natural History and Antiquarian Field Club.' It includes particulars of several northern country records.

FURTHER NOTES ON A SOLITARY WASP
(*Odynerus parietum*, Linn.).

W. M. EGGLESTONE,
Stanhope, Co. Durham.

I READ with some interest Dr. George's 'Notes on a Solitary Wasp' in the January number of the *Naturalist*, 1906, as I had a little experience with one of these interesting creatures during the same year.

On the 17th of August, 1905, I saw a Solitary Wasp on one

of the gate posts at the entrance to the offices of the Weardale Rural District Council at Stanhope, but on my nearer approach it made off.

This entrance consists of an iron gate and two cut freestone square pillars or posts, the columns of which, between the base plinth and the moulded capitals, are square, and have, on the front face, a panel formed by a sunk or incised line three-quarters of an inch wide and half an inch deep running up each side and across the top and bottom of the column. Thinking the black wasp I saw had some business on hand I examined the pillars and found on one of them two pairs of nests, built of mud and fixed in the angle of the incised line above mentioned, The cells are seen below the 'X' in the illustration.

These nests were nearly together, one of the double cells being fixed in one angle and the other in the opposite angle.

I dissected one pair of cells, and found in the upper one six green caterpillars rolled up very much like those illustrated on page 91 in Peckhams' book on Wasps.* Some of these caterpillars were still alive. In the lower cell I found two larger caterpillars. The nests were about three quarters of an inch long and joined together one above the other. I left the other double cells intact for further observation.

On the 22nd of August, on going to my office, I found the black wasp very busy at work on the top of the untouched double cell, so much so, that I got close to it and watched it for some time. Then the wasp dropped to the ground close to where I was standing and busied itself among some soft mud. Having evidently gathered up some of the mud, the wasp flew back to the cell, and set to work to strengthen the previously closed orifice. When it had accomplished its task, the little black worker flew overhead and I saw it no more. The top of the cell where it had been working was wet. I kept a watch on these mud cells, expecting some development in the spring, but one day in December I found that the cells were broken away, nothing being left but the ragged edges where the mud had stuck to the stone, resembling, but on a smaller scale, the remains of a broken swallow's nest. The cells were probably accidently destroyed by children. I saw no Solitary Wasps during the year 1906.

———◆◆———

Darwinism and the Problems of Life. A study of familiar animal life. By **Conrad Guenther,** translated by Joseph McCabe. London, 1906, Owen & Co. 436 pages, 12/6.

Both the translator and publisher of this work earn our gratitude for the facilities they have now given for enabling English readers to possess, in a convenient form, Prof. Guenther's most interesting volume. In it the author's aim is to vindicate the value and importance of Darwin's work, and the book is largely devoted to proving the truth of Darwin's doctrine. 'Every care has been taken to distinguish between facts and probabilities, and it has been clearly pointed out what general deductions may or may not be drawn from Darwinism. The ease with which the theory of evolution is grasped too readily, disposes people to regard Darwinism as the one true, natural, and sound view of the world-process. . . . The manner of presentation is simple, because the work is written for the general reader. No knowledge of science is pre-supposed, and the reader is briefly informed on all the questions that have a bearing on the theory of evolution.' From a perusal of the volume, it is obvious that the author has carefully consulted the now very extensive literature dealing with the subject discussed. The book, though massive, is very light, and the large size and clearness of the type makes its perusal particularly pleasant.

---

* "Wasps, Social and Solitary," by Mr. and Mrs. G. W. Peckham, 1905.

SPONDYLUS LATUS IN THE CHALK OF NORTH LINCOLNSHIRE.

H. C. DRAKE.
Hull.

I HAVE to record a fine specimen of *Spondylus latus* recently obtained from the larger chalk quarry on the Humber side at Barton, North Lincolnshire. It is 35 mm. in height, and is of the same width, and occurred in the *Holaster subglobosus* zone.

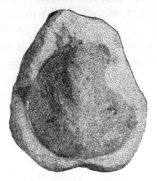

This species does not appear to have been previously recorded for the chalk of Lincolnshire. One specimen only is on record for Yorkshire, and was found by Mr. W. Hill in the railway cutting near South Cave station. This specimen, which measured 23 mm. each way, is now in the Woodwardian Museum at Cambridge. The Lincolnshire specimen can be seen in the geological gallery in the Hull Museum.

◆◆

A male Hoopoe was shot on the Moors near Whitby, and sent to Messrs. Rowland Ward & Co. for preservation, in November.

The recently issued 'Transactions of the Perthshire Society of Natural Science contains two admirably illustrated geological articles by Mr. G. F. Bates.

A puffin was caught alive in Springvale Road, Sheffield, in December. It was mistaken for a coot, and placed in a pond in one of the parks, where it died. It is being preserved.

In an article dealing with the Water-Pipit (*Authus spipoletta*) as a visitor to England, in the December *Zoologist*, Mr. M. J. Nichol refers to a record of the species at Tetney, Lincolnshire, in 1895.

It has been found that the oak beams supporting the picture galleries in the Bowes Museum are perishing of dry rot. An outlay of at least £15,000 will be necessary to replace them by iron girders, and as a consequence the Museum will remain closed until April, 1908.

Parts 4, 5, and 6, of 'The World's Commercial Products' have been received, and deal with sugar, tea, coffee, and cocoa. (Sir Isaac Pitman & Sons, 7d. net. each). Each is admirably illustrated by process blocks from photographs, and the coloured illustration of the Tea Plant in part 6 is really excellent.

LINCOLNSHIRE MITES.
RHYNCHOLOPHIDÆ.

C. F. GEORGE, M.R.C.S.
Kirton-in-Lindsey.

THAT these mites have been little recorded by English Acarologists is evident from the fact that when Professor Sig Thor of Christiana wrote his pamphlet on Norwegian *Rhyncholophidæ* in 1900, he records in his bibliography the names and works of no fewer than twenty-one writers, and only one of them is an Englishman! *viz.* Mr. O. P. Cambridge, the learned writer on British Spiders. His paper on ' *Calyptostoma Hardyi* ' may be found in ' The Annals and Magazine of Natural History,' vol. 16 (4th series), London, page 384. Yet these mites are by no means rare, even in Lincolnshire ; they are curious, and form, when well mounted, good and beautiful objects for the microscope. They are most frequently found under stones, or chips of wood, near plantations, and also in damp moss from tree trunks, ditches, etc. The first genus of *Rhyncholophidæ*, *Smaris*, was founded by Latreille in 1807. The Type species was figured and described by ' Hermann ' in 1804, under the name of *Trombidium expalpe.* It is perhaps not very common, as I have only one mount of this species, which I found many years ago. The mount is not a good one, but is sufficiently so to enable an arachnologist to name the species. When alive, the proboscis is so retracted that the animal appears to have no mouth organs ; hence the name '*expalpe.*' The pressure used in mounting, though not great in this instance, has been sufficient to show them partially ; as is well shown in Mr. Soar's drawing, fig. A, made from my slide. During life they can be protruded or retracted at the creature's will. The mite is darkish red in colour, long oval in shape, and rather thickly covered with scales, under which the skin is seen to be marked with circles, or pits, having a double contour (see fig. B), from the centre of which spring these bent scales. This skin structure requires to be carefully observed, as it distinguishes this mite from the next species ; the eyes are also very remarkable, being arranged in three pairs, as shown in fig. A. Mr. Soar gives the measurements as 2.34 mm. long, and 1.44 broad. Fig. C. represents the projecting mouth organs and palpi greatly enlarged ; the legs were too much doubled up and distorted, during mounting, for them to be figured ; the name of this mite is now *Smaris expalpis* Herman. The next mite of this

Genus *Smaris hardyi*, Mr. Soar has been able to figure with the
legs extended (Fig. D), and it will be seen that the tarsus of
the first pair of legs is considerably enlarged in club shape, the
second pair less so, the third pair only slightly, and the fourth

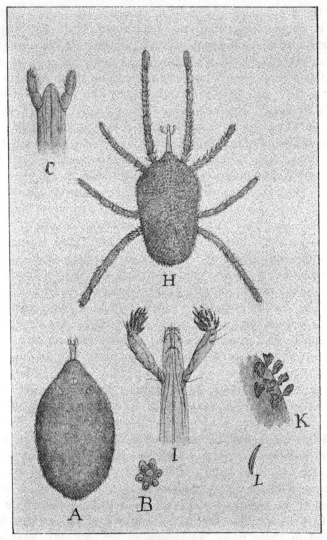

Water Mites.

scarcely at all thickened. They are all provided with powerful
double claws, and the penultimute. internodes are scarcely as
long as the tarsi, except in the last leg, where it is a little
longer than the tarsus. The legs are covered with strong and

pointed curved hairs, which are simple, *i.e.* not feathered. The body is rather lozenge-shaped, with rounded angles, and is broad at the shoulders. It is covered with rather short bent leaf-like scales, slightly pectinated on the surface, in at least some cases. The colour of the mite is dark orange red. The skin under the scales is marked out in a rather complex pattern, a small portion is shown much enlarged at fig. E, taking one of the button like bosses as a centre there radiates from it 3, 4, or 5 broad bands, sometimes looking almost like the spokes of a wheel. Professor Berlese in his *Acarus Nuovi*, 1905, Table XV., fig. V*a*, gives a similar figure, the mite from which it is taken (see fig. V), appears to be very much like *S. hardyi*, but is called by him *S. caelata*. I found *S. hardyi* in 1879, and gave an outline figure and some notes in the November number of *Science Gossip* for that year ; it will also be found mentioned by Murray in his 'Aptera,' page 140. The palpi and mandibles are shown much enlarged in fig. F, these last are seen to be lancet shaped, and carry a single barb almost at right angles to the blade. The mandibles of all the *Rhyncholophidæ* are more or less lancet shaped, and formed for piercing, and are mostly, if not always, more or less barbed. This is very different from the mandibles of the *Trombididæ*, which are sickle shaped, and fitted for tearing. The mandibles therefore are the great structural difference separating these two families. Fig. G represents what I take to be the larva of *S. hardyi*, it is remarkable in having three claws to each tarsus, the adult having two only.

The next Genus I have to mention is *Smaridia* Duges and the species, *S. ampulligera* Berlese. This is a very beautiful mite, of a rather brick red colour, and covered all over with dark cinnabar coloured scales, thickly planted all over the body, and also covering the legs, which are paler in colour than the body of the mite. The shape of the creature is well shown in Mr. Soar's drawing, fig. H. The eyes of the mite are not drawn ; even in life they are difficult to see, being covered with the scales ; but in a balsam mount made long ago, all colour has disappeared, and the eyes can not be made out. They are placed on the cephalothorax between the first and second pair of legs, there is also a line or furrow down the middle of the thorax, called the dorsal groove or furrow, not seen in the balsam mount, but well enough marked in the living creature ; and in this dorsal furrow there is a rod of chitin, enlarged almost like a battledore at either end, and having two very fine;

rather long, tactile curved hairs, springing each from the centre of a circular disc on the widened end of the rod. These can only be seen with great care under the microscope, in a balsam mount, but are much more evident in a freshly dissected

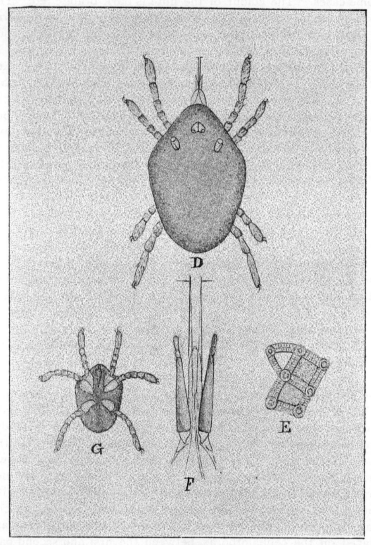

Water Mites.

mite. This dorsal groove and line are seen in many of the *Rhyncholophidæ,* and varies somewhat in detail, and so forms a help to diagnose certain species of the family. The tarsi of all the legs are more or less enlarged, the front ones considerably more than the others, the penultimate internodes are longer than the tarsi, especially those of the last leg, which are about double the length of the tarsus. The length of the mite is 1.60 mm., the length of the first leg 1.62 mm. Fig. I represents the mouth organs much enlarged. These when extended measure about 0.39 mm. The body is like a sac, and can be wrinkled at the will of the creature, which often causes the scales to appear irregularly placed. Fig. K shows the scales at the edge of the body, and L those on the legs ; the proboscis is retractile, but not to the same extent as in *Smaris*. To appreciate the mite properly it must be seen alive, as well as mounted and dissected.

The Victoria History of the Counties of England. Devonshire— Entomology. A. Constable & Co.

The Victoria Histories are apparently being issued rather more rapidly than they were at first, and now we have before us the Entomological portion of one of the three of our largest English Counties, Devonshire. We confess to a little disappointment in going through it, perhaps because Devonshire has for some years been one of our own favourite collecting grounds, and we cannot help thinking that more might have been made of it. No doubt nearly all the recorded species in all orders are included in the lists, but some of them, as the Orthoptera, show a want of acquaintance with recent literature, which otherwise should have enabled the compiler not only to add considerably more localities, but to indicate a number of species as common or even abundant, which he evidently considers as scarce. Then in the lists of Neuroptera, Trichoptera, Hemiptera, and Aphididæ, no localities whatever are given ; and even in such comparatively popular orders as the Coleoptera and Hymenoptera, most of the species are also without localities, or any indication as to their distribution. Personally, we regard such lists of species, in a large county like Devonshire, as little better than useless. Indeed, in the list of Neuroptera, old records are given of species which are not now regarded as British at all, and which probably never did form part of our fauna. The list of Lepidoptera is the most satisfactory, but even it leaves a good deal to be desired. We are specially surprised to find no mention made of the specimen of *Ophiusa stolida* taken by Mr. J. Jäger on the South Devon Coast, September 23rd, 1903. This very striking and beautiful noctua, though the first, and as yet the only one taken in Britain, was in such beautiful condition as to indicate that it had been bred on the spot where captured, its food plant being abundant there. There is no excuse whatever for its exclusion ; it is, in fact, far more entitled to be regarded as a native than is *Cucullia abrotani*, which the author includes in the list. This portion too has been badly edited. Some of the records have been incorrectly copied, and in one case a species (*Crambus uliginosellus*) is included, as recorded by ourselves, as having been captured by Mr. D'Orville. We know absolutely nothing of the circumstance ! Errors in spelling are plentiful, and evidently the proof sheets have been carelessly read. The part has been edited by Mr. Herbert Goss, F.L.S., and the following specialists are responsible for the compilation of the lists : Coleoptera, the Rev. Canon W. Fowler, M.A., F.L.S. ; Lepidoptera, the late Mr. C. G. Barrett, F.E.S. ; Hymenoptera, Orthoptera, Hemiptera, and Aphididæ, Mr. G. C. Bignell, F.E.S. ; Neuroptera and Trichoptera, Mr. C. A. Briggs, F.E.S. ; and the Diptera, Mr. Ernest E. Austen.—G.T.P.

NOTES ON YORKSHIRE BRYOPHYTES.

IV. *Frullania* and *Jubula*.

(Continued from page 16.)

F. CAVERS, D.Sc., F.L.S.

Professor of Biology, Hartley University College, Southampton.

THE sequence of the cell-divisions in the development of the sporogonium is extremely regular in *Frullania*,* and has been, on the whole, correctly described and figured by Hofmeister, Kienitz-Gerloff, Leitgeb, and Leclerc du Sablon. The first wall in the fertilised egg-cell is transverse to the long axis of the archegonium, and the lower cell (*hypobasal*) then divides by a vertical wall, while the upper cell (*epibasal*) divides transversely (Fig. 6, G). Each of the two upper cells then divides by two sets of vertical walls that cut each other at right angles. In the meantime, the two hypobasal cells have divided again by vertical walls, so that the embryo now consists of three tiers, each having four nearly equal cells. In some cases I have found one or more (as many as three in a few cases) further transverse divisions in the epibasal half of the embryo before the appearance of the vertical walls ; this is normally the case in other *Jungermanniaceæ*. The hypobasal cells soon project from the surface and divide rather irregularly, forming the blunt foot which presses into the tissue forming the stalk of the fertilised archegonium. The cells of the uppermost tier divide by tangential walls into four inner cells, which have denser protoplasm than the outer ones, and which form the archesporium, while the four outer cells produce the wall of the capsule. The tier of cells between capsule and foot divides by longitudinal walls, which separate eight outer cells from four inner ones ; this tier forms the seta or stalk of the capsule. The four primary archesporial cells divide repeatedly by longitudinal walls, resulting in the formation of a lens-shaped layer of about two hundred cells, the central cells being longer than the marginal ones (Fig. 6, G—L). These cells then become differentiated into two sets. Some of them divide transversely so as to produce rows of cells ; the others grow in length, but remain undivided, forming long cylindrical cells attached to the inner surface of the capsule wall above, and to the floor of the capsule below (Fig. 6, K L). The rows of

---

*. The details of development of *antheridium* and *archegonium* are omitted from this paper, since they do not differ from the normal process in other liverworts. See Scott, 'Structural Botany,' Part 2 ; Campbell, 'Mosses and Ferns,' 2nd Ed.

cubical cells alternate regularly with the long undivided cells ; the former gave rise to the spores, the latter to the elaters. The arrangement as seen in a cross section of the capsule is shown in the accompany- ing diagram ; S = spore-forming and e = elater- forming cells. According to Leclerc du Sablon,[*] the spore mother cells and elaters lose their cell walls at one stage, and afterwards acquire fresh walls, but this seems to be an error, due, probably, to imperfect methods of preparation. The walls are mucilaginous, and stain deeply in microtome sections, but there appears to be a definite wall at all stages, as in other liverworts.

S S S S
 e e e
S S S S
 e e e
S S S S

The spores of *Frullania* are roughly spherical (but rather irregular owing to their becoming flattened by contact in the longitudinal rows), and about 1.06 mm. in diameter ; the outer coat (*exospore*) is thick, brown in colour, and shows a number of shallow circular pits, the surface of which bears small tubercles.

Each elater becomes somewhat flattened at the lower end ; it contains a broad brown spiral band, passing into a ring at the trumpet-like lower end (Fig. 7, A D). In the developing elater the protoplasm becomes differentiated into an axial strand and a peripheral layer, which becomes spirally wound on the inner surface of the wall (Fig. 7, A). The elaters show this stage, and are still colourless after the spores have acquired their final form, and have the outer coat fairly thick and golden-brown, so that the elaters keep for a long time their primary function of conveying food materials to the developing spores. The cells in the lower part of the fertilised archegonium (*calyptra*) contain dense protoplasm with oil-drops, as do the cells in the tissue of the thickened archegonium stalk. The elaters alternate regularly with the longitudinal rows of spores, con-sequently, those nearest the central line of the capsule are the largest. There are from 80 to 100 in each capsule in *F. dilatata* and *F. tamarisci.*

The wall of the capsule at an early stage becomes two-layered. The cells of the outer layer have thick rod-like fibres on their lateral walls, especially at the angles between adjacent cells; these rods appear as dots in a surface view of the capsule (Fig. 7, C), and are especially marked, and deeply coloured brown at the two opposite points at the bases of the valves into which the capsule wall splits (Fig. 7, E). On the walls of the inner layer of cells the thickenings form an irregular network (Fig. 7, D).

---

[*] *Annales des Sci.*, Nat., 1885.

The divisions in the seta are very regular, and in cross sections there is seen a central group of four cells covered by three concentric layers of cells (Fig. 7, B). The seta is not so sharply marked off from the capsule as is usually the case in liverworts, for at the base of the capsule there is a mass of thin-walled tissue which may, as suggested by Spruce, be regarded as corresponding to the hypophysis or expanded upper part of the seta found in many mosses. Hence, when dehiscence occurs, the four valves of the capsule wall do not become free right down to the point of junction with the cylindrical seta, and *Frullania* capsules are usually described as opening for only about two-thirds of their length.

The calyptra, formed by the active growth of the fertilised archegonium, becomes five or six cells thick, and is raised on a massive stalk, leaving the unfertilised archegonium at the base (Fig. 6, E). The old archegonium neck at the top of the calyptr ι can be seen inside the perianth (Fig. 8).

The opening of the capsule has been accurately described by Goebel.* The capsule wall splits from above, downwards, into four valves, which spring backwards so as to lie almost horizontally at first, each valve carrying with it a share of the elaters and the rows of spores lying between them. Just before this, while the whole capsule is drying, the elaters contract, and their lower ends become free from the floor of the capsule, while the more firmly fixed upper ends remain attached to the inner surface of the capsule wall. The whole process is due to the shrinking of the membranes in the capsule wall and in the elaters ; when elaters from a dehisced capsule are placed in water they elongate again, and the spiral band becomes more loosely coiled. The spores are discharged with great force, and may be thrown out to a distance of three or four inches, so that the act of dehiscence might well be termed an explosion, quite different in degree, though not in kind, from the process observed in other liverworts, and in a few seconds from the rolling back of the valves the exploded capsule shows only a few spores that have remained entangled between the elaters. It is difficult to analyse this sudden action, but it is possible that the elaters themselves bend or twist on being set free at their lower ends, in addition to being passively moved outwards with the valves, and causing the spores to be jerked out.

The earliest divisions in the germinating spore take place

---

* ' Flora,' 1895, Heft 1.

before the rupture of the exospore or outer coat. The young plant at first grows in all directions, so that instead of forming a filament (germ-tube), as in most liverworts, it consists of a small oval mass of cells. At one end of this mass the superficial cells grow out to form rhizoids, while one of the cells at the other end becomes larger than the others, and divides by intersecting walls, forming a three-sided apical cell. The first few leaves are simple rows of cells, then come small bilobed cell-plates, the lower lobe of the side leaves only becoming pitcher-like after about a dozen of these simple leaves have been formed. The under leaves are at first represented only by small tufts of rhizoids, and when rhizoids occur on the mature plant they always grow out from the basal part of an under-leaf.

Isolated leaves of *F. dilatata* and *F. tamarisci*, or even small pieces including only a few cells, are capable of giving rise to new plants when cultivated. By growing leaves and leaf-fragments in culture solutions, the writer has several times obtained well grown plants, and can verify for both these species the account given for *F. dilatata* by Schostakowitsch,* who showed that the growth proceeds from a single leaf-cell; this cell divides so as to form two tiers with four cells in each, and the four upper cells then give rise to a small ovoid mass of cells, one of which forms the apical cell of the young plant.

Small disc-like cell masses (*gemmæ*) are sometimes formed on the leaves of *F. dilatata* and *F. tamarisci* by the growth and division of a single leaf-cell. These gemmae give rise to new plants, as also do the outgrowths on the perianth of *F. dilatata* when cultivated.†

For kind assistance in the supply of specimens of *Hepaticæ*, I am indebted to Messrs. W. Ingham, S. M. Macvicar, M. B. Slater, G. Stabler, and G. Webster, to whom I take this opportunity of expressing my thanks.

The majority of the illustrations in this paper have been reproduced from photomicrographs kindly taken for me by Prof. E. L. Watkin, M.A., Hartley University College, Southampton.

In the investigation of the *Hepaticæ*, I have been materially assisted by a grant allotted by the Government Grant Committee of the Royal Society.

---

* 'Flora,' 1894.

† Berggren (*Jakttagelser üfver mossornas könlosa fortplanting*, etc., Lund, 1895) says that the pitchers of *F. fragilifolia*, which readily become detached from the plant, give rise to leafy shoots on being set free. See the writer's paper 'On a sexual reproduction and regeneration in *Hepaticæ*,' 'New Phytologist,' vol. 2, 1903.

FUNGUS FORAY AT FARNLEY TYAS.

C. CROSSLAND, F.L.S.
Halifax.

THE sixteenth annual foray of the Yorkshire Naturalists' Union was held at Farnley Tyas, September 22nd to 27th, the object being the investigation of the neighbouring fields and woods. Permissions were granted by Lord Dartmouth and the West Riding County Council. The head-quarters were at the "Golden Cock." There were present Messrs. A. Clarke, Huddersfield; W. N. Cheesman, Selby; Harold Wager, Leeds; Thos. Gibbs, Derby; R. H. Philip, Hull; C. H. Broadhead, Wooldale; J. Needham, Hebden Bridge; H. C. Hawley, Boston, Lincs.; Thos. Smith, Alderley Edge; H. Humphrey, Stockport; J. W. H. Johnson, Thornhill; Rev. F. H. Woods, Bainton; and a Wakefield representative.

There were also representatives from the following Societies: Huddersfield, Milnsbridge, Slaithwaite, Berry Brow, Honley, Moldgreen, Primrose Hill, and Lindley. Each brought parcels of Fungi gathered in their own and other localities, such as Golcar, Kirkheaton, etc., too far away to be worked from this centre, though within the Huddersfield area.

This village has long been one of the occasional joint meeting-places for the many Field Naturalist Societies in the Huddersfield district. These Societies, especially the Huddersfield Botanical Society, have for nearly a quarter of a century included fungi in their investigatoins. It was partly on this account that the Committee decided to hold the foray here. There being a very suitable room at the Inn, with plenty of light and table space, it was thought an exhibition of the specimens collected might be made that would be of interest and value to all present.

Mr. A. Clarke provided a series of diagrams showing the order of classification, together with generic labels printed in large type, which were distributed on the walls and tables. These greatly facilitated the arrangement in proper order of the various species collected. Small wooden stands and pins were also made use of to show species in an erect and more natural position than if laid in a heap on the table.

A few uncommon species were brought from Hebden Bridge by J. Needham; others from near Wakefield by J. W. H. Johnson; and a consignment was sent from Luddenden Dean, near Halifax, by H. Waterworth, for exhibition purposes only, and are not included in the list of local species.

The foray opened by a short run out to Storthes Hall Wood on the Saturday afternoon. Evidence was not lacking that there must be no expectations of overflowing baskets at the end of each day's excursion. The gamekeepers had noticed the dearth of fungi in the woods. The delightful weather of the previous six or eight weeks had not been conducive to prolific fungus growths. However, observation having taught us that weather which retards the majority of species may be favourable to a few, there was no despondency. The elements were favour- able for collecting such as could be found, and it has often been noticed that when prospects are not specially good things altogether out of the ordinary line, worth perhaps a wagon load of old acquaintances, have been almost certain to crop up. It was so this time. There were one or two other compensations : the woods were near and easy to work.

In a delightful place in Mollicar Wood one party enjoyed themselves for a short while sitting on a log, as they could find no fungi by walking about. This proved to be the best way of searching at that particular place. Presently, after a little fun had partially subsided, one said, " I see something over there ; " another, " *And I see something over there.*" This caused a spurt in various directions, which led to the finding of one of the rarest hings collected during the foray—*Boletus parasiticus* in fine form attacking *Scleroderma vulgare*. This Boletus has only two previous Yorkshire records—Scarborough and Hebden Bridge— the latter too late to be included in the Yorks. Fungus Flora. The Hebden Bridge specimens were also on *S. vulgare*.

Varying fortune was experienced during the week's investiga- tions both in woods and pastures. While some pastures were rather barren, others yielded fairly abundantly. One in particular was remarkably good. This was an old pasture on the bank of the stream, about an acre and a half in extent, and partly surrounded by woodland. Here a party of five or six in a twenty minutes' search gathered thirty-eight species on the cattle dung and among the grass. Certainly, they were mostly species of common occurrence, yet there were three or four which have only a couple of previous Yorkshire records.

The pastures were all in good " heart," and showed no signs of suffering from lack of moisture : they were almost spring green. Pasture species, comparatively speaking, were more in evidence than denizens of the woodlands. At no pre- vious foray have thirteen species of *Clavaria* been found, which number is nearly half those hitherto recorded for the whole of

Yorkshire, while one—*C. incarnata*—is a new county record. One field, within a few hundred yards of head-quarters, produced no fewer than six species, including *incarnata*. This latter is not a very conspicuous *Clavaria* among grass, being only ½-1½ inches high, and was found partly by accident. Several of the company, desirous of having the exhibits as perfect as possible, cut out circular sods along with the fungus, and placed them on the tables in a growing condition. It was on one of these, under the pileus of a tall *Hygrophorus pratensis*, that *C. incarnata* was spotted. We were unable to find more in the field where the sod came from. In another field were some remarkably fine tufts of *C. fusiformis* and *C. coralloides*, tufts of the latter being 6-7 inches across. "Fairy rings" were scarce both in meadow and pasture, hence ring-dwellers, such as *Marasmius oreades* and one or two others, were conspicuous by their absence, which is quite unusual. The not common, *Hygrophorus ovinus*, was plentiful. Six of the twelve British *Panaeoli* were collected, while only one of the forty-three species of *Lepiota* was seen. At Maltby last year nine species were noted. There was almost a total absence of *Armillaria mellea*, although the district is fairly well wooded, and this species is usually one of the most common as a tree parasite.

In the woods there was a good variety of the genus *Russula*. Over one-third of the British species were seen, but only in small quantities, except the very commonest, as *ochrolenca emetica* and *cyanoxantha*. *R. rosacea* was one of the prettiest funguses at the foray.

A fine range of *Amanitopsis vaginatus* both in size and shades of colour was brought in, a specimen of the var. *fulva* being remarkably bright. *Collybia maculata* was the commonest woodland agaric. An extra effort made by a member of one of the local societies discovered the uncommon polypore, *Strobilomyces strobilaceus*, in Storthes Hall Wood. Mr. A. Denison, of the Milnsbridge Society, brought some remarkable dried specimens of a woody agaric, which he had collected about two years ago growing on the pine-board floor of a joiner's shop at Milnsbridge. They proved to be *Lentinus suffrutescens* Fr., and are the first recorded British specimens.

The spores of *Mucor mucedo* and the conidia of *Botrytis vulgaris* and *Cladosporium herbarum* had found a decaying banana skin by the road side in Storthes Hall lane, and were increasing a millionfold.

A specimen of *Hypholoma fasciculare* was seen growing from

inside a bracken stem, a peculiar habitat: its usual home is a dead stump.

Several of the members devoted Wednesday to the examination of material gathered for microscopic species.

At no previous foray has so much local interest been taken in the work both in collecting and in the classification of the exhibits. The number of people who visited the exhibition during the week cannot be less than a couple of hundred. There were numerous drawings and photographs of fungi on exhibit, and all necessary books and appliances were provided for working out the unfamiliar specimens.

It was scarcely possible under the circumstances to record the wood in which each species was gathered, but this matters little, as the woodlands in the neighbourhood are very similar in character. They consist principally of oak, with a slightly varying admixture of sycamore, wych elm, beech, ash, and birch; in some parts cf Storthes Hall Wood the latter is prevalent, along with its undergrowth companions, bracken and the flexuous hair-grass. Had the foray been held three or four weeks later the number of specimens collected would have been much greater.

With a view to keeping the list within as narrow limits as possible, all species previously recorded from the Huddersfield area are here omitted. This plan will also serve to show at a glance the additions made to the Huddersfield fungus flora during this meeting. One species is new to Britain, and seven species and varieties new to Yorkshire. Besides these there are numerous confirmations of hitherto single records which are quite as valuable as new ones, if not more so. The peculiar *Ptychogaster albus* was met with on a decaying stump. Some authorities think this may be the conidial condition of some Polypore.

On the Monday evening Mr. Massee was to have given an address on " Modern Mycologists," but being unable to be present, Mr. Wager kindly consented to give his lecture on " Recent Researches on Reproduction in Fungi " on that evening instead of Tuesday. There was a large and appreciative audience. On Tuesday evening Mr. Gibbs detailed some interesting observations, illustrated by drawings, on a series of South African micro fungi he had been investigating. These were collected by Mr. Cheesman in 1905 during the British Association Meetings.

The business meeting was held on the Wednesday evening.

Four new members were elected, and the Halifax Scientific Society affiliated with the Union.

Votes of thanks were unanimously passed to Lord Dartmouth and the West Riding County Council for the permissions to investigate their respective estates.

The following constitute the Mycological Committee for 1907 :—G. Massee, President; C. Crossland, Sec.; Rev. W., Fowler, Harold Wager, Alfred Clarke, W. N. Cheesman, Thos. Gibbs, C. H. Broadhead, J. W. H. Johnson, and R. H. Philip.

The place of meeting for 1907 is Grassington for Grass Woods, Bolton Woods, etc., September 21st to 26th.

All the species not otherwise located are from the woods and fields in the neighbourhood of Farnley Tyas, and are marked in the following list F.T. The other local initials used are :— M. = Milnesbridge; Sl. = Slaithwaite; H. = Honley; B.B. = Berry Brow; Sk. = Skelmanthorpe; W. = Wooldale; K. = Kirkburton.

   * = First Yorkshire record.

   ** = First British record.

BASIDIOMYCETES.
(GASTROMYCETES.)

Crucibulum Tul.
C. vulgare Tul.   F.T.
  On fallen twigs.
Lycoperdon Tournf.
L. echinatum Pers.   F.T.
  In woods among dead leaves.
L. depressum Bon.   F.T.
  On the ground in a wood.
L. spadiceum Pers.
Bovista Dill.
B. nigrescens Pers.   F.T.
  In pasture.
Mutinus Fr.
M. caninus (Huds.).   F.T., Ravensthorpe.   H. Parkinson.

(HYMENOMYCETES)
AGARICACEÆ.
Agariceæ.
Leucosporæ.

Tricholoma Fr.
T. sejunctum (Sow.).   F.T.
  On the ground in woods.
T. inamoenum Fr.   F.T.
  In pasture.
Clitocybe Fr.
C. dealbata (Sow.).   F.T.
C. tumulosa (Kalchbr.).   F.T.

C. infundibuliformis (Schæff).   F.T., M.B.   In woods.
Collybia Fr.
C. confluens (Pers.).   F.T.
  Among dead leaves.
Mycena Pers.
M. pullata (B. & Cke.).   Sl.
M. alcalina Fr.   F.T.
M. epipterygea (Scop.).   F.T., Sl.
  In pastures.
M. stylobates Pers.   F.T.
  On dead herbaceous stems.
M. hiemalis (Osbk.).   F.T.
  On decaying bark.
Omphalia Fr.
O. integrella (Pers.).   F.T.
Pleurotus Fr.
P. tremulus (Schæff.).   F.T., Sl.
  On bare ground.
       Rhodosporæ.
Pluteus Fr.
P. phlebophorus (Ditm.).   Sl.
  Among rotting logwood chips.
Entoloma Fr.
E. prunuloides Fr.   F.T.
E. nidorosum Fr.   F.T.
  Both in pastures.
Clitopilus Fr.
C. carneo-albus Wither.   Sl.
       Ochrosporæ.
Pholiota Fr.
P. togularis (Bull.).   F.T.
  On the ground in a wood.

P. mutabilis (Schæff.). B.B.
On dead stump.
Hebeloma Fr.
H. sinapizans Fr. F.T.
* H. nudipes Fr. F.T.
Naucoria Fr.
N. semiorbicularis (Bull.). F.T.
In pastures.
Galera Fr.
G. tenera (Schæff.). F.T., Sl.
* Var. pilosella Pers. F.T.
Tubaria Sm.
* T. cupularis (Bull.). F.T.
Bolbitius Fr.
B. titubans (Bull.). F.T., B.B., Sk.,
M.B. In pastures.
Cortinarius Pers.
C. (Ino.) violaceus (L.). F.T.
C. (Ino.) hircinus Fr. F.T.
C, (Tela.) paleaceus Fr. F.T.
C, (Hygr.) acutus (Pers.). F.T.
All on the ground in woods.

Melanosporæ.

Stropharia Fr.
S. merdaria Fr. F.T.
On cow dung.
Hypholoma Fr.
H. pyrotrichum (Holmsk.). F.T.
Among grass.
Panæolus Fr.
P. retirugis Fr. F.T., M.B.
P. sphinctrinus. F.T.
P. fimicola Fr. F.T.
Psilocybe Fr.
P. ericæa (Pers.). F.T.
P. coprophila (Bull.). F.T.
On dry cow dung.
* P. canobrunnea Fr. F.T.
On manure heap.
Psathyra Fr.
P. corrugis (Pers.). Golcar.
In cornfield.
Coprinus Pers.
C. deliquescens (Bull.) F.T.
Among grass.
C. Gibbsii, Mass. & Crossl. F.T.
C. stercorarius (Bull.). F.T.
The preceding two on dung.
C. hemerobius Fr. F.T.

Paxilleæ.

Hygrophorus Fr.
H. Clarkii (B.&Br.) F.T.
H. ovinus (Bull.). F.T.
H. laetus Fr. F.T., Sl.
H. miniatus Fr. F.T.
H. calyptræformis Berk. Sl.
All in pastures.

Lactarieæ.

Lactarius Fr.
L. minimus W. G. Sm. Sl.

Russula Pers.
R. adusta Pers. F.T., Sl.
R. densifolia Secr. F.T.
R. semicrema Fr. Sl.
R. furcata Pers.
* Var. ochroviridis Cke. Sl.
R. sanguinea Fr. F.T.
R. rosacea Fr. M.B., F.T.
R. purpurea Gillet. (=R. *Queletii*
Fr., var. *purpurea*). F.T.
R. lactea Fr. F.T.
R. cutefracta Cke. M.B.
R. rubra DC. F.T., Sl.
R. granulosa Cke. F.T., H., W.
R. integra Fr. Sl.
R. puellaris Fr. Sl.
R. ochracea A.&S. F.T.
All the Russulæ were found on
the ground in or near woods.

Cantharelleæ.

Cantharellus Adams.
*C. Friesii Q.
On woodwork of coal-frame,
in garden, Spa Baths, Slaith-
waite.
C. infundibuliformis Fr. F.T.

Marasmieæ.

Marasmius Fr.
M. sclerotipes Bres. (*Collybia*
tuberosa). F.T.
Growing from sclerotia on dead
stump.
Lentinus Fr.
** L. suffrutescens Fr. M.B.
Growing from the floor of
joiner's shop.

POLYPORACEÆ.

Boleteæ

Boletus Dill.
B. parasiticus Bull. Mollicar Wood,
on *Scleroderma vulgare.*

Polyporeæ.

Polyporus Mich.
P. rufescens Fr. F.T.
On dead stump.
P. adustus F. Huddersfield.
On tree stump.
P. chioneus Fr. F.T.
Poria Pers.
P. mollusca Fr. F.T.
P. sanguinolenta (A.&S.). F.T.
Last two on rotting branches.

HYDNACEÆ.

Hydnum L.
H. argutum Fr. F.T.
On dead wood.

THELEPHORACEÆ.
Stereum Pers.
S. sanguinolentum Fr. F.T.
Corticium Fr.
C. calceum Fr. F.T.
Peniophora Cke.
P. velutina (Berk.). F.T.
On decaying ash wood.
Cyphella Fr.
C. capula Fr. F.T.
On dead herbaceous stems.
Thelephora Ehrh.
T. laciniata Pers. F.T., Sl.
On the ground in woods.

CLAVARIACEÆ.
Clavaria \ ahl.
C. muscoides L. F.T., B.B.
C. coralloides L. F.T., B.B.
C. umbrinella Sacc. F.T.
All three in pastures.
C. cinerea Bull. F.T.
C. cristata Holmsk. F.T., M.B.
C. rugosa Bull. F.T.
All three in woods.
C. fusiformis Sow. F.T., M.B.
* C. incarnata Weissm. F.T.
C. tenerrima Mass & Crossl. F.T.
All in pastures.
Typhula Pers.
T. erythropus Fr. F.T.
On dead herbaceous stems.
Pistillaria Fr.   .
P. tenuipes Mass. F.T.
In pastures.
P. puberula Berk. F.T.
On dead leaves in a wood.

TREMELLACEÆ.
Calocera Fr.
C. cornea Fr. F.T.
On rotting branch.

UREDINACEÆ.
Coleosporium Lév.
C. sonchi (Pers.). H., F.T.
On *Tussilago farfara.*
Uromyces Link.
U. polygoni (Pers.). F.T.
On *Polygonum aviculare.*
Puccinia Pers.
P. menthæ Newsome.
On *Mentha viridis* in garden.
P. rubigo-vera (DC.). F.T.
Uredospores on grass.
P. poarum Niels. F.T.
Æcidiospores on *Tussilago far-fara.*
P. glomerata Grev. F.T.
On *Senecio aquatica.*

ASCOMYCETES.
(PYRENOMYCETES).
Cordyceps Fr.
C. militaris (L.). F.T.
On dead beetle among grass.
Hypocrea Fr.
H. rufa (Pers.). F.T.
On dead wood.
Xylaria Hill.
X. polymorpha (Pers.). M.B.
On dead stumps.
Poronia Willd.
P. punctata Fr. F.T.
On dry horse dung in pasture.
Hypoxylon Bull.
H. rubiginosum (Pers.). F.T.
Eutypa Tul.
E. lata Tul. F.T.
On dead branches.
Sordaria C.&DeN.
S. coprophila (Fr.). F.T.
On horse dung.
S. decipiens Wint. F.T.
S. curvula DeBy. F.T.
Sporormia DeN.
S. intermedia Awd. F.T.
The last three sp. on rabbit dung.
Sphærella C.&DeN.
S. fragariæ (Tul.). F.T.
On the leaves of garden strawberry.
Sphærotheca Lév.
S. pannosa (Wallr.). F.T.
On cultivated rose bushes.
S. castagnei Lév. F.T.
On garden peas.
Erysiphe Hedw.
E. communis (Wallr.). F.T.
On *Polygonum aviculare.*

(DISCOMYCETES.)
Peziza Dill.
P. badia Pers. F.T.
On the ground in a wood.
Dasyscypha Fr.
D. virginea Fckl. F.T.
On dead herbaceous stems.
D. Soppittii Mass. F.T.
D. ciliaris Sacc. F.T.
Both on dead oak-leaves.
Ciboria Fckl.
C. luteovirescens Sacc. F.T.
On dead leaf-stalk.
Helotium Fr.
H. claroflavum Berk. F.T.
On rotting branch.
H. citrinum Fr.
Var. pallescens Mass. F.T.
On decaying wood.
H. aciculare Pers. F.T., Sl.
On moss-covered stumps.

H. cyathoideum Karst.   F.T.
 On dead plant stems.
Mollisia Fr.
M. lignicola Phil.   F.T.
 On decaying wood.
Ryparobius Boud.
R. argenteus.   F.T.
 On rabbit dung.
Ascophanus Boud.
A. minutissimus Boud.   F.T.
 On rabbit dung.
A. equinus F.T. .
 On horse and rabbit dung.
Saccobolus Boud.
S. neglectus Boud.   F.T
S. Kerverni Boud.   F.T.
 On cow and rabbit dung.

PHYCOMYCETES.
Pilobolus Tode
P. crystallinus Tode.   F.T.
P. Kleinii Van Teigh.   F.T.
 Both on cow dung.
Mucor Mich.
M. mucedo L.   F.T.
 On decaying banana skin laid
  on the road side.
Spinellus Van Teigh.
S. fusiger Van Teigh.
 On *Boletus.*
Cystopus Lév.
C. candidus Lév.   F.T.
 On shepherd's purse in garden.
Phytophthora DeBy.
P. infestans DeBy.   F.T.
 On potatoes.

DEUTEROMYCETES.
Sphæronemella Karst.
S. fimicola Marchal.   F.T.
 On rabbit dung.

HYPHOMYCETES.
Cylindrium Bon.
C. flavovirens Bon.   F.T.
 On dead oak-leaves.

Penicillium Link.
P. glaucum Link   F.T.
 On dead plant stems.
P. candidum Link.   F.T.
 On *Polyporus squamosus.*
Botrytis Mich.
B. vulgaris Fr.   F.T.
 On decaying banana skin laid
  on the road side.
B. fascicularis Sacc.   F.T.
 On dead stems of cow parsnip.
Sepedonium Link.
S. chrysospermum Fr.   F.T.
 On dead *Boletus.*
Arthrobotrys Corda.
A. superba Corda.   F.T.
 On rabbit dung.
Torula Pers.
T. herbarum Link.   F.T.
Cladosporium Link.
C. herbarum Link.   F.T.
Stilbum Tode.
S. erythrocephalum Ditm.   F.T.
 On rabbit dung.
Ægerita Pers.
Æ. candida Pers.   F.T.
Fusarium Link.
F. roseum Link.   F.T.
 On decaying plant stems.

MYXOMYCETES.
Stemonitis Gled.
S. fusca Roth.   F.T.
Trichia Haller.
T. varia Pers.   F.T.
Chondrioderma.
C. difforme Rost.   F.T.
 The last three on rotting wood.
Craterium Trent.
C. confusum Mass.   F.T.
 On dead leaves.
Fuligo Hall.
F. varians Somm.   F.T.
 On decaying wood.

Where the Forest Murmurs. Nature Essays. By **Fiona Macleod.**
London, 1906. 389 pages. Price 6/- net.

In this little volume are gathered together several articles by this well-known writer, which have appeared in *Country Life,* and which her admirers will be glad to get in a more permanent form. Undoubtedly the authoress has every sympathy with Nature, and with all that is beautiful. She is also able to express her thoughts in beautiful language. Here and there, however, the desire for 'fine' writing, with exceeding short sentences following one another in rapid succession, and the constant repetition of the same words over and over again, grows just a little wearisome. On a single page we learn 'The last enchantment of mid-winter is not yet come. . . The forest-soul is no longer an incommunicable mystery. It is abroad. It is a communicable dream. In that magnificent nakedness it knows its safety. . . It is not asleep as the poets feign.' It is 'chronic'!

THE YORKSHIRE NATURALISTS' UNION IN 1907.

At the Annual Meeting of the Yorkshire Naturalists' Union, held at York on December 15th, the following officers were elected :—

President—C. Crossland, Halifax.

Secretary—T. Sheppard, The Museum, Hull.

Divisional Secretaries—York, S. W.—H. H. Corbett, 9, Priory Place, Doncaster ; A. Whitaker, Savile House, Worsborough Bridge, Barnsley. York, Mid. W.—Riley Fortune, Lindisfarne, Dragon Road, Harrogate. York, N.W.—W. Robinson, Greenbank, Sedbergh. York, N.E.—J. J. Burton, Rosecroft, Nunthorpe, R.S.O., Yorks. York, S.E.—J. W. Stather, Brookside, Newland Park, Hull.

Sectional Officers.

Vertebrate Zoology :—

President—Riley Fortune, Harrogate.

Secretaries—T. H. Nelson, Redcar. A. White, Leeds. E. W. Wade, Hull. H. B. Booth, Shipley.

Entomology :—

President—Wm. Hewett, 12, Howard Street, York.

Secretaries (for Lepidoptera)—A. Whitaker, Barnsley ; and T. A. Lofthouse, Middlesborough. (Hymenoptera, Diptera, and Hemiptera) W. D. Roebuck, Leeds. (Neuroptera, Orthoptera, and Trichoptera) G. T. Porritt, Huddersfield. (Coleoptera) E. G. Bayford, Barnsley.

Conchology :—

President—J. E. Crowther, Elland.

Secretary—W. Denison Roebuck, Leeds.

Botany :—

President—W. G. Smith, Ph.D., Headingley.

Secretaries—H. H. Corbett, Doncaster ; J. Fraser Robinson, Hull.

Geology :

President—Cosmo Johns, Sheffield.

Secretaries—A. J. Stather, Hull ; E. Hawkesworth, Leeds.

Committees of Research.

Yorkshire Micro-Zoology and Micro-Botany Committee:—
Chairman—M. H. Stiles, Doncaster.
Convener—R. H. Philip, Hull.

Yorkshire Coleoptera Committee:—
President—M. L. Thompson, Saltburn-by-Sea.
Convener—E. G. Bayford, Barnsley.

Glacial Committee:—
Chairman—Prof. P. F. Kendall, M.Sc., Leeds.
Conveners—J. H. Howarth, J.P., Halifax; J. W. Stather,
Hull.

Yorkshire Bryological Committee:—
Chairman—M. B. Slater, Malton.
Convener—J. J. Marshall, Beverley.

Mycological Committee:—
President—G. Massee, Kew.
Convener—C. Crossland, Halifax.

Yorkshire Fossil Flora and Fauna Committee:—
Chairman and Convener—R. Kidston, F.R.S., Stirling.

Geological Photographs Committee:—
Chairman—Prof. P. F. Kendall, Leeds.
Convener—A. J. Stather, Hull.

Yorkshire Coast Erosion Committee:—
Chairman—F. F. Walton, Hull.
Convener—E. R. Matthews, Bridlington.

Yorkshire Marine Biology Committee:—
Chairman—Dr. H. C. Sorby, F.R.S., Sheffield.
Convener—Rev. E. H. Woods, B.D., Bainton Rectory, nr.
Driffield.

Wild Birds' and Eggs' Protection Committee:—
Chairman—W. H. St. Quintin, J.P., Rillington.
Conveners—R. Fortune, Harrogate; T. H. Nelson, Redcar.

Yorkshire Botanical Survey Committee:—
Chairman—Dr. T. W. Woodhead, Huddersfield.
Convener—Dr. W. G. Smith, Leeds.

Committee of Suggestions for Research:—
Chairman—Prof. P. F. Kendall, M.Sc., Leeds.
Convener—Dr. W. G. Smith, Leeds.

Hymenoptera, Diptera, and Hemiptera Committee:—
Chairman—G. T. Porritt, Huddersfield.
Convener—W. Denison Roebuck, Leeds

NOTE ON A VARIETY OF *LIMNÆA* *STAGNALIS.*

J. W. TAYLOR.
Leeds.

THE interesting specimen of *Limnæa stagnalis,* found by Mr. Hutton (see p. 61), exhibits clearly and distinctly a character which is liable to occur in almost any of our Mollusca.' Analogous specimens are occasionally found amongst the terrestrial species, but the peculiarity is much more frequently met with amongst the freshwater shells ; *Limnæa peregra* and *Physa fontinalis* being especially subject to this mode of ornamentation. White spiral banding of this kind is always adventitous, never commencing *ab ovo,* but in every case originating during the free life of the animal. It is probably an effect arising from injury to, or laceration of the outer margin of the mantle by fish or some other predatory creature, by which the secretory glandules are injured or destroyed, and as the particular glands secreting the outer or epidermal layer are the most externally placed, they are the most liable to injury and destruction.

Injury to the secretory cellules results in a deficient secretion of the protective outer epidermis, which in the Limnæidæ gives the colour to the shell, so that its absence or unusual delicacy and thinness exposes the white calcareous stratum beneath, and enables it to become strikingly perceptible and in strong contrast to the horny-brown colour of the adjacent uninjured shell surface, each injured gland or group of glands giving rise to a slender or broader white or whitish line, which revolves spirally around the whorls in strict correlation with the direction of the coiling, while the rate of increase in its breadth is in correspondence with the general enlargement of the shell itself.

If the laceration of the marginal glands is not so severe as to totally destroy, but merely tears the glandular fringe, the injury may in process of time become healed, and the white revolving lines on the shell, due to the injury, will be gradually obliterated by the overgrowth of epidermis of normal density and appearance.

If, however, the injury to the pallial margin be more serious, then the more deeply seated lime cells may also be destroyed, in such case, not merely is the outer epidermal layer involved ; but the substance of the shell itself may show its effects by the shell wall being distinctly cleft.

FIELD NOTES.

BIRDS.

The Hoopoe in Lincolnshire.—On my way from the Black Hut, which stands on the bank of Kelsey Beck, in the parish of Cadney, on December 17th, I saw a pair of Hoopoes. Though it was mid-winter, I was surprised to see that they had what I took to be fully developed crests. I found, on getting home and looking into Saunder's Manual, that it is generally after stormy weather that these birds are observed on the east coast. We had certainly experienced enough rough weather just before I saw them.—D. WOODRUFFE-PEACOCK, Cadney Vicarage, Brigg.

—: o :—

SHELLS.

Variety of *Limnæa stagnalis* found near Leeds.—The specimen figured herewith I took in a pond south-east of Leeds,

on November 23rd last. It will be seen that it has a single white narrow band in the centre of the whorl. The shell is $1\frac{5}{16}$ inches in length.—W. HARRISON HUTTON, Leeds.

—:o :—

DIPTERA.

Note on *Volucella pellucens* at Worksop.—On July 27th, 1906, whilst working amongst my insects between 9 and 10 p.m., I was surprised to see a fine specimen of this Dipteron fly into the room attracted by the gas which was burning. The night was a very hot and still one, and the windows were wide open.

Oddly enough, exactly the same thing happened just a year previously on July 28th, 1905, under exactly similar circumstances.

It seems strange that such a sun-loving species as *Volucella pellucens* should be on the wing at night.—E. MAUDE ALDERSON, Worksop, October, 1906.

At the recent annual meeting of the Lancashire and Cheshire Entomological Society, Mr. S. J. Capper was elected President, and Dr. J. H. Bailey, Mr. E. J. B. Sopp, Prof. E. B. Poulton, Messrs. J. R. Charnley, H. H. Corbett and W. Mansbridge, Vice-Presidents.

REVIEWS AND BOOK NOTICES.

Upper Nidderdale, with the Forest of Knaresborough.*
Certainly no county has such a wealth of valuable books dealing
with its history, topography, etc., as has Yorkshire. This is
largely due to the charms of the county, but is to some extent
attributable to the number of able writers which the county
has been able to boast. Amongst the living authors who have
done so much in praise of Yorkshire, Mr. Harry Speight holds
a prominent place. His most recent book, ' Upper Nidderdale,'

Old Lead Mines, Merryfield Glen.

is before us, and should do much to make this part of the
county more appreciated even than it is at present. A little
over a year ago his ' Nidderdale, from New Monkton to
Whernside,' was issued, and had a ready sale. The present
work is practically a reprint of that book, with the exception of
that part relating to the district below Knaresborough, which
is now omitted. Whilst it appeals particularly to the antiquary,

---

* Being a record of the History, Antiquities, Scenery, Old Homes,
Families, etc., of that Romantic District. By Harry Speight. Elliot
Stock, 1906. Pages 368+lxxii. Price 5/- net.

there is much in the volume of general interest, and no one visiting the neighbourhood, and anxious to learn what is to be known about it, can afford to be without 'Upper Nidderdale.' The volume is very full in so far as it relates to the history of the various old families living in the area. There are numerous illustrations, mostly of an excellent character. One of these we are kindly permitted to reproduce.

Transactions of the Hull Scientific and Field Naturalists' Club for the year 1906. Vol. III., Part IV., with Title page and Index. Edited by **Thomas Sheppard, F.G.S.** Price 2/6 net. A. Brown & Sons, Ltd., 1907. 8vo.

We have once more to congratulate our Hull friends on the production of the account of another year's excellent work, and on the part now before us, with its wealth of illustrations, which this time runs to an admirable coloured plate of the only known British Eggs of Pallas' Sand Grouse, four eggs, of which Mr. T. Audas is the proud possessor.

The part opens with a detailed account, by Mr. Sheppard, of a collection of Roman Antiquities from South Ferriby, in North Lincolnshire. The place is outside the East Riding, but only just on the other side of the Humber, and being his birthplace, the Editor is fully justified in preserving so complete an account of the interesting Roman remains, so fully described and illustrated in this paper, six plates being devoted to it.

The next paper, by John Nicholson, deals with 'Some Holderness Fighting Words,' which would convey the idea that the East Yorkshireman is a particularly quarrelsome person, so great is his wealth of fighting words. The author might have added another word, ' Snappers,' the by-name of the 15th or East Yorkshire Regiment of the Line. Pure natural history follows, Mr. Boult's pessimistic account of East Yorkshire Lepidoptera (hardly Entomology as he phrases it) in 1906, and Mr. T. Dobbs' report on East Yorkshire Conchology in 1906, are succeeded by a Bibliography and List of East Riding Hymenoptera, in which the sparse total of 23 species is recorded as against nearly 600 in the Yorkshire County List.

An excellent memoir, with portrait, of the great Hull Entomologist, William Spence, joint author of the famous Introduction to Entomology, is full of interesting detail.

Mr. R. H. Philip—a speaking likeness of whom, in conjunction with other four ex-presidents of the society, forms the frontispiece—describes the work done in Diatoms in 1906;

Mr. Arthur R. Warnes does the same for Fungi, and Mr. T. Stainforth for Coleoptera. In connection with the latter it is satisfactory to note that it is proposed to devote a cabinet in the Hull Museum to a type collection of British Coleoptera.

Mr. Sheppard's very appropriately written Presidential Address is worthy of one who combines with that office the Curatorship of the splendid museum Hull can now boast of, and who is this year honoured with one of the awards of the Geological Society of London.

Mr. J. Fraser Robinson treats of Botanical Notes of 1906, rounding off the series of local reports of investigation, and the Report of the Committee gives a record of unabated enthusiasm and well-deserved success.

We look forward for more and similarly satisfactory records of future work.—R.

NORTHERN NEWS.

In the journal of the Quekett Microscopical Club (Vol. 9, No. 57), Mr. C. D. Soar has an interesting paper entitled 'Notes and Observations on the Life-History of Fresh-water Mites."

A 'Knowledge' Book Club, in connection with 'Knowledge and Scientific News,' formed in December, was discontinued owing to lack of support in January, and the books are to be offered for sale in February.

'Ornithological Notes from Derbyshire,' for the year 1905, is one of the many interesting papers in the 'Journal of the Derbyshire Archæological and Natural History Society' for 1906, recently published.

Under the heading 'Myrmacophilous Notes for 1906,' in the December 'Entomologist's Record,' Mr. H. St. J. K. Donisthorpe enumerates several captures in Northumberland, Durham, and Cumberland. The same journal also contains particulars of 'Additions to the Coleoptera of the Northumberland and Durham district,' 1906, by Mr. R. S. Bagnall.

The recently issued 'Proceedings of the Geologists' Association' (Vol. 19, Pt. 10) contains a record of the Association's 'long excursion' to the Yorkshire Coast in 1906, illustrated by some excellent photographs by Mr. Godfrey Bingley. The same Proceedings also contain a paper on 'The Geology of the Yorkshire Coast between Redcar and Robin Hood's Bay," by Mr. R. S. Herries.

The Geological Society of London will this year award its medals and funds as follows :—The Wollaston medal to Prof. W. J. Sollas, M.A., F.R.S. ; the Murchison medal to Mr. Alfred Harker, M.A., F.R.S ; the Lyell medal to Dr. J. F. Whiteaves of Ottawa ; the Wollaston fund to Dr. Arthur Vaughan, B.A. ; the Murchison fund to Dr. Felix Oswald, B.A. ; and the Lyell Fund to Mr. T. C. Cantrill and Mr. Thomas Sheppard.

It is gratifying to find that the study of nature is spreading ! At a meeting of Suffolk guardians recently, it was reported that in consequence of workhouse fare an inmate known as the 'human hairpin' had become as 'fat as a mole.' Possibly the above paragraph will help us to understand what was meant by the item, a 'human flea,' on the list of exhibits at the recent excellent conversazione of the Doncaster Scientific Society.

SPECIAL SPRING NUMBER.
ONE SHILLING NET.

No. 602
(No. 380 of current series)

MARCH 1907.

THE NATURALIST.

A MONTHLY ILLUSTRATED JOURNAL OF

NATURAL HISTORY FOR THE NORTH OF ENGLAN

EDITED BY

T. SHEPPARD, F.G.S.,
The Museum, Hull;

AND

T. W. WOODHEAD, Ph.D., F.L.S.,
Technical College, Huddersfield.

Contents :—

LONDON :
A. Brown & Sons, Limited, 5, Farringdon Avenue, E.C.
And at Hull and York.
Printers and Publishers to the Y.N.U.

(*Continued on page 3 of Cover.*)

British Eggs of Pallas' Sand Grouse.
(Actual Size).

NOTES AND COMMENTS.

ROSEBERRY TOPPING.

WHEN the Yorkshire Naturalists' Union visited Guisborough last year (see 'Naturalist,' Nov. 1906, pages 393-394), Mr. J. J. Burton pointed out that the levels and thicknesses of several of the zones appearing on Roseberry Topping are incorrectly given by the Geological Survey. He therefore kindly prepared the

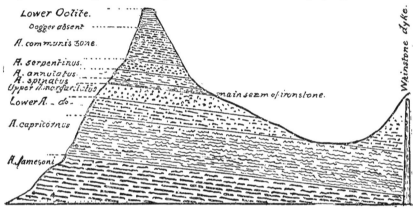

Section through Roseberry Topping, N.N.W. to S.S.E.
Horizontal scale, 880 feet to 1 inch. Vertical scale, 200 feet to 1 inch.

section, reproduced herewith, which may be taken as approximately correct. Mr. Burton is engaged in investigating this curious hill, respecting which there are a number of interesting problems, and we hope to give our readers the results of these investigations later.

BRITISH EGGS OF PALLAS' SAND-GROUSE.

By the permission of the Editor of the 'Transactions of the Hull Scientific and Field Naturalists' Club,' we are able to present our readers with a coloured plate, representing, in actual size, the only four British eggs (two clutches) of the Pallas' Sand-grouse known to exist. From a note in the Transactions just referred to, we gather that both clutches were obtained on the high wolds west of Beverley, one on June 15th, and the other on July 5th, 1888. They are by no means the only eggs laid during that extraordinary immigration of 1888, as some of the broken egg-shells were found on the fields. The eggs herewith figured (Plate XIII.) came into the possession of the late Johnson Swailes of Beverley, whose fine collection (minus these four, which are owned by Mr. T. Audas) is now in the Hull Museum.

It is gratifying to find that the sand-grouse eggs are in Hull, and are likely to remain there, though at present they are several hundred yards away from the museum !

DISTORTED STRATA IN THE LITTLE DON VALLEY.

In the recently issued 'Transactions of the Manchester Geological and Mining Society' (vol. 29, part 9), Mr. Wm. Watts gives some useful geological notes on 'Sinking Lansett and Underbank Concrete-Trenches in the Little Don Valley.' During this work many interesting geological sections

were exposed, which are reproduced in the paper by means of photographic blocks. One of these, by permission of the Institution of Mining Engineers, we are enabled to reproduce. This shows the distorted strata in the discharge-tunnel, the beds in the foreground evidently forming the arch of an anticline. An account of the author, and his work, with portrait, appears in 'Yorkshire Notes and Queries' for February.

NOTTINGHAM NATURALISTS.

The 'Fifty-fourth Annual Report and Transactions of the Nottingham Naturalists' Society' is to hand, and contains a useful account of the year's work, together with one or two notes of distinct local value. The report contains obituary notices of Prof. J. F. Blake and Prof. H. Marshall Ward, both of whom were at one time prominent members of the Nottingham Society. We are permitted to reproduce the portrait ot Prof. Blake, which will be interesting to readers of this journal,

The late Professor Blake.

having regard to his contributions to the geology of various parts of the north of England. Amongst the other items in the report are—' Weather Charts and Weather Forecasts' (the Presidential Address of Mr. H. Mellish), and ' The Stickleback : its Personal and Family History ' (with plate), by Dr. H. H. Swinerton. The Hon. Secretary of the Society, Prof. J. W.

Carr, gives a list of new Nottinghamshire spiders and false-scorpions, together with some notes on new and rare Nottinghamshire plants. Our contributor, Mr. H. Wallis Kew, describes a recent addition to the British false-scorpions.

BRADFORD NATURALISTS.

No. 11 of the *Bradford Scientific Journal* has appeared, and we are pleased to notice that with the twelfth issue there will be an index to the parts already published. In the present part is an account of the Ice Age in Wharfdale, by Mr. F. Hall, in which some sections in the drift, etc., are described ; and there are notes on Local Cockroaches by Mr. J. W. Carter ; Local Flies, by Mr. J. H. Ashworth ; and some interesting notes concerning the Nightjar by Mr. A. Badland, in which he describes the sound of the bird as 'a weird, long-drawn KrrrrrrrrrrrrOrrrrrrrrrrrr.' This paper is illustrated by a plate containing photographs by Mr. J. W. Forrest, showing the eggs, young, and the adult bird on the nest. A useful feature is the summary of natural history observations during the year 1906, which includes the various reports of the recorders of another Bradford society, the Bradford Natural History and Microscopical Society. The *Bradford Scientific Journal* gives evidences of continued enthusiasm on the part of the Bradford naturalists. There is one little point, however, that we do not quite like with regard to the journal, and that is the nature of the advertisements. That on the cover, in which we are told to drink Blank's tea, which may be blended 'scientifically,' being particularly ugly.

RECORDERS' REPORTS.

Another Bradford Society, the Bradford Natural History and Microscopical Society, has favoured us with a copy of its recorders' reports for 1906, which occupy a pamphlet of sixteen pages, and are principally reprinted from the *Bradford Scientific Journal*. Mr. J. Beanland contributes Phanerogamic Botany ; Mr. Malone, Cryptogamic Botany ; Mr. H. B. Booth, Vertebrate Zoology ; Mr. F. Rhodes, Conchology ; Mr. J. W. Carter, Lepidoptera, and a note on Local Cockroaches ; Mr. J. Ashworth, Diptera ; Microscopy by Mr. F. C. Sewell ; and Geology by Mr. J. H. Ashworth. We notice that the Society has resolved to print the whole of the records made since its formation, and that the Society's financial position warrants it. Mr. Badland, who is largely responsible for the *Bradford Scientific Journal*, is the President of the Natural History and

Microscopical Society, and this possibly accounts for the apparent co-operation of the two societies as regards printing. We do not know the local circumstances, but we should have thought that it would have been advisable for the two societies, covering a somewhat similar ground, to have amalgamated. It is quite possible that many of the members are paying subscriptions to the two societies.

'FOSSIL MUSHROOMS.'

Visitors to the well-known chalk quarries on the south Humber shore, near South Ferriby and Barton, will be familiar with the small circular 'sea-urchins' which occur in some

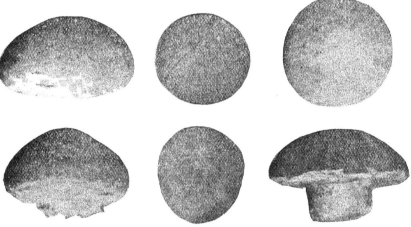

'Fossil Mushrooms.'

numbers in the lower part of the pits. They are best seen protruding, mushroom-like, from the bed of chalk immediately below the 'Black-band,' or zone of *Belemnitella plena*. This 'Black-band' is cleared away for convenience in blasting operations, leaving a shelf of exposed chalk, below which is an excellent collecting ground. The quarrymen, who secure these specimens, are evidently not believers in the deep-sea origin of the chalk; the pointed teeth occurring there, are, by them, known as 'fossil birds' tongues,' the Ammonites were once snakes, and the echinoderms were mushrooms. As 'proof' of the last, specimens are produced with the 'stalk' still adhering, this having been cut from the solid by the workmen. Two such specimens are shown on the bottom row in the above illustration, the centre specimen being *Echinoconus castanea* from another part of the pit; all the remainder are *Discoidea cylindrica*.

1907 March 1.

BRITISH TUNICATA.*

The Ray Society is well known for the excellence of the work it accomplishes in connection with its memoirs relating to the more neglected branches of natural History. In recent years these have been produced with remarkable regularity, largely as a result of the energy of the society's honorary secretary, Mr. John Hopkinson, who, it need hardly be said, is a Yorkshireman. Its latest publication is vol. 2 of 'British Tunicata,' an unfinished monograph, by the late Joshua Alder and the late Albany Hancock, which should be of particular interest to northern naturalists. Mr. Hopkinson has edited the memoir, and, with the addition of a wealth of beautiful plates, has produced a handsome and useful volume, the first part of the work having been issued two years ago. Of additional interest is an account of the life of Alder, by the Rev. Canon Norman, and the life of Hancock, by Dr. D. Embleton, with an addendum by Canon Norman. A portrait of Hancock appears as frontispiece to the volume.

THE ILLUSTRATIONS.

To the plates, too much praise cannot be given; they are such as will be of the very greatest service to students of this neglected order. In addition there are numerous illustrations in the text. Both sets of illustrations are photographically reproduced from the drawings of Alder and Hancock, under the supervision of Mr. Hopkinson. By the courtesy of the society and its editor, we are able to present our readers with one of the plates (plate XV.), which will speak more for their excellence than will any words of ours. Some of the specimens will probably be familiar to those who are in the habit of collecting amongst the flotsam and jetsam of our coasts. On the plate—Figs. 1-4 = *Styelopsis grossularia* (Van Ben.) Traust. : 1, a group, natural size ; 2, a single individual from this group, twice natural size ; 3, an individual, probably a variety of this species, with a young one attached, three times natural size ; 4, the same, natural size. Fig. 5 = *Styelopsis glomerata* (Alder): a cluster, natural size. Figs. 6-8 = *Thylacium aggregatum* (Rathke) V. Carus: 6, a group, one half natural size ; 7, a single individual, natural size ; 8, a group, probably a variety of this species, of the size and form of var. *maculatum*,

---

* xxviii. + 164 pp. and 50 plates, 1907. Issued to the members of the Ray Society for 1906. 25/- net.

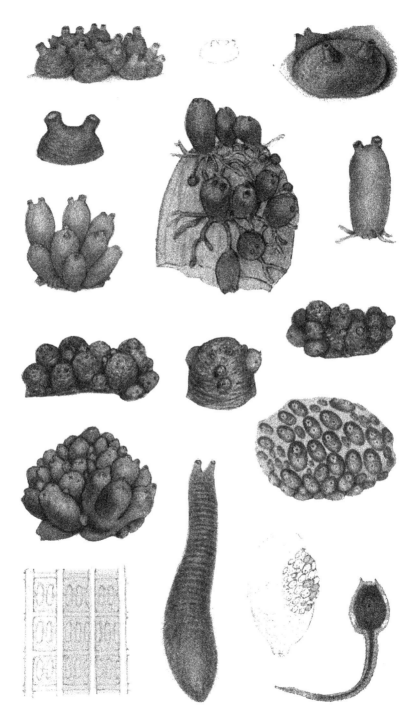

British Tunicata

but apparently not spotted, natural size. Figs. 9-11 = *Thyla-cium Sylvani* V. Carus : 9, a group of small ones ; 10, a group of rather larger ones ; 11, a single individual, much larger ; all with young ones attached, and natural size. Figs. 12-14 = *Thylacium variolosum* (Gaertn.) : 12, a mass, twice natural size; 13, showing mode of growth on an Ascidian, natural size ; 14, a larva, magnified. Figs. 15 and 16 = *Pelonaia corrugata* Forb. and Goods. : 15, Test, natural size ; 16, part of branchial sac, magnified.

THE SLAUGHTER OF KINGFISHERS.

We heartily agree with the following note which appears in a local magazine in Hull, and can only regret that Hull is not the only place where this senseless slaughter is allowed. When will the police authorities understand that their duties do not end in seeing that the placards are duly posted in more or less prominent places in their districts ? The attention of the police has been called to this case. We have also been advised of similar slaughter of kingfishers near Huddersfield. As we have previously pointed out, however, in Yorkshire at any rate, the ' Wild Birds' Protection Act,' so far as the police are concerned, is very largely a farce. ' Hull is not a lovely city. The objects of beauty are not very many, and the praiseworthy endeavours of some of its citizens to multiply them, only make slow progress. But sometimes, even within the boundaries of its smoky fields, nature is kind. Flowers still open in summer, and sometimes from brighter scenes come birds which bring us a message of grace. So it has been this winter in the parish of Newland. Wanderers in our fields might see the glorious vision of the little kingfisher, with its tropical luxuriance of colour, winging his way along the waterside, for just now he finds more food here than inland. Such a thing should be the pride of a great city, better than an art gallery in himself. Alas for our civilisation ! This is the barbarian's chance. Parliament has passed the Wild Birds' Protection Act, the Hull Corporation and the East Riding County Council have scheduled the kingfisher for protection all the year round, what does it matter ? Men go out with their guns and slaughter every one they can find. We have reason to believe that no fewer than ten of these birds have been killed in this way during the last few weeks. Such a thing is a disgrace to the community, and anyone who could give evidence for the conviction of the perpetrator of it would be a public benefactor.'

1907 March 1.

A YORKSHIRE VARIETY OF A RARE BRITISH TARDIGRADE.

G. S. WEST, M.A., F.L.S.

Birmingham University.

DURING the recent examination of some old algal material from Penyghent, W. Yorkshire, I came across a dilapidated specimen of a Tardigrade, the like of which I had no recollection of having seen before. Further search through the material fortunately resulted in the finding of several individuals in a much better state of preservation.

The animal appears to be a variety of *Macrobiotus papillifer*

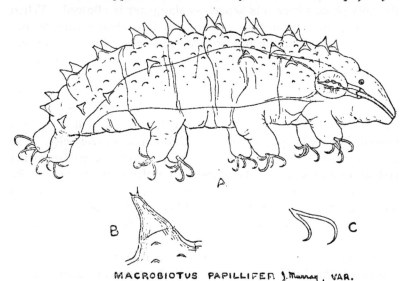

MACROBIOTUS PAPILLIFER J. Murray, VAR.

A.—Animal seen from the side, x 500.
B.—Single dorsal process showing the two minute apical spines, x 1000.
C.—Pair of claws, x 1000.

James Murray ('The Tardigrada of the Scottish Lochs,' *Trans. Royal Society, Edinburgh*, vol. xli., part iii., No. 27, 1905, p. 692, t. iii., f. 15 *a-c*). The specific name '*papillifer*' is in reference to the conical papillæ or processes along the dorsal surface and sides of the body. These projections are disposed in transverse and longitudinal rows, each one having an acuminate apex.

The Yorkshire animals were collected in April, 1896, and occurred among mosses, and the submerged portions of sedges and rushes in a small rivulet on the western slope of Penyghent at an altitude of about 1400 feet. They were of rather small dimensions, having a maximum length of $\frac{1}{133}$ - inch ($=$ 190 μ). They do not agree very strictly with the Scottish specimens, differing chiefly in the fewer processes along the sides of the body and in the presence of numerous small rounded warts or tubercles between the processes. These small tubercles are much scattered, but cover the whole dorsal surface and upper parts of the sides of the body not occupied by the processes, diminishing in size, and ultimately disappearing towards the ventral region (fig. A). The five median pairs of processes, which form part of the double series which extends down the back of the animal, were not acuminate, but slightly truncate at the apex, and furnished with two minute spines (*vide* figs. A and B). The cuticle of the processes was also noticed to be much thinner than the rest of the body-cuticle (fig. B).

Each leg was furnished with two equal pairs of strongly-hooked claws, one claw of each pair being somewhat larger than the other (fig. C).

The pharynx and teeth differed in no respects from the Scottish specimens described by Mr. James Murray. No eggs were seen.

Macrobiotus papillifer has only previously been observed from Loch Ness and Loch Morar in Scotland. It is closely related to *M. tuberculatus* Plate.

———◆◆———

FLOWERING PLANTS.

The montane form of *Myosotis sylvatica*.—Referring to Mr. Pickard's *Botanical and other Notes at Arncliffe* in the December number of the 'Naturalist' (pp. 425-8), the small montane form of *Myosotis sylvatica* therein mentioned was, I believe, first placed on record by myself as Phanerogamic Secretary to the Botanical Sect. of the Yorks. Nat. Union in the 1883 Report (see 'Trans. Miscellaneous Botanical Papers,' Vol. I., p. 204).

This form is given in 'Speight's Craven Highlands' as one of the noticeable plants of the Malham district, doubtless on the authority of the above report. I was climbing Gordale Scar one sunny day in June, 1883, and about half way up my attention was arrested by a patch of blue colour on a shelving bank. This I found to be a much dwarfed mass of *Myosotis* in full bloom, and I at once thought of *M. alpestris*, the true alpine of Mickle Fell. On examination, however, the plants proved to be but the montane form of *M sylvatica*.—P. Fox Lee, Dewsbury.

NOTES ON THE BREEDING HABITS OF BATS.

ARTHUR WHITAKER,
Worsbrough Bridge, Barnsley.

In the 'Naturalist' for November, 1905 (pp. 325-330), I gave a few notes on this subject relating principally to the Noctule Bat (*Pterygister noctula*), and expressed a hope that I might be able to supplement them in the future. This I am now in a position to do.

I entered upon 1906 with the firm intention of rearing some of our British Bats in captivity. I find, however, that the difficulties of doing this are very great, and once again have to record partial failure, though I have succeeded in carrying my observations a degree further than was the case last year. As this article deals with a number of incomplete observations, it will facilitate matters to enumerate these at once, and therefore I here give a table of the instances which have come under my personal observation, during the past few years, of the breeding of bats.

(1). Pipistrelle.
    (*Pipistrellus pipistrellus*). Specimen accidently killed during last week of July, 1901, contained a fully developed embryo, which would probably have been extruded within forty-eight hours.

(2).   ditto   Specimen netted May 28th, 1905 (confined along with two males and two other females). Died on July 14th, 1905, and was then found to contain a very small embryo, probably not more than half developed.

(3).   ditto   Specimen caught June 30th, 1906, gave birth to a single young one, July 2nd, 1906.

(4).   ditto   Specimen caught June 7th, 1906, in the New Forest (confined along with one other female) gave birth to a young male on July 18th, 1906.

(5).   ditto   Specimen caught at Barnsley June 16th, 1906 (confined along with last mentioned female) gave birth to a male on July 19th, 1906.

(6). Noctule.
    (*Pterygister noctula*). Naked young male about a week old, taken from hole in a tree occupied by a colony of adults, Barnsley, June 29th, 1905.

(7).   ditto   Adult female taken from above-mentioned colony gave birth to a young male, June 30th, 1905.

(8). Lesser Horseshoe.
    (*Rhinolophus hipposiderus*). Specimen taken Wells (Somerset), July 20th, 1906. Died July 24th, 1906. Found to contain a fully developed embryo ready for extrusion.

(9). Daubenton's.
    (*Myotis daubentoni*). Specimen taken Barnsley, June 19th, 1906. Gave birth to a young female same night.

In every one of the nine instances above mentioned, it will be noticed that a single young one only was (or would have been) produced. This substantially confirms the statement often made that in this country it is a rare occurrence for a bat to give birth to ' twins,' as they are affirmed to do on the continent.

The dates given above indicate that July is the month when the majority of bats are born, though it will be noticed in the cases of Daubenton's and the two Noctules, the latter part of June was the time. In most cases, I should think, the young will not be able to fly by themselves until the latter part of August, but this is more or less conjecture.

One of the most interesting conclusions to be drawn from the above table (see No. 4) is that in the case of the Pipistrelle the period of gestation is not less than forty-one days. Although the exact period is not ascertained, it is probably about six weeks, and this agrees with the only other information we seem to have on this point, that given by Mr. G. Daniell (Proc. Zool. Soc. 1834), who ascertained that in the case of the Noctule the period of gestation exceeded thirty-eight days. Further and more definite information on this point is wanted, however.

Of Nos. (1), (2), (6), and (7) on the above list a full account was given in my previous article, and nothing further need be said about them.

The Pipistrelle referred to in No. (3) was never in my own possession, but was caught by a lad in Barnsley, who managed to knock it down with his cap on the evening of June 30th, whilst it was flying. He told me that the bat gave birth to a young one sometime during the second night after he obtained it. He also stated that on the day following the birth of the young one the mother escaped from him whilst he was playing with it in the daytime, and crawled up the house wall out of his reach, creeping behind the spout, where it remained hidden during the day, the newly-born young one being beneath its wing at the time. At dusk it came out and commenced to fly about the yard, but it flew so heavily and slowly that he re-captured it again without difficulty. The following day it again made good its escape from a rabbit hutch in which he had placed it, but he could see it squeezed in its former position behind the spout, and he promised me he would re-capture it at dusk and let me have it. Unfortunately, it rained heavily, and the bat did not move at all in the evening, but the following morning it was missing, and no further traces of it were seen.

The Daubenton's Bat referred to (No. 9 on the list) was one of a party of four, which were feeding at dusk, on June 18th, 1906, over the surface of the Serpentine in Stainbrough Park. I succeeded in netting it as it approached near to the side. This was at about ten o'clock at night.

The following morning, at about eight o'clock, I found it hanging in the corner of the cage in which I had placed it the previous evening, and under its right wing was a 'baby,' to which it had given birth during the night. The young one was of a dark purplish flesh colour, the wing membrane being only very slightly darker in colour than the skin of the rest of the body. It was blind, and naked save for a few fine, straggling hairs on the muzzle. It clung tenaciously to its mother during the whole time it lived (except when I separated them to photograph), and was calling continually with a very soft, 'sucking' kind of chirrup, scarcely audible at a distance of a few feet, and very diffierent indeed from the loud, deliberately repeated call of a baby Noctule.

The mother quite failed to give her young one any attention, and when, at dusk of the night following her capture, I endeavoured to feed her, she obstinately refused to take the least particle of food or drink. Upon examination, I found the reason for this was that her tongue was inflamed and much swollen, from what cause I do not know. The bat was evidently in a 'bad way,' and I found I could do nothing for it, and consequently was not surprised, though much disappointed, to find next day that both the young bat and its mother were dead.

The relative sizes of the young female and its mother (given in inches and decimals of inches), measured carefully immediately after death, were as follows :—

Length of head and body ...   Adult 1.70   Immature 1.15
Length of tail ....   Adult 1.3   Immature .6
Expanse of wings ....   Adult 9.5   Immature 3.3

On Plate X. are reproduced four photographs of the bat with its one-day-old youngster by its side, taken by my friend Mr. Wakefield.

The most interesting observations which I have been able to make during the past season were made in connection with two Pipistrelles referred to on the foregoing list as Nos. 4 and 5.

During the early part of June I took a short holiday in the New Forest, Hampshire, and on the day previous to my return I met a gamekeeper, who, in answer to inquiries of mine, informed me of a colony of bats that occupied the roof of his cottage, which was situated in the middle of the forest, and to which I arranged to pay a visit that evening.

Arriving there about half-an-hour before dusk, I was somewhat disappointed to find that, although the roof was evidently inhabited by a very large colony of bats, as their squeakings plainly testified, the fact that it was an old tiled roof, affording numberless openings for the egress of its occupants, would inevitably prevent me from securing many specimens with the single small butterfly net, which was all I had to capture them with. On the keeper's advice, I decided to try the back slope of the roof, from which he said most of the bats emerged, but I was again disappointed to find that the only available ladder was so short that I had to stand on its topmost rung (which necessitated my holding on to the spout with one hand) in order to reach to the roof ; in addition to this, I had to splice my net to a broom handle before it would reach to the exit holes, which were all in the vicinity of the ridge. On the whole, under these adverse circumstances, I was not at all surprised at only securing three specimens, which were all that emerged from the hole I had elected to guard. From other holes in the roof the bats, which were all small in size, came pouring in little streams, from about fifteen minutes after sunset to some three quarters of an hour later, but for long after that time the noisy squeaking proceeding from the inside of the roof proclaimed the fact that some individuals had not come out. Apparently they had no intention of doing so, for I heard them squeaking inside fully an hour and a half after the bats had quite ceased to issue from the roof. If I were to make a guess at the strength of the colony, I should put it down at between three and four hundred, and I think the bats were probably all Pipistrelles ; certainly most of them were. The three specimens I obtained were all females of this species.

The following day I returned to Yorkshire, and, owing to an unfortunate accident, the label got scrubbed off the trunk in which these bats were packed, and the box was consequently missing when I reached Sheffield. That was on the Friday afternoon. In spite of the fact that every effort was made to recover the box, I did not receive it until the following Wednesday evening, and I was consequently not at all surprised to find the bats in the last stage of exhaustion. One or two Whiskered Bats, which I had also obtained in the Forest, were, in fact, dead on arrival, and although the three Pipistrelles lived for a short time, I only managed to nurse one of them round into proper health again.

On June 16th I netted a female Pipistrelle as it was flying at dusk by Worsbrough Reservoir, and this I placed in the same cage as the female still living, of whose capture I have just given an account.

These two bats were both fed exclusively on mealworms, which they would pick up for themselves from the floor of their cage after the first week of captivity. The New Forest femele was, for the first two months, in much better health than her companion, and would consume nearly twice the quantity of food, managing to dispose on the average of about forty meal-worms per day, a greater quantity than I have known any other individual of this species to consume.

Sometime between 9 a.m. and 4 p.m. on July 18th, forty-one days after its capture, the New Forest Pipistrelle gave birth to a young male. On looking into her cage at the latter hour, I saw her crawling about on the bottom with the youngster under her right wing.

The following day, sometime in the afternoon, the other female gave birth to a young one, also a male.

Both the young Pipistrelles at birth were of a dull flesh colour, blind, and naked save for a few slight hairs on the muzzle. The wing membranes and ears were decidedly darker than the rest of the skin in colour. They were very small at first, so small, that when tucked under the maternal wing their presence would not be detected unless one were either looking for them or happened to notice an occasional suspicious undul-ation of the membrane of the mother's wing near her shoulder, as the little one squirmed about underneath.

The difference in the rate of growth between these two young bats was really remarkable, and undoubtedly due to the fact that at birth, and for some time afterwards, the New Forest parent was in much better health than the Worsbrough one. The baby first born began to grow darker in colour day by day, especially the wings, ears, interfemoral membrane, muzzle, etc. I first saw its eyes open on the eighth day after birth, but it did not seem to use them much, for it would only open them when handled, and often crawled about with them closed as long as it lived.

The fur began to show at the end of the first week, and as it became more noticable, imparted a silky, golden apperance to the back, and a more silvery one to the chin and breast, and was especially noticeable if the bat was viewed sideways. The skin of the bat, as well as the wings, etc., grew steadily darker and darker in colour, and at the end of three weeks it was almost black, except on the belly, which was very dark purplish flesh colour (see Plate XI). The hair first commenced to grow on the shoulders and back, then on the head and chin, and lastly on the breast. The belly was still almost naked at the time of its death, which occurred when it was thirty-three days old.

The other baby bat which, it will be remembered, was only born one day later, lived until it was forty-three days old, but it hardly grew at all, and at the time of its death was blind, naked, and almost unchanged from the day it was born. That it should survive so long when evidently deriving insufficient nutriment from its mother, who was in bad health during the whole of the time, is a wonderful indication of the extraordinary amount of vitality possessed by these creatures. Both mothers were fed almost exclusively on mealworms during the whole of the time, so that it was evidently not in any way a question of diet.

This great difference in the rate of growth between the two young ones in captivity indicates that much depends upon the state of the parent's health, and it is quite possible that, in a state of nature, growth may be even more rapid than it was in the case of the more healthy of these two young ones.

I found the association between the young bats and their parents not nearly so close as I had anticipated. When they were but a few days old, I not infrequently found them hanging quite alone several inches away from their respective mothers, and this applies to the youngster which was quite well and growing fast at the time, as well as to the one which was certainly from the first somewhat neglected by its mother. After they were a couple of weeks old I often found them at the opposite side of the cage to their mothers. This separation did not seem to cause the young ones any trouble or uneasiness, for they would sleep thus contentedly for many hours without showing any signs of anxiety. Often I should find the two young ones asleep touching one another.

When touched or disturbed, or when wanting their mothers, they would lift themselves well up on the wrists, and raising the head very high, turn it anxiously about from side to side, uttering a deliberate chirrup resembling the soft smacking of one's lips, which was very faint when the creatures were young, but steadily grew in power as the days went by. This noise

was made by the bat with widely open mouth ; and after calling
for a time, it would set off on a searching expedition for the
mother, crawling slowly, but with a firm grip of anything to
which it could cling, and keeping up the search with great
perseverance until it was successfully ended. The youngsters
seemed to recognise their own mothers easily, and would take
little or no notice of the wrong parent. When one was looking
for its mother, and got near to her, it would grab hold with its
mouth of any part of her it could catch. If the mother happened
to be busy feeding at the time, she would often take no notice
of it, but drag it carelessly about with her, whilst it clung with
a kind of dogged perseverance to the fur of her back, her
interfemoral membrane, or any part of her which it happened to
have got hold. Whenever the adult bat paused, the youngster
would try to improve its grip, and work into a safer and more
comfortable position, and eventually it would manage to squirm
either over the mother's shoulder or under her interfemoral
membrane, and so get under her wing ; when this was the case,
the mother would nearly always bend her head under her wing,
and apparently tuck the young one into a mutually comfortable
position, and at such times the young one could be heard
making a very soft, but rather musical, twittering.

After the mothers had fed, they always used to suckle their
respective young ones immediately. If 'baby' were under the
wing of one of them when she came out to feed, it did not seem
to hamper her movements seriously until it came to 'pouching'
a mealworm, but this operation—always difficult for a bat to
perform on *terra firma*—the presence of the young one seemed
to render ten times more difficult.

In case some readers of this article are unaware of what is
meant by the term 'pouching,' a short digression may be
pardoned, in order to make the term clear.

Bats secure the insects they feed upon in a natural state
whilst they are on the wing, and as the 'gape' of a bat is
comparatively small, and many of the insects fairly large and
strong, it is not the easiest matter for a bat to secure a firm
grip of its prey until its struggles are overcome. To avoid the
risk of losing their captures owing to this difficulty, these
creatures have acquired a curious habit. When they have
seized an insect, and whilst they are still flying, they bend the
hind legs and tail forward under the body, and then bend their
heads down into the bag of skin thus formed by the membrane
connecting the legs, tail, and wings. The insect is then in a

Daubenton's Bat and Young one.

FIG. 1.—One day old.

FIG. 2.—Seven days old.

FIG. 3.—Fourteen days old.

FIG. 4.—Fourteen days old.

FIG. 5.—Twenty-one days old.

FIG. 6.—Twenty-one days old.

Young of Pipistrelle or Common Bat.

kind of trap, and is pressed against the interfemoral membrane, as this portion of the skin is termed, until the bat has overcome its struggles and secured a good grip of it. Excellently though this little manœuvre works whilst the creatures are flying (as is probably always the case in a natural state), it will at once be seen that there are difficulties in the way when the bat is on the ground; and in captivity, its food is, of course, always given to it as it crawls about on the floor of its cage. Under these altered circumstances it is *standing* upon its feet, and when it endeavours to pouch an insect, it curves its tail under the body, raises itself upon its wrists, brings the feet as far forward as it dare, tucks down the head, and then generally discovers too late, as it goes topping over on its back, that it is unable to go through the performance without losing its balance. Bats are soon 'rigwelted,' and its prey is almost always dropped in its struggle to regain a normal position. After a bat freshly introduced to captivity has done this a few times, it seems to fully realise the difficulties of its new life, and its excitement and annoyance become greater every time the performance is repeated. So that usually, after a short time, every mealworm given to a bat results in the little creature losing its centre of gravity, and the observer losing his gravity altogether. After a few weeks of captivity they become a little more expert; but feeding from the ground is always a difficult matter to some species, especially the Pipistrelle.

Now, my two female Pipistrelles, which had young ones, had been feeding themselves from mealworms, which I simply threw into the cage for them, so long that I never anticipated they would require assistance; but after a time I noticed that they seemed not to consume much of the food that I put in for them, and both bats were getting into a rather weak condition. This was about a month after the birth of the young ones, and from that time I paid a great deal more attention to them, and found that the reason of their illness was difficulty in feeding, owing to the young ones often clinging to them and hampering their movements, and so preventing them from getting sufficient food. I was careful after finding this out to feed them again by hand, and also to make them take at least thirty minutes exercise every day, flying round the room *without* their young ones. By these means I managed to bring them both back again into good health. But their illness resulted in their giving insufficient nutriment and attention to their offspring, so that I was bitterly disappointed, just when I was beginning to congratulate myself that I should safely rear the young ones, to lose them both. The one which was most advanced died first, on the 21st of August, at the age of thirty-three days, and the other lingered on without developing until Sept. 2nd, when it also died.

At the time of death the former of these immature bats was carefully examined and measured, and its size, compared with that of an average adult, was as follows : —

	Adult.	Young.
Head and body......	1.55	Aged 4 weeks, 1.28
Tail....................	1.11	,, .62
Wing expanse	8.50	3.64
Ear (length)	.42	.28
Tragus (length) ...	·19	
Tibia	.50	·35
Forearm	1.21	,, .65

The above comparative measurements should convey a good idea of the relative sizes of the two ; but it will be noticed that I give the age of young at four weeks. The specimen my measurements were taken from was, it is true, nearly five weeks old ; but seeing that during the last week of its life it had not developed, much allowance should be made, and probably a young one in good health would easily attain the development indicated in four week's time.

Many of the most interesting points still remain unsolved, and require careful observations in the future. During the thirty-three days through which the most interesting of these two young Pipistrelles was in my keeping it subsisted entirely on its mother's milk, and took no solid food. It became strong enough to crawl quite briskly about, but showed no sign of any inclination towards flight, save that during the last week or ten days I observed it frequently open and stretch its wings, and once or twice beat them in the air in a 'flipperlike' manner, but only one at a time. When young birds begin to do this, it is a sure indication that they will soon make more serious attempts at flight, and doubtless the analogy holds goods with regard to bats.

I several times allowed my adult Pipistrelles to fly in the same room, with the young ones attached, when the proceeding was always the same. The mother would crawl up my jacket sleeve with the young one under her wing, and then, even when realising that she was at liberty to go, would hesitate for some time, turning her head about from side to side. After this she

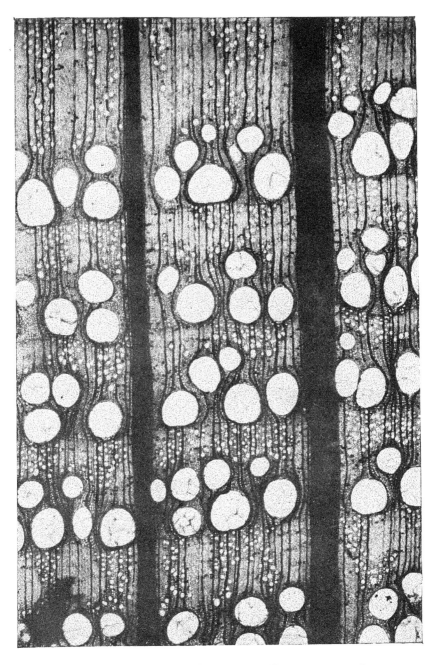

Transverse Section of Wood of Oak (x 30 diameter). (See p. 83.)

would extend both wings fully and flatly, and lift them gently up and down six or eight times, evidently to feel that 'baby' was not clinging in a position which would incur risk to either of them ; often she would not be quite satisfied, but would tuck her head down, and move it slightly. After these preliminaries had been gone through, she would launch off, but seldom flew more than five or six lengths of the room, and that with a slow, straight, and heavy flight. It was so obviously hard work for her that I did not often encourage the proceedings, especially as she often had difficulty in turning after flying the length of the room, several times being unable to do so, and colliding with the wall and falling to the ground, which I feared might result in injury to the young one. Sometimes, also, she would fly round the room four or five times, gradually falling towards the floor, like a bat only half awake.

When the young ones were very small, however, I once or twice got their mothers to fly pretty briskly with them, and as this was in the daytime, I had a good view of the procedure, and found that the 'baby' clung to the mother's nipple with its mouth, and allowing its body to hang quite down, merely brought up its feet and clung to her fur with them also. In this position the youngster was quite conspicuous, as its mother flew about, at close quarters, in the daylight, hanging down like a little ball attached to her breast.

Mr. Wakefield very kindly took photographs of one of my young Pipistrelles week by week, and a series of these are reproduced on Plate XI., showing it at one, seven, fourteen, and twenty-one days old. The gradual increase in size and darkening in colour is well shown in these photographs ; but the hairiness which should be slightly apparent even in fig. 2 does not become noticeable until fig. 5.

———◆◆———

Familiar Trees. By **Prof. G. S. Boulger.** Cassell & Co., London.
Messrs. Cassell are bringing out, in fortnightly parts at 6d. net each, Prof. Boulger's 'Familiar Trees.' The present edition, which has been revised throughout and enlarged, is being produced in 29 parts, several of which have already been issued. The author's name above is a guarantee for the accuracy of the descriptions given. Each species is carefully described and figured, the photographs of the characteristic trees of each type being all that can be desired. There are also no fewer than 114 coloured plates, mostly admirably done, illustrating the fruit, flowers, foliage, or the entire tree. Of great value also are the 114 full-page plates, showing the structure of the different woods, from micro-photographs. One of these the publishers have kindly enabled us to reproduce (see plate XVIII.). The work should have a very large sale, and will doubtless do much to popularise the study of the trees which form so picturesque and prominent a feature in almost every British landscape.

1907 March 1.

NOTES ON *CHRYSOPA PERLA* AND *C. FLAVA.*

By E. MAUDE ALDERSON, F.E.S.,
Worksop.

IN the following notes I have endeavoured to compile a summary of the observations which I jotted down in my entomological diary during the breeding of the above two species of *Chrysopidæ.*

The most interesting points seem to me to be the existence of what I have regarded, perhaps incorrectly, as a sub-imaginal stage in the life history of the genus, and also the presence in each species of some very characteristic markings on the head. These are so striking and so diverse (in the species which I have seen and in the few drawings I have been able to find in books) that I should think they must form a very easy, if superficial, means of identification of species in the larval stage.

The presence of the sub-imaginal stage, if correct, seems to form a strong link between this group and the *Ephemeridæ.*

C. perla.—On June 29th of last year (1905) I obtained a ♀ of *Chrysopa perla,* and found, on opening the chip box in which I had placed her, that she had deposited several ova. She remained alive until July 1st, and at the end of that time had laid about a dozen ova. Each ovum was deposited separately. They were placed at random all over the box; and of a bright, shining bluish-green colour, almost exactly the shade of the green parts of the body of the parent. The ova were on long foot-stalks, 5-6 mm. in length, and were very beautiful objects.

As I was leaving home within the next few days, I took the ova with me in order to observe them closely. By July 3rd they had lost their brilliance, and assumed a greyish tinge. This gradually deepened until the period of hatching, when they appeared wholly grey, the change in colour being evidently due to the young larvæ showing through the transparent shells. The young larvæ all emerged after just a week—July 7th-8th. They were grey in colour, and soon became very active. I fed them with 'green fly,' and they grew surprisingly quickly. I could not observe any change of skin, but at the rate at which they increased, I should think they must have moulted two or three times at least.

The young larvæ were most active at night; and as I kept

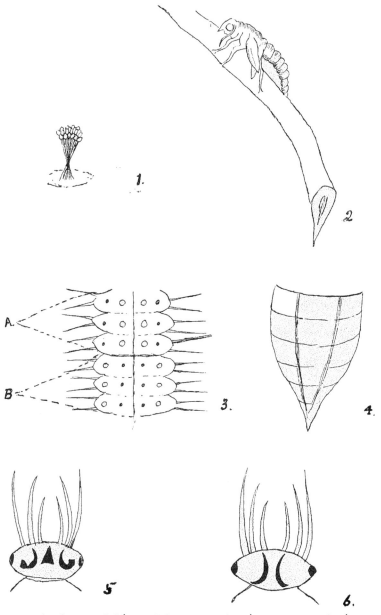

1. Ova of C. flava. 2. Sub-imago C. flava. 3. A. Thoracic ⎫ Segments—showing warts
                                                 B. Abdominal ⎭ C. perla--larva.
4. Abdomen—underside—C. perla. 5. Head of C. perla. 6. Head of C. flava.

them in a small tin box with a glass lid, it was most interesting to watch them hunt their prey. They seemed most voracious little creatures, and would move restlessly about at a great pace until they met with an aphis. They did not appear to be guided by sight in finding food, but rather by sense of feeling. When they came across an aphis they would strike with great rapidity, fixing their sucking spears into the body of the unfortunate victim, which never, so far as I could see, would offer the slightest resistance. It was a wonderful sight to see the subsequent proceedings. Both insects would remain perfectly still, and by the aid of an ordinary lens it was quite possible to see the vital juices of the aphis pass through the sucking spears into the body of the *Chrysopa*, until in a very short time the one became an empty sack of skin and the other a full-fed gourmand. After a short rest the same process would be repeated, many times a day, and also during the night, for I frequently found the larvæ still feeding when I looked at them after dark. I think it probable, too, that they sometimes mistook one another for aphides, as their numbers certainly decreased, and they seemed to have little powers of discrimination. In the case of *C. flava*, one larva was actually seen devouring a larva of *Dictyopteryx bergmanniana*, which had been introduced with the rose leaves. The larvæ became much more sluggish as they grew older, probably owing to their increased bulk, as they grew with great rapidity.

The following is a rough description at seven days old :—

Head.—Pale greenish, with three characteristic jet-black marks.

Antennæ.—Madder brown.

Sucking Spears.—Pale madder brown ; darker at the tips.

Eyes.—Jet-black.

Thorax.—Whitish green at sides and beneath, brown madder above ; two rather crescent-shaped marks at sides, jet black.

A black dividing line runs all down the centre of the back, and on each segment are placed four warts (two larger and two smaller), two on each side of the dividing line.

The thoracic warts are *black*, and the *larger* ones are placed nearer to the central line.

The abdominal warts are *whitish green*, and the positions are reversed, the smaller one being placed next to the dividing line.

The underneath parts of the larvæ are whitish green, with two madder lines running down the abdomen.

The larvæ pupated July 18th-19th, the larval stage lasting just ten days. The cocoons were of white silk, like small pellets of cotton wool, about 4 mm. × 3 mm. One spun up on the muslin of the cover, the others amongst the debris at the bottom of the jar. I kept the jar in the house during the whole of the winter months, and on May 24th, 1906, I found that one imago had emerged. Unfortunately I missed seeing the actual emergence, and with the exception of one crippled specimen, which I found afterwards, no further emergences took place.

C. flava.—On July 14th of the same year I obtained ova from a wild *C. flava.* The eggs were laid in a group on the lid of the box, the foot-stalks being united by their middles into a bundle. The ovum was elliptical and of a bright green —the exact colour of the body of the parent. They began to hatch on July 20th, emergence apparently taking place from the apex of the ovum. There were thirty-nine ova in the group, but only a small proportion of these, some eight or nine, emerged. The day before hatching two conspicuous brown spots appeared, one on each side of the ovum, which were evidently the eyes of the embryo. The eggs did not change colour nearly so much as in the case of the *C. perla.* The apex of each egg became a yellowish green shortly before the brown spots appeared, and both ends and sides became transparent as the embryo became more fully formed. After hatching, the larvæ remained perfectly inactive on the ova, and continued so for some time. I could not discover what they did during this period, but I do not think that they devoured their egg shells. On touching them with a camel's hair brush they at once became very active, and once removed from the empty eggs they did not return to them, but ran about very quickly.

The larva is about 2 mm. in length, of a shining transparent white, of a pearly lustre; the eyes large and black; the sucking spears and legs white. The most striking feature in the appearance of this larva, in all stages of growth, is that it appears divided into three portions : (1) the thoracic segments ; (2) first four abdominal segments ; (3) last five abdominal segments. When very young (1) appears blood-red ; (2) much darker red, owing, I suppose, to the intestines showing through the transparent skin ; (3) wholly transparent and of a yellowish tinge. After a few days the distinguishing marks on the head began to appear. They are black, the same as the eyes, and are rather like two crescents, placed back to back.

There are also two similar lines, one on each side of the head, running to the eyes. The tibiæ appear to be fuscous in shade, and give the legs the appearance of having dark bands round them. I am unable to say how many changes of skin are effected, but I should think, judging from the rate of growth, about four or five at the least. At five days old some of them appeared to be in their third skin. They are extremely active larvæ, and use the last segment of the body as claspers to walk with.

Rough description at ten or eleven days old :—

Head.—Yellowish. (The characteristic marks had disappeared).

Eyes.—Black.

Antennæ and Sucking Spears.—Madder brown.

Body.—Yellowish white ; central line and markings crimson ; warts on thorax black ; side lines crimson.

The claspers at the extreme end of the abdomen seemed very prehensile. Not only did the larvæ walk by means of them, but they also used them as a means of attachment to some substance on a change of skin. During this process the larva hangs head downwards, attached by the tail, and by repeated efforts gradually frees itself from its old skin, the feet being disengaged last. In one instance, through my interference, the larva became detached, and for a whole day rolled about helplessly at the bottom of the box, unable to extricate its legs. It fed, whilst in this condition, whenever an aphis came sufficiently near to be seized ; but after some time, as it seemed unable to free itself, I came to its assistance, and with a fine pair of forceps liberated each leg separately, when it was at once able to stand.

On August 5th I left home, so on the 3rd I confided the larvæ to the care of Mr. J. T. Houghton, of this town, who made the following notes for me during my absence :—

Aug. 3rd—Received larvæ.

,, 5th—One seen devouring a larva of *Dictyopteryx bergmanniana.*

,, 6th—First one pupated in a rose leaf.

,, 8th—Two more pupated amongst debris at bottom of the jar.

,, 9th—Last one spun up without cover of any kind.

I kept the jars containing the pupæ indoors all the winter, and on May 24th, 1906, I discovered at 1-15 p-m. that two perfect imagines had emerged, and that a third was half out of its

cocoon. This last remained in the same position until about
2 p.m., when it freed itself, and to my surprise, instead of
gradually developing into a perfect insect, remained enveloped
in a thin, transparent pellicle. Soon after this it apparently
lost its hold of the cocoon, and began rolling about at the
bottom of the jar in its efforts to find some object to which to
attach itself. I placed it on a piece of twig, but it seemed
quite unable to cast off its sub-imaginal skin, and after two or
three days, as it gradually got weaker, I dropped it into a tube
of formalin for preservation.

One other imago emerged after this, but the time elapsing
between the emergence from the pupa case and the casting of
the sub-imaginal skin is evidently very short in a healthy specimen, and I was not fortunate enough to see it. The pupa case
appears to open by means of a small lid at one end. The cast
skins are perfectly transparent, and consist of the covering of
the body, legs, and head. The thorax splits, in order to allow
the insect to emerge. In the sub-imago the wings are only
3 mm. in length, the fore wing appearing slightly shorter than
the hind wing, and resting above it. The antennæ are quite
short, and are folded *round* the eyes like a ram's horns. They
appear to lengthen rapidly, and when fully extended remain
curled underneath the body. I could not help wondering if
this lengthening of the antennæ might not throw some light on
the subject with regard to the genus *Adela* in the Micro-
Lepidoptera. It has always puzzled me, and perhaps others,
how the long antennæ in this genus can be folded in the pupa
case. Might not this lengthening out be a possible solution of
the mystery?

The eyes of the sub-imago are of the same bright green as
in the perfect insect, the body also of the same shade, and the
bright yellow line on the thorax as distinct as in the full
emergence. The wings also show their iridescence through the
thin membrane that covers them.

I feel uncertain if this stage ought to be regarded as a part
of the pupal existence or as a true sub-imaginal one. I can find
no information about it in the books I have, so I have preferred
to call it the latter, especially as it seems to me to represent a
distinct stage in the life history. It is evidently of very short
duration in a healthy subject, as except in the case of the insect
which I preserved, I never saw any of the others except as
imagines.

Emergence seemed to always take place in the morning.

Unfortunately only four of the pupæ produced insects. One or two of the cocoons showed signs of a coming emergence, the lids turning yellowish, but nothing further resulted.

I am afraid these few notes are very imperfect, but I found the study of these *Chrysopidæ* so interesting that I hope at some future time to be fortunate enough to again have the chance of investigating their life history.

Should any readers of these pages obtain ova which they do not require, I should be very glad if they would send them to me, particularly those of other species than the above.

———◆◆———

The editor of the *Museum Gazette* does not mind confessing that thus far the journal threatens to involve a larger loss than is pleasant.

A marine laboratory is to be erected at Cullercoats, at a cost of £3000, by Mr. Hudleston, who has agreed to let it to the Armstrong College at a yearly rental of 3 per cent. on his outlay.

Prof. E. B. Poulton has a lengthy and valuable paper on ' Predaceous Insects and their Prey,' in the ' Transactions of the Entomological Society of London,' recently published. It occupies over 80 pages.

We are pleased to note that Dr. W. E. Hoyle, of the Manchester Museum, is to be the President of Section D (Zoology) at the next meeting of the British Associations at Leicester. Prof. J. W. Gregory will preside over Section C (Geology).

In a recent East Yorkshire paper it is recorded that two ducks were shot lately with one barrel. One of the ducks was found to have a trout in its bill which weighed one and a half pounds ! It is not stated what was in the bill of the other bird. Probably a fib !

Mr. G. W. Lamplugh, F.R.S., favours us with a reprint of his ' Notes on the Occurrence of Stone Implements in the Valley of the Zambesi around Victoria Falls,' which have recently appeared in the Journal of the Anthropological Institute. The specimens in question were collected in the Zambesi district by the author in 1905.

The Board of Agriculture and Fisheries has received information that the American Gooseberry mildew (*Sphærotheca mors-uvæ*) has been discovered in more than one place in England, and as there is reason to believe that the disease, in at least one case, is of some years' standing, they think it desirable to warn all fruit-growers of the dangers involved. Particulars of methods to be adopted to eradicate the pest can be obtained free on application to the Board of Agriculture and Fisheries, 4, Whitehall Place, London, S.W.

The Rev. J. Conway Walter of Langthorpe sends us a pamphlet containing two poems, (1) ' The Destruction of St. Peter's Church, Mablethorpe, by the Sea, in 1287,' and (2) ' The Old Black Oak, or a Fenland Record.' As regard the first, it is an attempt to represent in rhyme a catastrophe in local history, of which no full and connected account has yet been published. We learn that on January 1st, 1287 :—

> ' The night, it was dark ; and the wind, it howled,
> On Malbertoft's desolate shore.
> Above the storm-ridden heaven's scowled ;
> Below was the breaker's roar.' etc.

Mr. W. J. P. Burton has an interesting paper on ' The Ancient Volcanoes of Derbyshire ' in the ' Transactions of the Burton-on-Trent Natural History and Archæological Society,' recently issued. The same publication contains a paper dealing with the ' Nests and Eggs of Local Birds ' (Burton-on-Trent), by Mr. C. Hanson.

LIFE ZONES IN BRITISH CARBONIFEROUS ROCKS.

Part II.—The Fossils of the Millstone Grits and Pendleside Series.

(*Continued from page 23.*)

(PLATE XIV.)

WHEELTON HIND, M.D., B.S., F.R.C.S., F.G.S.

Salter's original specimen came from the Grits of Pule Hill; unfortunately the zone was not stated. The place where *G. bilingue* is most common is on spoil heaps on Pule Hill and at Marsden Station. It occurred in small nodules in a band of shale passed through when driving the L. & N. W. Railway tunnel. The exact position of the band in the series is therefore uncertain, but in the Valley of the Noe, Derbyshire, it occurs well down in the Pendleside series. Messrs. Barnes and Holroyd collected the following fauna at Pule Hill :—

Cælonautilus quadratus.	*Nuculana stilla.*
Glyphioceras diadema.	*Schizodus antiquus.*
,, *bilingue.*	*Sanguinolitos tricostatus.*
,, *reticulatum.*	*Posidoniella lævis.*
Gastrioceras listeri.	*Pterinopecten papyraceus.*
,, *carbonarium.*	*Aviculopecten fibrillosus.*
Euphemus urei.	*Rhizodopsis* sp.
Macrocheilina gibsoni.	*Strepsodus saurordes.*
,, *reticulata.*	*Elonichthys aitkeni.*

So much, then, for the Pendleside series, which passes above insensibly into the Millstone Grit, and with this passage *Gastrioceras listeri* gathers strength, both in numbers and size. Of this species Spencer remarked : (Proc. York. Geol. and Poly., 1898, Vol. XIII, p. 390) ' In the Yoredales (Pendleside series) this shell is of small form, and good specimens are somewhat rare. . . . . In the Yoredale Shales of Todmorden small limestone nodules occur, in which small specimens of this species are found in a good state of preservation ; all the specimens I have seen are of small size, but as we have their crushed forms through the shales of the Millstone Grit, they seem to gradually increase both in numbers and size.' I have never found this species myself below the Grits, but I was fortunate enough to acquire Mr. Spencer's fine collection, in which undoubted specimens of *G. listeri*, small in size, are labelled Horsebridge Clough.

G. carbonarium, which is associated so abundantly with

G. listeri in the lower portion of the Coal Measures, does not seem to come in till late on in Millstone Grit times.

Some years ago a collection was made from the Shales of Eccup, presumably underlying the 3rd Grit, on behalf of the Carboniferous Zone Committee of the British Association, before I became Secretary to the Committee. I, however, worked over a good deal of the material, and the list fairly represents the fauna :—

Lamellibranchiata.

Aviculopecten gentilis.
Pterinopecten papyraceus.
Posidoniella lævis.
  ,,   *kirkmani.*
Nucula æqualis.
Nuculana stilla.
Ctenodonta lævirostris.
Schizodus antiquus.
A large sulcate *Edmondia.*
Myalina peralata.

Cephalopoda.

Cælonautilus subsulcatus.
Glyphioceras reticulatum.
Orthoceras morrisianum.

Brachiopoda.

Productus, a scabriculate form.
Lingula mytiloides.
Orbiculoidea nitida.

The late James Spencer, of Halifax, watched the excavations and shafts sunk for the Halifax waterworks under Wadsworth Moor.

He describes (Trans. Manchester Geol. Soc., Vol. XIII., pp. 209-212) the sections, and I make out the following species in his collection from that horizon, which he determines as the shales between the 3rd and Kinderscout Grits.

Glyphioceras reticulatum.
  ,,   *phillipsii.*
  ,,   *bilingue.*
Dimorphoceras gilbertsoni.
  ,,   *loonyi.*
Temnocheilus sp.
Solenocheilus cyclostomus.

Cælonautilus quadratus.
Orthoceras aciculare.
Posidoniella lævis.
  ,,   *minor.*
Pterinopecten papyraceus.
Ptychomphalus ip.
Productus sp.

Spencer states that *Gastrioceras listeri* and *Nomismoceras spirorbis* occur here, but I cannot find specimens in his collection.

Barnes and Holroyd have described a fossiliferous grit on the flanks of Pule Hill, near Marsden, which occurs in Netherley Quarry on Pule side. I visited the quarry on one occasion with them. They consider the bed to be either high up in the Kinderscout Grit or low down in the 3rd grit. The list of fossils contains no species typical of the horzion. Cephalopoda and Brachiopoda are conspicuously absent. Gasteropoda in the form of casts are referred by them to eleven species, of which the

most common are two forms of *Euphemus*. The following
lamellibranchs also occur :—

Myalina verneuillii.	*Schizodus antiquus.*
,, *flemingi.*	*Posidoniella ?* sp.
Sedgwickia attenuata.	

Such a fauna has been met with nowhere else in the grits,.
and should be looked for in other localities.

I found a large angular boulder, some years ago, in a small
stream at the head of the Elkstone Brook, near the Mermaid
Inn at Morredge, N. Staffordshire.   This was a grit containing
casts of *Schizophoria, Productus, Orthotetes, etc.*   Judging from
its condition, and from the fact that there is little or no drift on
Morredge, which is more than 1000 feet above sea level, I don't
think this rock was far from its parent bed, which I hope to find
one day in the neighbourhood of the Roches.

A most important shell bed occurs in the neighbourhood of
Pateley Bridge and Harrogate.   It is well seen at Cayton Gill,
Clint Quarries, at several places near Pateley Bridge, and Hazel
Hill near Sawley.   The Grit series here is estimated at 1900 feet
thick, and the shell bed is 1000 feet up in this series.

The fauna is large and peculiar, totally different from that
which is generally regarded as characteristic of the Millstone
Grits.   Brachiopoda and Cephalopoda are rare.

At Clint Quarries I obtained the following species :—

Brachiopoda.

Chonetes, cf. *laguessiana.*
Derbya sp.
Productus cora of late mutation.
 ,,   *longispinus.*
 ,,   *scabriculus.*
Spirifer bisulcata.
Spiriferina cristata.
Seminula ambigua.
Schizophoria resupinata.

Lamallibranchiata.

Aviculopecten dissimilis.
Leiopteria laminosa.
Parallelodon [cast.], sp. (*P. obtusus*).
Sanguinolites sp.

Cephalopoda.

Stroboceras sulcatum.
Orthoceras sp.

Pisces.

Petalodus acumenatus (tooth).   .
Phizodopsis sauroides (scale).

In a small quarry, one mile from Pateley Bridge, on the
Ripon Road I obtained :—

Derbya sp.
Productus cora, a late mutation.
 ,,   *longispinus.*
Schizophoria resupinata.
 ,,   (*Rhephidomella*) *michi-*
           lini.
Seminula ambigua.
Spiriferina cristata.
Aviculopecten sp.

Amusium concentricum.
Leiopteria squamosa.
Cypricardella sp.
Parallelodon obtusus ?
Edmondia sp.
Loxonema rugifera.
Euphemus sp.
Stroboceras sulcatum.
Fenestella.

The *Derbya* is very large and very common in the lower beds of the quarry. Casts show very beautifully the internal structure of the shell and the centre plates which distinguishes this genus from *Orthotetes*.

At Hazel Hill, in the parish of Sawley, near Ripon I obtained

Productus cora, a late mutation.
,, *longispinus*.
Chonetes, sp. *laguessiana*.
Derbya sp.
Schizophoria resupinata.

Lamellibranchiata.
Aviculopecten dissimilis.
,, *stellaris*.
,, *semicostatus*.
Pterinopecten whitei.
Edmondia maccoyi.
,, *rudis*
Leiopteria sp.
Lithodomus jenkinsoni.

Mytilimorpha sp.
Parallelodon sp.
Palæolima sp.
Protoschizodus curtus.
Sanguinolites sp.
Tellinomorpha cuneiformis.

Gasteropoda.
Macrocheilina sp.
Naticopsis sp.
Euphemus sp.

Cephalopoda.
Stroboceras sulcatum.
Epipphioceras bilobatum.
Orthoceras sp.

This fauna is a remarkable one to find so high up in the Carboniferous series. Possibly the shell bed on Pule Hill may represent it. A very large number of species are of Lower Carboniferous forms which do not appear in British strata, as far as we know, in the 2000 feet of rocks immediately below. It would be interesting to know where they were living in the meanwhile. The fauna indicates much clearer waters during the period in which the shell bed was laid down than the characteristic fauna of the Millstone Grit.

In North Staffordshire a marine band with *Gastrioceras listeri, Pterinopecten papyraceus*, and *Posidoniella laevis* always underlies the rough rock or Roches Grit. Workings to reach the coal on the Grit below turn out a shale rich in compressed specimens and fragments of these fossils in the neighbourhood of Ipstones and Oakamoor. The fauna is found in shales below the first Grit at Knypersley.

Mr. Spencer's experience in Yorkshire is very similar, and we may consider it a fact that the abundance of *Gastrioceras listeri* increases at higher horizons in the Millstone Grit until a maximum is reached in the Bullion, Mountain-mine, or Hard Bed Coal of the Halifax district.

At Caton Green, a few miles from Lancaster, in the Lune Valley, a series of shales below some grits have been worked for brick making.   Several pits are opened showing sections of sandstones and shales, cut in the most westerley a dark grey shale worked containing many round bullions or concretions. Most of the bullions yield no fossils, but I obtained there *Pleuronautilus nodosocarinatus, Solenocheilus,* Sp. nov. a body chamber of which I think I have from Congleton Edge Quarry, 500 feet below the third Grit.   *Posidoniella laevis* and *Elonichthys aitkeni.*

Dr. A. H. Foord, (Geol. Mag., Dec. 3, Vol. VIII, 1890, p. 481), has recognised in addition *Solenocheilus latiseptatus, Pleuronautilus armatus,* and *Actinoceras sulcatulum* from these beds.   This fossiliferous horizon occurs somewhat nearer the top than the base of the Millstone Grit series.

P. nodosocarinatus has been found high up in the Yoredales of Swaledale, and in the Arden Limestone series of the West of Scotland.   From this locality it was described by Armstrong under the *N. nodiferus,* but Roemer has previously described the shell and the name *N. nodosocarinatus.*   A sharp look out should be kept for traces of this very fine and characteristic species in the Millstone Grits of the Midlands.   A fine specimen lately found during excavations by the Corporation at Harrogate is now in the Natural History Museum, South Kensington.

At Holt Head, near Slaithwaite, near Saddleworth, Mr. Barnes found a bed of shale crowded with the valves of a shell which I figured and described as *Sanguinolites ovalis.*   (Pal. Soc. Brit. Carb. Lamell, Pt. V., p. 411, Pl. 46, figs. 14-17.)   The horizon is probably below the 3rd Grit, and the fauna which accompanies it consists of :—

Nuculana stilla.	*Posidoniella lævis.*
Nucula gibbosa.	*Lingula mytiloides.*
Schizodus antiquus.	*Ostracoda.*

This, then, is the extent of our knowledge of the molluscan fauna of the Pendleside series and its distribution, and I can only hope it may be useful in stimulating local geologists to take up the subject of life zones in the Millstone Grit rocks. Taking it as a whole, the fauna is a fairly extensive one, and in certain localities is most prolific.

Before terminating this paper, I propose to give lists of the fish and flora which are found in the Pendleside series.   These lists are comparatively meagre, but I have no doubt that they can be considerably increased by research.

Dr. Wellburn compiled the following table showing the

distribution of fish remains in the Pendleside series for the Report of the British Association Committee for Life Zones in British Carboniferous rocks : to this I have added a column for the Pendleside series of North Wales.

Pisces.	Pule Hill, Marsden.	Crimsworth Dean.	Todmorden.	Derne Valley.	Whitewell.	River Hamps, Staffs.	Burnsall and Thorpe Fells.	Astbury, Cheshire.	N. Derbyshire.	N. Wales.
Cladodus mirabilis		*								
„ sp.							*	*		
Orodus elongatus	*							*		
Acanthodes sp.	*	*	*					*		
Marsdenius summiti	*									
„ *acuta*	*									
„ sp.				*						
Strepsodus sauroides	*									
Rhizodopsis sauroides	*									
Cœlocanthus hindei						*				
„ sp.									*	
Radinichthys circulus									*	
„ sp.	*									
Elonichthys aitkini	*	*		*	*					*
„ sp. nov.	*									
„ *obliquus*	*			*	*					
Acrolepis hopkinsi	*	*								*
„ *wilsoni.*									*	*
Platysomus sp.					*					

The following table shows the distribution of the flora of the Pendleside series, all of which have been determined by Mr. Kidson. There is abundant room here for local collectors to extend the list and localities.

Plantæ.	Polvash, I. of Man.	N. Staffs.	N. Derbyshire.	Teilia, N. Wales.	Pendle Hill.	Congleton Edge, 300 feet below 3rd grit.
Adiantites antiquus	*	*		*		
„ *tenuifolius*		*				
„ *machaneki*	*					
Archæopteris tschermaki		*				
„ sp.						
Asterocalamites scrobiculatus						
Dactylotheca aspersa		*				
Lepidodendron branchlets		*				
„ *veltheimianum*					*	
Lepidophyllum n. *L. lanceolatum*		*				

1907 March 1.

Plantæ.	Polvash, I. of Man.	N. Staffs.	N. Derbyshire.	Teilia, N. Wales.	Pendle Hill.	Congleton Edge, 300 feet below 3rd Grit.
Lepidostrobus sp.		*				
Neuropteris antecedens		*	*			
Rhabdocarpus, n. sp.		*				
Rhodea, n. sp.		*				
Phacopteris glabellata						
,, *inequilatera*						
Sphenopteris quercifolia		*				
,, *pachyrachis*	*			*		
,, ,, var. *stenophylla*	*			*		
,, ,, ,, *affinis*				*		
,, ,, ,, *subgeniculata*				*		
,, ,, ,, *schlebani* ...				*		
Stigmaria ficoides						*

EXPLANATION OF PLATE XIV.

FIG.

1. *Posidonomya becheri,* a left valve, from shales in the Mixon Hey Brook, Staffordshire, one mile below Mixon Hey Farm.
2. *Posidonomya membranacea,* a right valve, from shales in the same stream, about half a mile above Onecote Grange.
3. *Chaenocardiola footii,* 2 left valves, from shales in the River Dove, near Glutton Bridge.
4. *Aviculopecten fibrillosus,* the convex valve, from the Quarry N.W. of Holly Wood, Congleton Edge.
5. *Aviculopecten fibrillosus,* the smooth flattened valve, same locality.
6. *Glyphioceras bilingue,* an enlarged view, showing the mouth of the specimen, from shales in Wild Moor Bank Hollow, about two miles N.W. of the Cat and Fiddle Inn.
7. *Glyphioceras bilingue,* a partially decorticated specimen, but uncrushed.
8. *Glyphioceras reticulatum,* a young specimen, from shales in the River Dane.
8a. *Glyphioceras reticulatum,* a section showing the chambers of the shell, same locality.
9. *Glyphioceras spirale,* a crushed specimen, from the Quarry on Congleton Edge.
10. *Lingula scotica,* a very rare Brachiopod, hitherto only found in Scotland and Northumberland; with (a) *Orbiculoidea nitida,* same locality.
11. *Prolecanites compressus,* showing the characteristic suture lines, from the old Limestone Quarry, Astbury.

All the specimens are in my own collection.

———◄·◊►———

Mr. G. T. Porritt has a note on Hereditary and Sexual dimorphism in *Abraxas grossulariata,* var. *varleyata,* in the January 'Entomologist': Monthly Magazine.'

The Belfast Municipal Art Gallery and Museum is following the example of another institution in publishing quarterly notes. Nos. 3 and 4 are before us, and have been reprinted from the local press.

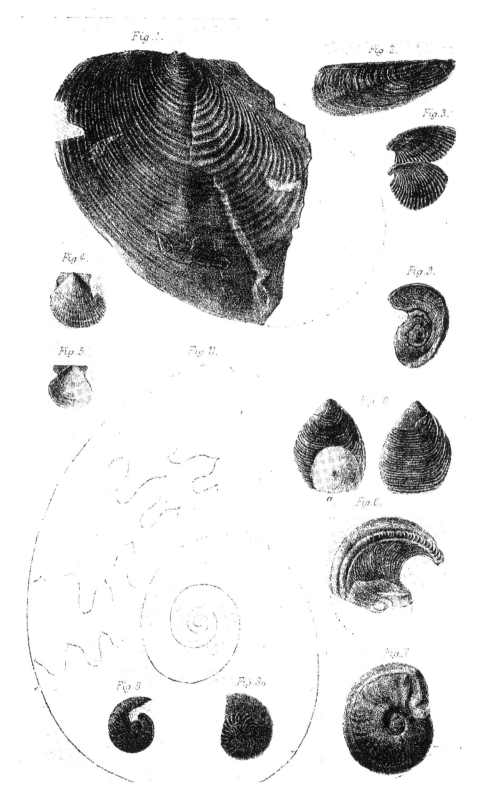

RECENTLY DISCOVERED FUNGI IN YORKSHIRE.

C. CROSSLAND, F.L.S.
Halifax.

THE present is considered a suitable time to bring together, under one heading, the Fungi discovered in Yorkshire since the publication of the Yorkshire Fungus Flora in 1905. This will enable mycologists to post their copies up to date without much trouble. Many species have been recorded in the pages of the 'Naturalist' in the interval; in all such cases references to pages and date are given. To facilitate the placing of these records in their proper sequence, the number in the flora which each must follow is added. Two are included which were accidentally omitted when the work was compiled.

It will be seen that two are new to science, 7 new to Britain, and 46, and 3 vars. new to Yorkshire. This brings the present total of known Yorkshire species to 2681.

The number of new vice-county records has also been considerably added to:—S.W., 46; MID W., 8; N.W., 69; N.E., 32; S.E., 57. These, however, can easily be inserted in the Flora by aid of the 'Naturalist,' 'Transactions,' and the Trans. of the Hull Sci. and F. N. Club.,' '05-6, by any one sufficiently interested in the subject.

NEW SPECIES.

Clavaria gigaspora Cotton n. sp.

Cæspitose but distinct at the base, or solitary, greyish with tinge of yellow, whitish at base of stem, small, up to 3 c.m. high, branched, flesh tough, smell and taste absent; branching irregular, sometimes almost palmate, branches erect, occasionally forked, often wrinkled, solid, terete or compressed, much compressed at the acute axils, ultimate branches attenuated, apices blunt; stem 1 c.m. long, or shorter, slender, not very distinct; internal structure of densely packed hyphæ 4-4.5 μ diam. forming a firm, tough tissue, rather horny when dry; basidia large, 60-70 × 15 μ, contents finely granular, sterigmata four, rather stout, 8-10 μ long, spores broadly elliptical, slightly oblique, ends somewhat narrowed, average 12-16 × 7.5-8 μ, very variable (10-20 × 7-9μ), guttulate, then granular, hyaline, smooth.

S.W.—Habitat. Amongst moss on rocky, heathy slope, road side between Flappit and Crossroads near Cullingworth. —C. Crossland and Thomas Hebden, Nov. 1906.

To follow No. 1218.

A small dingy yellowish-white plant, scarcely over-topping the moss in which it grows. It appears to resemble certain forms of *C. cinerea* and *C. cristata*, but is readily distinguished from either by the large spores. The structure is also somewhat exceptional, being composed of very fine, densely matted hyphæ, which run out into unusually large basidia.

Clavaria gigaspora Cotton sp. nov.

Planta tenax, irregulariter ramosa, alba v. pallido-alutacea, ad 3 c.m. alta. Caulis tenuis, circiter 1 cm. longus. Rami breves, erecti, teretis, saepe rugosi, infra ramulos compressi ; ramuli ultimi sensim attenuati, apicibus sub-obtusis. Basidia majuscula 60-70 × 15 μ, intus granulosa. Sporae late ellipsoideae, grandis, 14 × 7 μ (10-20 × 7-9 μ) hyalinae, laeves.

Ad terrum muscosam, prope. Cullingworth, Yorks. Brittanniæ.

Verticicladium Preuss.

Sterile hyphæ creeping, fertile erect, septate, verticillately branched, branchlets subulate ; conidia continuous, solitary at the tips of the branchlets, soon falling away.

Verticicladium Cheesmanii Crossl. n. sp.   Plate IX.

Effused in pale red-brown patches, sterile hyphæ creeping, septate, contents granular, deep red-brown, 4-5 μ thick, fertile hyphæ erect, pale red-brown, 2-3 times branched, 4 μ thick, slightly wider at base of branches, branches spreading, ultimate branchlets mostly in pairs, opposite, occasionally solitary, subulate, slightly inflated at base, erect, 13-15 × 3.5 μ ; conidia broadly elliptical, ends obtuse,. pale red-brown, 6-8 × 3.5-5 μ.

Mid W.—Hab.   On decorticated wood.   Stainor Wood near Selby.—W. N. Cheesman, Nov. 1906.

To precede Stachylidium.

Verticicladium Preuss.

Hyphæ steriles repentes, hyphæ feraces erectae, septatae, verticillatim ramosae, ramuli subulati, conidia continua, solitaria ad apices ramulorum.

Verticicladium Cheesmanii Crossl. sp. nov.

Effusum, stratis pallido-spadiceis, hyphis sterilibus repentibus, septatis, cytoplasma granulari, rufo-spadiceis, 4-5 μ crassitudine, hyphis feracibus erectis, ramosis bis vel ter, 4 μ crassitudine, leviter latioribus, ad basin ramorum, pallido-spadiceis, ramis patentibus, ramulis ultimatis generaliter binis· et oppositis, subinde solitariis, subulatis, leviter inflatis ad basin, erectis,. 13-15 μ × 3.5 μ, conidiis late ellipticis, polis obtusis, pallido-spadiceis, 6-8 μ × 3.5-5 μ, cito disjunctis.

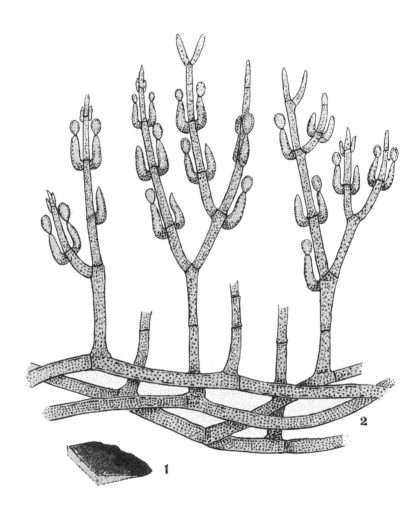

Verticicladium Cheesmanii Crossl.

FIG. 1.—Portion of patch on decorticated wood, natural size.
FIG. 2.—Enlarged about 640 diams.

NEW TO BRITAIN.
Geaster triplex Jungh.

S.W.—Hebden Bridge, on the ground in Pecket Wood, June 1905 (J. Needham).

'Unexpanded plant acute, exoperidium recurved (or when not fully expanded, somewhat saccate at base), cut to the middle, or usually two-thirds, to five or eight segments. Mycelial layer adnate. Fleshy layer generally peeling off from the segments of the fibrillose layer, but usually remaining partially free as a cup at base of inner peridium. Inner peridium sub-globose, *closely sessile.* Mouth *definite,* fibrillose, broadly conical. Columella prominent, persistent, elongated (see fig. 49). Threads thicker than spores. Spores globose, roughened, 3-6 m.c. ('The Geastrae,' C. G. Lloyd, June 1902, pages 25-27, figs. 47-49).

To follow No. 12.

Hebeloma subsaponaceum Karst.

S.E.—Allerthorpe Common, near Pocklington, on the ground, under beeches (* 'Nat.' Sept. '05, p. 267 ; l.c. Jan. 6, '06, p. 7).

To follow No. 460.

Cantharellus hypnorum Brond., Rev. Myc., 1892, p. 65 ; Sacc., Syll. ii. p. 32, 1895.

S.W.—Ferrymoor near Cudworth (* 'Nat.' Oct. 1905, p. 295). For description and note see l. c., Jan. '06, pp. 7-8.

To follow No. 905.

Lentinus suffrutescens Fr.

S.W.—Milnsbridge near Huddersfield, growing from the wood flooring of a joiner's shop (F. F. '06, 'Nat.' Feb. '07, pp. 52 and 54). Spores probably on the timber when imported. Certe A. Clarke.

To precede No. 948.

Lachnea gilva (Boud.). Sacc. Syll., n. 747. *Peziza gilva* Boudier, Icon. 37.

S.W.—Hebden Bridge, on sandy ground, among moss, by the river side. For description and note see ' Nat.' Jan. '06, pp. 8-9.

To follow No. 1840.

Zygodesmus fulvus Sacc., Michelia ii., p. 147.

Var. *olivascens* Sacc., Mich. ii., p. 585.

Mid. W.—Stainor Wood, Selby, on decaying wood, '06 (W. N. C.).

To follow No. 2386.

Graphium xanthocephalum Sacc. Certe. G. Massee.

N.W.—Masham, on bamboo canes used as plant supports in greenhouse. Jan. '06 (W. A. Thwaites).

To follow No. 2455.

NEW TO YORKSHIRE.

Lycoperdon cruciatum Rostk.

S.W.—Maltby and Stubbins Woods, among moss and dead leaves. (F.F. '05, 'Nat.', Nov. '05, p. 338, where *L. velatum* Vitt. is suggested, and l.c. Jan. '06, pp. 6-7, with photo. and diagnosis. Some authorities consider this to be *L. velatum* Vitt.

To follow No. 23.

Lycoperdon Cookei Mass., Mon. Lycop. n. 52, pl. xiii., figs. 24-26.

S.W.—Farnley Tyas, on the ground in woods (F.F. '06, 'Nat.', Feb. '07, p. 54, as *L. spadiceum* Pers., which is supposed to be a synonym).

To follow No. 24.

Lycoperdon depressum Bon.

S.W.—Farnley Tyas, on the ground in woods (F.F. '06, 'Nat.', Feb. '07, p. 54).

To follow No. 25.

Amanita cariosa Fries. Hym. Eur. p. 24.

S.W.—Huddersfield, Sept. 1895 (Dr. Cooke, Tr. Brit. Myc. Soc., 1903, p. 13).

To follow No. 48.

Lepiota granulosa var. **rufescens** B. & Br., Ann. Nat. Hist., n. 1834; Cke., Illustr., pl. 213A.

S.W.—Firbeck (F.F., '05, Tr. 33, 1907).

Clitocybe subinvoluta (Batsch.).

S.W.—Firbeck, on the ground among grass (F.F., '05, 'Nat.', Dec. '05, p. 369; Tr. 33, '07).

To follow No. 159.

Mycena lineata (Bull.).

S.W.—Roche Abbey Valley (F.F. '05, Tr. 33, '07),

To follow No. 217.

Volvaria Taylori Berk., Outl., p. 140.

S.W.—Slaithwaite, near Huddersfield. Typical specimens found on grass plot in front of his house by E. J. Walker, Nov. 1906, Certe. A. Clarke.

Differs from *V. violacea*, which it somewhat resembles, by the gills being remarkably attenuated behind, and by the small, brown volva.

To follow No. 317.

Pholiota heteroclita Fr.

S.W.—Hewenden Bridge near Cullingworth, Sept., '05, on dead poplar.

To follow No. 403.

Pholiota tuberculosa (Schæff.).

S.W.—Heaton Wood near Bradford (Bradford Sci. Jour., Jan. 1907, p. 341).

To follow No. 409.

Hebeloma longicaudum (Pers.).   Var. **radicatum** Cke., Hdbk., p. 164.

S.W.—Firbeck (F.F. '05, Tr. 33, '07).

Hebeloma nudipes Fr.

S.W.—Farnley Tyas, on the ground in a wood (F.F. '06, ' Nat.', Feb. '07, p. 55).

To follow No. 463.

Tubaria cupularis (Bull.).

S.W.—Farnley Tyas, among grass in pasture (F.F. '06, ' Nat.', Feb. '07, p. 55).

To follow No. 515.

Cortinarius (Phleg.) **triumphans** Fr.

S.W.—Cullingworth, on the ground under birch trees. (Thos. Hebden).

To precede No. 537.

Cortinarius (Phleg.) **scaurus** Fr.

S.W.—Maltby Wood, on the ground among moss, decaying twigs, etc. (F.F. '05, ' Nat.', Nov. '05, p. 340 ; Tr. 33, '07).

To follow No. 551.

Cortinarius (Ino.) **arenatus** (Pers.).

S.W.—Firbeck, on the ground in a wood (F.F. '05, ' Nat.', Nov. '05, p. 340 ; Tr. 33, '07).

To follow No. 570.

Cortinarius (Tela.) **helvelloides** Fr.

S.W.—Maltby, on the ground in a wood (F.F. '05, ' Nat.', Nov. '05, p. 340 ; Tr. 33, '07).

To follow No. 597.

Cortinarius (Hygr.) **uraceus** Fr.

S.W.—Firbeck, on the ground in a mixed wood (F.F. '05, 'Nat.', Nov. '05, p. 340; Tr. 33, '07).
To follow No. 615.

Hypholoma leucotephrum (B. & Br.).

S.E.—Holmpton, Withernsea, near base of tree trunk, among grass on road side, Oct. 1905 (Hull S. & F. N. C. Tr.; '06, p. 292).
To follow No. 666.

Psilocybe canobrunnea Fr.

S.W.—Farnley Tyas, on dry manure heap (F.F. '06, 'Nat.', Feb. '07, p. 55).
To follow No. 692.

Hygrophorus (Hygr.) **spadiceus** Fr.

S.W.—Firbeck, in pasture (F.F. '05, 'Nat.', Nov. '05, p. 339).
Differs from *H. conicus* in virgate pileus, and thicker gills not narrowed behind.
To follow No. 801.

Russula furcata Pers.   Var. **ochroviridis** Cke.

S.W.—Slaithwaite (F.F. '06, 'Nat.', Feb. '07, p. 55).
To follow No. 860.

Cantharellus Friesii Quel.

S.W.—Slaithwaite, in cold frame in garden (F.F. '06, 'Nat.' Feb. '07, p. 56).
To follow No. 904.

Corticium populini Fr.

S.E.—Hull, on pine pit-props, probably in this case imported (T. Stainforth, Jan. 1907).
To follow No. 1156.

Clavaria incarnata Weissm.

S.W.—Farnley Tyas, in pasture ('Nat.', Feb. '07, pp. 52 and 55).
To follow No. 1242.

Uromyces junci (Desm.).

S.W.—Askern Bog, Æcidium stage on *Pulicaria dysenterica* (* 'Nat.', Oct. '06, p. 374).
To follow No. 1307.

Sporormia pascua Niessl.
S.W.—Sheffield, on rabbit dung, April, 1903 (T. Gibbs),
Distinguished from *S. octomera* Phil. and Plow. by the short
foot and broad subtruncate apex of the ascus,
To follow No. 1617.

Humaria Phillipsii Cke.
N.W.—Masham. For description, synonyms, and notes,
see 'Nat.', Jan. '06, pp. 9-10.
To follow No. 1811.

Lachnea cinnabarina (Schw.).
S.W.—Hebden Bridge. For description and notes see
'Nat.', Jan. '06, p. 8.
To follow No. 1824.

Helotium Hedwigii Phil.
S.W.—Hardcastle near Hebden Bridge, on fallen elm-twigs,
June, 1897 (J. Needham). Accidentally omitted from the Flora.
Distinguished from allied species by the lower half of stem being
white, and by the remarkably swollen, woolly base.
To follow No. 1952.

Myxotrichum deflexum Berk.
S.W.—Copley near Halifax ('Nat.', Aug. '05, p. 254).
To follow No. 2150.

Penicillium hypomycetis Sacc.
S.W.—Maltby Wood, spreading over a group of sporangia
of *Trichia fragilis* (F.F. '05, 'Nat.', Dec. '05, p. 371).
To follow No. 2320.

Rhinotrichum Bloxamii B. & Br.
Mid W.—Stainor Wood near Selby, Nov. 1906 (W. N.
Cheesman).
To follow No. 2328.

Stemonitis laxa Mass. [*Comatricha laxa* Rost.].
S.E.—Hull, very small sporangia on a pine log, West Dock
Reservation, Sept., 1903 (Trans. Hull S. & F. N. C., 1905,
p. 202, T. Petch).
To follow No. 2500.

Stemonitis flavogenita Jahn.
S.E.—Hedon; Thearne; Hornsea; Snake Hill, South Cave,
(l.c. 1905, p. 202, T. Petch).
'This species was, till recently, designated *S. ferruginea*

Ehr., and East Riding specimens have been distributed under this name. The discovery of Ehrenberg's type specimen proves that his *S. ferruginea* is the modern *S. Smithii* and necessitates a change of nomenclature' (Jour. of Botany, vol. 42, p. 194).

To precede No. 2502.

Amaurochæte atra Rost.

S.E.—On pit props just landed ex steamer from Norway, Fish Dock Extension, Hull, Sept., 1903 (Tr. Hull S. & F.N.C., 1905, p. 203, T. Petch).

To precede No. 2503.

Lamproderma irideum Mass.

S. E.—Thorp Garth, Aldborough, on twigs in a stick heap, Aug. 1903 ; Hornsea, plantations north of the Mere, May, '04' (l.c. 1905, p. 203, T. P.).

To follow No. 2505.

Margarita metallica Lister.

S.E.—Abundant on hawthorn branches in the winter in the neighbourhood of Hedon, Dec. 1902 and 1903 ; Rose Hill; Newton Garth; Aldborough; Thearne; Tansterne; and Hambleton (l.c. p. 208, T. P.).

To follow No. 2512.

Dianema depressum Lister.

S.E.—Hambleton ; Tansterne ; Bale Wood, Aldborough, Jan. '04 (l.c. p. 208, T. P.).

To follow No. 2513.

Trichia lutescens Lister.

S.E.—Dryham, North Cave, Aug. 1903, on sticks, probably hawthorn, forming a fence whose base rested in swampy ground by the roadside (l.c. p. 295, T.P.).

To follow No. 2530.

Didymium clavus Rost.

S.E.—Thearne ; Aldborough, in dead grass and leaves (l.c. p. 201, T. P.).

To follow No. 2543.

Physarum calidris Lister.

S.E.—Thorp Garth, Alborough, Aug. '06 (l.c., p. 200, T. P.).

To follow No. 2562.

Physarum Phillipsii Balf. fil.

S.W.—Halifax, on rotting rope in warehouse, July 1895, collected by the late H. T. Soppitt.

To follow No. 2563.

Badhamia decipiens Berk.

S.E.—Tansterne fox cover, on wood, moss, etc., in dry ditch, Aug. '03 (Tr. Hull S. & F. N. C., '05, p. 199, T. P.).

To precede No. 2567.

Badhamia verna Rost. (*Physarum vernum* Sommf.).

S.E.—Hedon, abundant on dead hawthorn branches, Dec. '03 (l.c., p. 200, T. P.).

To follow No. 2568.

Badhamia foliicola Lister.

S.E.—Tansterne fox cover, in abundance on dead hawthorn twigs in a dry ditch, Aug. '03 (Jour. of Botany, vol. 42, p. 129).

To precede No. 2569.

The discovery in Yorkshire of twelve of the last thirteen species is due to the excellent field investigations in quest of Mycetozoa, or Myxomycetes, carried on by Mr. T. Petch, B.A., B.Sc., during 1903-4. A few were new to Britain. Altogether Mr. Petch found in the East Riding upwards of sixty species of this group, *Vide.* Trans. of the Hull S. and F. N. Club, '05, pp. 196-208. Here we have another example of what may be accomplished in field research by persistent work as opportunity affords.

———◆◆———

In a paper upon ' Local Birds,' by Mr. C. F. Innocent, read recently before the Sheffield Naturalists' Club, the following classification was suggested as more detailed than the usual meagre division into residents and migrants.

(A) *Residents:*—
  1. Eu-residents: resident all the year through, *e.g.* house sparrow.
  2. Pen-residents: resident all the year through, except for a few weeks, *e.g.* starling.

(B) *Migrants:*—
  3. Pseudo-residents : resident as species, but migratory as individuals, *e.g.,* robin.
  4. Winter residents : *e.g.* fieldfare.
  5. Summer residents : *e.g.* swallow.
  6. Eu-migrants : which only visit on autumn or spring passage to winter or summer residence, *e.g.* ringed plover.

(C) *Erratics:*—
  7. True waifs and strays: *e.g.* puffin.
  8. Archæo-residents: which formerly lived in the district and of which individuals occasionally return to former haunts of the species, *e.g.* eagle.

BIRD NOTES.—YORK DISTRICT.

SYDNEY H. SMITH.
York.

I cannot but comment on the large numbers of Redwings that have visited us this winter. They literally swarmed all over the country during December, but I am afraid their flocks were sadly decimated by the severe weather in January. Fieldfares do not appear as plentiful as in previous years, and for some weeks I have only remarked these handsome immigrants in small batches of four to twelve birds. Hooded Crows, locally termed Grey Backs, appeared about the middle of October, and were in full force by November, many thousands roosting nightly in Crompton Wood, their usual winter quarters, sharing the branches with parties of immigrant and local Carrion Crows and a tremendous body of Rooks and Jackdaws. At dusk the immense circling flock of dusky birds made all the din they possibly could, and was apt to leave a lasting impression on the mind of a student of nature. A small party of Grey and Pied Wagtails frequents the shallows along the River Foss. Thanks to the Birds' Protection Order, the brilliantly plumaged Kingfisher is more often seen on both Ouse and Foss, sometimes I notice one right in the heart of the city. A few Siskins have been caught on the Malton Road, and several Bramblings seen in private gardens during the recent hard weather, when they fed along with Sparrows and Chaffinches. On the flooded meadows at East Cottingwith duck appeared in their usual number (about 400 birds), chiefly Mallard and Wigeon, with a few Teal, Scaup (occasional), Pochards, Tufted, and Golden-eye. A gaggle of geese pitched one night, but it was too dark to distinguish the species (probably brent or grey), and half-a-dozen handsome Whooper Swans spent two days on the fresh water, departing in the night to other parts. Mr. Snowden Sleights, the local fowler, sent me two female Goosanders early in January; every year a few females turn up at Cottingwith, but no males. According to Messrs. Booth and Riley Fortune, a small party of males appears in the Washburn Valley every year. The question to be settled is, are these birds all of one immigrant party, the sexes mutually agreeing to separate during their stay in Yorkshire?

THE PROTECTION OF BIRDS IN THE WEST RIDING.

R. FORTUNE, F.Z.S.

FOR some time the Wild Birds' and Eggs' Protection Committee of the Y.N.U. has been endeavouring to get the existing Wild Birds' Protection Act of Yorkshire into line, so that the same order should apply to the three Ridings. It has not been found practicable to do this. The East Riding has met us in a fair spirit, the North Riding has quite ignored our suggestions, up to date. The West Riding authorities have dealt with our suggestions in a very broad manner, and it is very pleasing to record that this authority seems particularly alive to the requirements of wild birds.

A new Schedule or Order has just been issued, and it may be interesting to note the additions to this. Practically we have got all the birds we wished to have scheduled, and it is particularly gratifying to note that both the Raven and Peregrine Falcon are now absolutely protected during the whole of the year.

All birds, with the exception of our familiar friend the House Sparrow, are protected from the last day of February to the 12th day of August; but with the exception of those specially mentioned in Schedules A and B, owners or occupiers of land, or persons authorised by them, are at liberty to destroy them should they so desire. The birds named in Schedules A and B are protected absolutely, not only against the general public, but also against owners or occupiers, who must not lift their hands against them.

Schedule A gives protection to the species included from the last day in February to the 12th day of August. There are four additions to this list, viz., the Great Northern and Red Throated Divers, Gadwall and Quail. There is absolutely no necessity for the inclusion of the first two species, as they are not likely to be seen in the West Riding between the dates named, if at all. The same remark might also apply to the Gadwall; but as it is a species which is increasing as a resident in Norfolk, there is no reason why it should not extend its range to Yorkshire. At any rate no harm can be done by anticipating this event somewhat. The Quail nests in some part of the county every summer, and every inducement should be given it to continue to do so.

There are about sixteen omissions from the previous list,

but as they have been moved up and placed amongst the birds
protected all the year round we cannot complain.

Schedule B is the most interesting. The birds in this list
are absolutely protected against everyone all the year round.
To this Schedule no fewer than twenty-six species have been
added. The Peregrine Falcon and the Raven, both practically
extinct as breeding species, have now a chance to regain their
lost ground. Other nesting species included for the first time
are the Corn Crake, which, without doubt, has decreased in
numbers very considerably of late years. The Dotterel, of
which only about a single pair attempts to nest in the Riding.
These birds are shot on migration in spring for the purpose
of obtaining feathers for dressing flies for fishing, but as
feathers may be obtained from the Starling answering quite as
well, if not better than those of the Dotterel, it is scandalous
that the rare bird should be brought under contribution when
the slaughter of a few Starlings can do no harm. The Black
Headed Gull, a much maligned species, which observation has
certainly proved that, in the West Riding at any rate, they
do no harm, but a great deal of good. I have investigated
several cases of alleged interference with grouse eggs by this
bird, but in no case was the bird guilty. The Kittiwake Gull
is also included. This bird does not nest or frequent the West
Riding, but occasionally an individual may be blown inland
from the coast. He may now visit us safely, at least according
to law. The Hedge Sparrow and Tree Sparrow are bracketted
together under the head of Sparrows ; the first is of course not
a Sparrow but a Warbler. It is pleasing to know that our
unassuming little friend has been placed upon the list. The
gentlemanly Tree Sparrow is such a very local bird, nowhere
very abundant, that he deserves to receive sanctuary. He may
be readily distinguished from his vulgar relation by his chestnut
head, two white bars on the wing, and his more musical note.
The Spotted Crake and Water Rail, two very rare nesting
species, may now attend to their household affairs in security.
Many have been shot every autumn, but this is now prohibited.
The Stone Curlew, nearly extinct as a nesting bird, may now
possibly increase in numbers. The Turtle Dove, which is extending its range in the county, may now do so in security.
The Twite, an interesting and very local species, is placed on
the list for the first time. I know of one place in the fell district
where the manager of a local bank, for several mornings in
succession, sallied out to shoot Twites, killing from twenty

to forty each time ; it is pleasant to know that such 'sportsmen' are now answerable to the law for their dastardly conduct. Of non-breeding species the Bee Eater is included, and if we bear in mind the Bentham episode, we cannot but feel that it is only just that it should be so. The slaughter of Rough Legged Buzzards, which sometimes visit us in numbers, will for the future be prevented, and in like manner the young Sea Eagles which visit us are to have every protection. The Golden Eagle was on the list before. The wanton destruction of Pallas' Sand Grouse, should they again visit us, will probably be prevented, and the inclusion of all the British breeding Terns may save some senseless destruction, should any of them by chance pay a visit to the West Riding.

Professional bird catchers are to be prevented from following their nefarious calling, for those favourite cage birds, Bullfinch, Goldfinch, Linnet, and Chaffinch, must not now be caught at any time of the year. They have all suffered considerably in the West Riding, the first three especially, and while not being against the keeping of cage birds, the cruelties I have seen practised in the wholesale capture of wild birds, makes one extremely glad that they are now thoroughly protected. The Heron too is safe against the selfish angler who cannot bear anyone but himself to catch a fish, probably the antipathy to the Heron arises in many cases because *he can* catch fish ; the fact that he feeds considerably upon other fare seems to be entirely overlooked. The bold little Merlin, the Lady's Falcon, must not now be molested at any time, nor must our useful friend the Kestrel.

In addition, the eggs of all the birds mentioned in Schedules A and B, ninety-three of which at any rate nest in the county, are also protected, formerly there was a separate schedule for the eggs of certain species, but it certainly simplifies matters when it is understood that the eggs of all the birds scheduled are not to be taken. Probably no naturalist is opposed to egg collecting when conducted reasonably, but the senseless manner in which rare local species are harried merits the condemnation of every true ornithologist.

There are many species scheduled in A for protection during the nesting season—as, for instance, the Fulmar, Avocet, Smew, etc.—which it seems absurd to place upon the list. This I pointed out to the West Riding authorities, but they were determined not to omit any species which had been scheduled before. Hence the unnecessary inclusion of certain names. The same applies in Schedule B, where birds are protected all the year round which are not with us in the winter months, as, for instance, migratory species like the Corncrake, Dotterel, Nightingale, Sandpiper, Terns, &c. But this we need not cavil at, so long as the species really necessary are included. On the whole, the Union and its Committee dealing with these matters are to be congratulated in accomplishing what has been done.

The eggs of the Lapwing, which, in the previous order, were not allowed to be taken after the end of March, may now be taken up to and including the 15th day of April. The first date was absurd, as there are no eggs, or at least very few, in the West Riding until the first few days in April, and as in some parts of the Fell districts a considerable trade is done in Lapwings' eggs, it was felt that some hardship was entailed by not allowing the farmers a little opportunity for reaping a profit. The birds will not suffer from this extension, for, as a rule, there is a great percentage of loss amongst the early eggs from early frosts, want of cover, and the harrowing, etc., of the fields.

Special attention is drawn to the fact that the setting of Pole Traps is illegal, and also to the fact that the police have instructions to take proceedings against all persons offending against the Order. This is pleasant news, as last year the police had to be nearly goaded into taking action against some persons who endeavoured to destroy the Peregines at Ingleborough.

There is, I think, some idea that in the near future the whole of the Bird Protection Orders throughout the kingdom will be overhauled and brought into line, so that the present existing confusion may be avoided. The simplest plan in my mind would be to protect ALL birds; but allow the various authorities to withdraw the protection from certain species (with the sanction of the Home Office) which have become a nuisance, as we know many species are likely to be if they become too numerous.

—◆◆—

A paper on 'The Boultham Well at Lincoln,' by Wm. McKay, with details of the strata passed through, to a depth of 1561 feet, appears in part 1. of vol. 30 of the 'Transactions of the Manchester Geological and Mining Society.'

We learn from 'The Museum News' that 'The success of Dr. Hyatt's lecture on "Jumbo's Teeth and other Teeth" *needs no further proof* than the fact that 102 children in the lecture room (*which seats only 60 persons comfortably*), listened with *close* attention and interest, for forty minutes.'

NOTES ON THE COMMON SWIFT IN THE BRADFORD DISTRICT.

HARRY B. BOOTH, M.B.O.U.

I HAVE had many opportunities of watching the habits and peculiarities of this rushing, dashing bird, that makes light of distance and space, and disdains to set foot on earth. I have been rather well situated, having been able to easily watch one particular colony which annually visits Heaton Grove, at Frizinghall, a suburb of Bradford. During my residence of nearly thirteen years at Frizinghall, they were all the summer constantly before my eyes (even from my bedroom window), and oftimes when I could not see them they were heard, and I loved to hear their harsh screaming notes in concert. Since leaving Frizinghall six years ago, I have, on the top of the tramcar, usually passed the same colony several times each day.

Up to last season (1906), the colony has generally consisted of thirty to forty pairs, but last summer there would not be more than twenty pairs present. This decrease is rather singular, because Swifts are steadily increasing in this district. Their arrival and departure at this breeding place during the last nineteen years has been extremely regular ; with the single exception of the cold wet spring and summer of 1903, they have always arrived at the breeding quarters on the 10th, 11th, or 12th of May, and as a colony they have departed on the 18th, 19th, or 20th of August. Taking the arrivals and departures of this colony which I have chronicled, I find that they have spent exactly an average of 101 days each year around this nesting site. In 1903 they did not put in an appearance until May 18th, and they stayed until September 1st, thus not only coming a week later, but increasing their stay with us to 107 days. Possibly this extra week's delay at their nesting quarters was occasioned by the difficulty of obtaining sufficient food in order to bring the nestlings forward.

The above dates refer to the arrivals at and departures from this nesting place. A single bird, or a pair, will usually arrive in the surrounding district between the 1st and the 7th of May, and on two occasions they have been noted on the 30th of April. After the arrival of the first harbingers they gradually increase almost daily. If the weather should be fine and clear then, many of the new arrivals spend a good deal of time over the surrounding hills and moorlands, but if very cold, wet, or 'muggy,' or with a strong north-east wind blowing, they repair to the valley near to the river. During the early part of their arrival I have never seen any near to the nesting colony at Frizinghall, if, indeed, they are the actual birds which frequent it later. It may be, these early Swifts pass on to other districts to breed. Anyhow, about the 11th of May, the birds (about half the number that will eventually occupy it) will be seen circling round and flying about the nesting site. After their departure about the 19th of August, it is usual for several birds to return to the vicinity of their breeding quarters a few days after, and sometimes they will remain for several days, but they rarely stay later than the 26th or 27th of August. By their appearance I usually take these latter to be birds of the year.

It is rather a curious fact that immediately the Swifts have finally left their feeding area at Frizinghall, several House Martins take their place. These latter do not breed, so far as I know, within about a mile away, and I have never seen them there when the Swifts were about. I mention that as a curious fact in connection with *this* colony, and not because I wish to impute any antagonism between the Swifts and House Martins, which are usually good friends, and will be seen flying and feeding together in many places. For the last two or three weeks before their departure, if the weather be fine, many of the Swifts again visit the higher moorlands, as the first-comers did in the spring.

Last season I was greatly surprised to see a Swift hawking for food by the river, and evidently quite at home, on Oct. 7th. The weather was mild, with plenty of insect food about, and I watched it for half-an-hour in the morning, but it had apparently disappeared in the afternoon. This is more than a month later than I had previously seen the species in this district. Last year the remaining moiety of the Frizinghall colony departed on August 19th, but a few birds returned on the 24th and 25th, from which dates I had not seen any until the one on Oct. 7th. Almost every year very late occurrences of the Swift are reported from some part of the country—generally near to the coast—and they are most often erroneously chronicled as ' Late *Stay* of the Swift.' The observer usually states that none have been seen for some time previously, and no doubt this habit of the Swift in occasionally returning misled Gilbert White into believing that they were temporally coaxed out of their hibernating quarters by the mild weather.

My observations seem to show that with this species the date of their departure is fixed more by the forwardness of the young brood, and by their ability to undertake the long journey, than by the state of the weather, or of their food supply at the time of leaving. I find that in the finest summers, and consequently when there is the largest supply of winged insect food, this colony usually breaks up a day or two earlier than in the colder and wetter seasons, and they will leave sometimes when there is apparently an unlimited supply of food about. Nesting appears to be their sole object here, and as soon as this is completed their restless and active spirits fall an easy prey to the migration 'fever.'

Several writers have stated that during fine, warm, clear, and still nights in June, the male Swifts remain the whole night on the wing, while the females are sitting. Certain it is that often on such evenings and just before dark, several birds will gradually soar and circle higher and higher until they are no longer visible. Several years ago I spent some time in watching this curious vesper flight with a good field-glass, but each time I lost the birds in the darkness before any apparent descent had been commenced. However, about a quarter of an hour later, by standing beneath the eaves where a part of the colony nested, I frequently, quite distinctly, heard the fluttering in the darkness of one or more birds against the wall up above me, which convinced me that it was the return of the birds that I had been watching, to their nesting holes for the night.

Gilbert White states definitely that he has seen Swifts pairing in the air. This statement has been greatly ignored by most later writers, and some have even doubted it. I have never witnessed this myself, but Mr. Fred Jowett, a careful observer, reports the following occurrence : At Saltaire, on May 26th last year, at 8-20 a.m., he was noticing a pair of Swifts which were flying and sailing at a good elevation. Suddenly they appeared together as if one bird, slowly descending in a vertical line during this time, and shortly after they separated and flew about as when first noticed. Mr. Jowett afterwards heard from a gentleman who lives near that he had witnessed exactly the same action two mornings before, at the same place, and at about the same hour. It may be that this habit is better known than I suspect, but that it is not often recorded.

At this season of the year it is interesting to note with what eagerness Swifts will pick up, and take away, any small feather, straw, or light substance that is carried into the air by the wind.

It is easy to discover where they are nesting. During the

screaming, tearing, rushing flights of the males in large circles on fine mornings and evenings, it will be noticed that each bird will pass quite close to the entrance of the hole wherein its mate is sitting, sometimes even going out of its way to do so.

———◆◆———

THE PROTECTION OF THE BIRDS AT SPURN, ETC.

An Appeal.

At a meeting of the Wild Birds and Eggs Protection Committee, held at York on Saturday, Feb. 16th, the president, Mr. W. H. St. Quintin, J.P., M.B.O.U., etc., in the Chair, it was resolved to appeal to the naturalists of Yorkshire for subscriptions towards a fund, to be employed in effectually protecting the birds of Spurn Point, etc., by keeping watchers there during the nesting season.

Spurn Point is the only nesting place in Yorkshire of the Lesser Tern, and one of the only two places where the Ring Dotterel nests. The Oyster Catcher and Sheld Duck are also to be found nesting there, and it is feared that unless adequate protection is afforded, these birds may soon be wiped out as nesting species.

If funds will allow, the committee would also like to place a watcher at Hornsea Mere, where many interesting species nest, and also give rewards for the protection of isolated nesting birds, to ensure their safety, as was done with the Bempton Peregrines last year.

The following sums have already been promised—

	£	s.	d.
W. H. St. Quintin	3	0	0
T. H. Nelson	1	0	0
Oxley Grabham	1	0	0
C. E. Elmhurst	1	0	0
H. B. Booth	1	0	0
R. Fortune	1	0	0
T. Sheppard	1	1	0
Left from last year	1	0	0

Further subscriptions, which will be duly announced in this journal, are urgently needed, and may be sent to Mr. T. Sheppard, the Secretary of the Union, or to Mr. T. H. Nelson, Redcar, or Mr. R. Fortune, Harrogate, Secretary of the committee.

1. Pipistrelle Bat about to take flight.
2. Long=eared Bat asleep on Bracken. (See p. 115.)

It will be necessary to start the watchers about the middle of April, therefore the committee trust to have a speedy and generous answer to their appeal. Naturalists' Clubs in the County are specially asked to assist.

In view of the difficulties the police have to contend with in administering the Wild Birds' Protection Acts, from the fact that very few of them are able to recognise either the birds themselves or their eggs, the committee propose to appoint referees in various districts in the county, whose duties will be to assist the authorities by identifying any birds or eggs which may be submitted to them.

———◆◆———

Every Boy's Book of British Natural History. By W. Percival Westell. London : The Religious Tract Society, 1906. 279 pages, numerous plates. Price 3/6.

Mr. Westell knows how to produce a book. A short time ago we noticed his 'Country Rambles,' which had an introduction by Mr F. G. Aflalo. Then followed 'British Bird Life,' with an introduction by the Rt. Hon. Sir Herbert Maxwell, Bart., M.P. The present volume has an introduction by Lord Avebury. We are wondering who will write the introduction to Mr. Westell's next ! On the cover of 'Every Boy's Book of British Natural History' we find the name of Lord Avebury in larger type than that of the author, and we hasten to peruse his Lordship's contribution. We find *nearly two pages* from his Lordship's pen ! and nearly two pages of very general information too. We look for an opinion of the book—one paragraph after another—and at last, in the last line but one, we find it :—' *The photographs are charming*'*!* and on the very next page Mr. Westell informs us that ' *Mr. Sedgwick* is responsible for the whole of the photographs, and also for chapters 2 and 3 dealing with the camera and the uses to which it can be put.' The book is largely devoted to the birds ; and, whilst we have not checked every entry, the items in the present work are apparently entirely copied—perhaps just a little 'boiled down'—from 'British Bird Life.' The chapter on Mammals and Fish are admittedly largely drawn from Aflalo's 'Natural History of the British Isles'; Messrs. Arnold have given great assistance by their Life Histories of Fish, Insects, and other forms of British Wild Life ; Mr. Lucas' books on Butterflies and Moths have provided information on these subjects, and Mr. A. E. Burgess has given valuable assistance in connection with the botanical section. In this way, and with the help of the photographs and chapters by the Rev. S. N. Sedgwick, and the Introduction by Lord Avebury, has been produced *Westell's* 'Every Boy's Book of British Natural History.' It is, nevertheless, very attractive in appearance, most of the illustrations are really very fine (two of which we are kindly permitted to reproduce, see Plate XVII.), and there is no doubt it will appeal largely to the young naturalists for whom it has been prepared. There is an index occupying nearly three pages, and the price is very reasonable.

Messrs. Watts & Co. are to be congratulated on being able to produce Haeckel's 'Evolution of Man,' in two volumes at the phenomenally low price of 6d. each. The volumes form Nos. 26 and 27 of the Rationalist Press Association's cheap reprints. The first, dealing with Human Embryology, or Ontogeny, contains pages 1-178, and 209 illustrations ; the second is devoted to Human stem-history, or phylogeny, and contains pages 179-364, and 199 illustrations. We are not surprised to find on the cover the words 'Second impression, completing 50,000 copies.' At the price, no one ought to be without ' Haeckel's greatest work.'

ECOLOGY OF WOODLAND PLANTS NEAR HUDDERSFIELD.

"THE Ecology of Woodland Plants in the Neighbourhood of Huddersfield," which recently appeared in the *Journal of the Linnean Society*,* has won for its author the degree of Doctor of Philosophy of the University of Zürich. This is a guarantee of the quality of the work, but it must not be thought that the paper is of 'Swiss manufacture.' Even a casual perusal will reveal much careful investigation which could only be done on the spot, and the facts which form the basis of this paper were already recorded before Dr. Woodhead went to Zürich early in 1905. There, we have good reason for saying it, his work

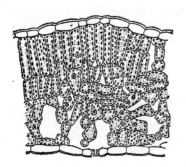 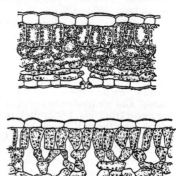

received the hearty approval of Professor C. Schröter, one of the leaders in Plant Ecology, and one can see the impress of this great master on the finished publication.

Readers of 'The Naturalist' have already had a sample of the kind of investigations in 'Notes on the Bluebell' ('Naturalist,' 1904, February and March). The present paper opens almost where the 'Bluebell' paper left off, the first study being Birks Wood, the one which was figured in 1904. Reduced figures are given of maps originally prepared on the Ordnance Survey '25 inches to one mile' sheets. One of the maps is a soil-map of the wood; another shows the distribution of oak, beech, Scots elm, sycamore, and conifers in the same wood. The other three maps show common plants of the undergrowth. A comparison of these excellent maps shows at a glance that the Bracken (*Pteris*) occurs mainly under the trees

---

* Vol. XXXVII., 1906, pp. 333-406.

with open canopy (oak and birch), and is almost absent under shade-trees (beech, elm, and sycamore). The Bluebell (*Scilla*), as already described in ' The Naturalist,' is more influenced by soil, being most abundant where there is a few inches of humus over loam, and almost absent on light sandy soil and clay. Two Grasses (*Holcus mollis* and *Deschampsia* or *Aira flexuosa*) are also abundant, *Holcus* mainly on the moist humus soils, and *Aira* on the lighter sandy parts. The author has already shown the common occurrence of *Scilla*, *Pteris*, and *Holcus* as the chief constituents of the undergrowth of the Yorkshire valley wood, and that they occupy three distinct zones in the soil (see fig. 11 in ' Bluebell,' ' Naturalist,' March, 1904). These three plants are thus non-competitive in the soil, and also to some extent non-competitive in their aerial parts, since *Scilla* is almost finished before *Pteris* comes. To describe this kind of plant association Dr. Woodhead suggests the term ' complementary association.' Birks Wood is taken as an example of a typical mixed deciduous wood of the Coal Measure area of this part of Yorkshire. The second study is a typical mixed deciduous wood of the plateau and slopes of the Millstone-Grit area. The woods actually investigated were those near Armitage Bridge, which lie on the slopes of the Netherton Plateau   The two maps showing the dominant trees and the principal plants of the undergrowth respectively are somewhat disappointing. The difficulty of making charts of woods on steep slopes is here exemplified, and it is one which occurs in all steep-sided valleys. It seems to us that here the cartographer will be forced to abandon the horizontal representation of these woods on the Ordnance Maps, and in the case of studies of particular woods, must resort to some convention by which the wood will be represented as flat. The ecological results obtained in the Millstone Grit woods again show the influence of light and soil. The higher slopes have a shallow, sandy soil covered with thin peaty humus ; the lower slopes have a deeper and moister soil, resulting from weathered shales. The Bracken favours the ground under open canopy and shuns the shade : on the lower slopes the conditions resemble a Coal Measure wood, and the undergrowth is somewhat similar ; on the dry, sandy upper slopes the Bracken is less favoured and occurs in patches, while its rhizomes, being unable to penetrate deeply, are found interlaced with and competing with those of the more abundant Ling and Bilberry. The Bracken thus occurs as one of the dominant plants of two plant associations, the distinction of which is emphasised in this paper. We summarise here the distinctive features :—

Name of Association...	Meso-pteridetum	Xero-pteridetum
Character of plants ...	Mesophytic (little adapted to drought)	Xerophytic (adapted to drought)
Soil...	Deep, moist, with humus	Shallow ; liable to dryness ; peaty humus.
Distribution in time and space	Complementary	Competitive.
Characteristic species	Scilla festalis	Calluna Erica
	Holcus mollis	Vaccinium Myrtillus.
	Lamium Galeobdolon	Deschampsia flexuosa.

This exact definition of these two plant associations is a distinct step in the right direction.

The next study is a comparison, by means of two maps, of the dominant woodland trees with the plants of the undergrowth. The two maps are a disgrace to the *Journal of the Linnean Society*. They include sixty-six square miles to the south of Huddersfield, and we know that the author prepared them with great care on large-scale maps. In the publication they have been reduced to a single-page demy octavo, and are also badly printed. The symbols are almost illegible, and the features intended to be shown can only be made out with great difficulty. The wood map indicates where trees have been found buried in the peat, and one sees how much higher the old woodland area has been. The Huddersfield district is shown from these maps to consist of three zones : (1) Moss-moor on peat, with Cotton Grass dominant and Bilberry, etc., on the more elevated and better-drained ridges ; (2) Millstone-Grit Plateau, an ericaceous zone, with Ling, Bilberry, etc., on shallow peat ; (3) Lower Coal Measure area, with deeper and moister soils, and a meso-pteridetum as the undergrowth of the woods. The names used are not quite aptly chosen. Geologically, the Millstone-Grit Plateau includes the Moss Moor, and in a vegetation study one would rather see terms used which indicate the nature of the vegetation.

The influence of geological formations is shown by two maps (p. 364), from which it is evident that the xerophytes are present on the drier soils of the Millstone Grit, while the mesophytes frequent the moister soils of the Coal Measures.

The second part deals with the anatomical structure of woodland plants which grow sometimes under the shade of trees, sometimes on dry or moist soils in the open. The nature of the changes can be seen from the figures which the Linnean

Society has permitted us to reproduce. Other plants examined and figured are Bluebell (*Scilla*), Hair Grass (*Deschampsia flexuosa*), Yorkshire Fog (*Holcus mollis*), Hogweed (*Heracleum*), Dog's Mercury (*Mercurialis perennis*), and Yellow Dead Nettle (*Lamium Galeobdolon*). These figures indicate that while most species of plants occur in definite plant associations, and are limited to certain conditions of ærial and soil conditions, they have considerable powers of adaptation to change of environment. This fact cannot be lost sight of in ecological botany, and the details shown by Dr. Woodhead emphasise the importance of careful examination.

The bibliography appended to this paper is a long one, and is well worth careful perusal, if only to gather some idea of the complex nature of these ecological studies.

The paper as a whole is an excellent example of the results which may be expected from careful ecological survey, although the author himself states that much still remains to be investigated regarding the relation of plants to their environment. Under the guidance of the painstaking and careful author of this paper, these more exact studies are in progress, and the results will be looked for with keen interest.—W. G. S.

The Science Year Book: Diary, Directory, and Scientific Summary. King, Sell, and Olding. Price 5/- net.

This indespensable volume increases in usefulness each year. No working scientific man can afford to be without it. The present issue has many improvements. To astronomers particularly is it invaluable, and contains much information. There is a review of 'Science in 1906,' in which the more important discoveries, etc., are enumerated; a useful but apalling glossary of scientific terms that have been recently introduced; a directory of Periodicals, Public Institutions, Societies, Universities, etc.; a Biographical Directory, and a daily diary. As a frontispiece is a portrait of Lord Rayleigh. An abridged edition, without the diary, but containing all the articles, is on sale at 3/-.

The Romance of Animal Arts and Crafts. By H. Coupin and John Lea. London: Seeley & Co., Ltd., 1907. 356 pages and 27 plates. Price 5/-

This is a companion to 'The Romance of Plant Life,' noticed in these columns for December last. It is a substantial book, and has a gaily coloured cover. This volume, too, is for the young naturalist, and contains much interesting information. The fact that it is a 'Romance' prevents criticism so far as scientific accuracy is concerned. The sub-title perhaps explains its scope 'being an interesting [!] account of the spinning, weaving, sewing, manufacture of paper and pottery, aëronautics, raft-building, road-making, and various other industries of wild life.' The animals are classified according to whether they are 'excavators and miners,' 'makers of mounds,' 'masons,' 'carpenters,' 'trappers,' 'harvesters,' etc., etc.; and the authors have been able to gather together quite a wonderful series of stories which should not fail to instruct the individual who is so handsomely catered for now-a-days—'the young naturalist.' The book is cheap, and is indexed.

OLD HALIFAX.*

In this handsome volume the Hon. Curator of the Bankfield Museum, Halifax, has published an admirable piece of work. He has permanently placed on record much information relating to old Halifax, and particularly to the part played by the Halifax coiners. This information could only have been obtained by years of patient research, and is of such a nature that, had Mr. Roth not secured it whilst opportunity offered, it would have been exceedingly difficult, if not impossible, for a future worker to have gathered the facts together. There is much in the book which we must pass by, as it hardly comes within the scope of this journal. There are illustrations and descriptions of what some Hull—and Halifax town councillors might call 'rubbish,' but which we are delighted to see Mr. Roth has secured for the Bankfield Museum. Many of the

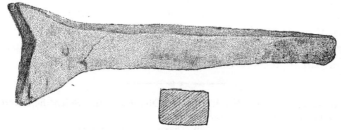

Fig. 1.—Object found in Shibden Park.

objects he figures and describes were likewise once familiar objects in other districts, and are preserved in various museums; a few, however, we make acquaintance with for the first time. All interested in bygone times, and all collectors of relics of the past, should see the numerous illustrations in the book under notice. At least one of the 17th century tobacco pipes figured on p. 281 appears to have been made in Hull,† and some years ago a large hoard of Portuguese and other counterfeit silver coin was found in Hull whilst excavating for one of the docks. These are very similar to and of about the same date as the Portuguese counterfeit coin made in Halifax, so fully described by Mr. Roth. From these and

* 'The Yorkshire Coiners, 1767-1783, and notes on Old and Prehistoric Halifax.' By H. Ling Roth. F. King & Sons, Halifax, 1906. Pages xxvii.+322, with numerous illustrations. Price 21/- net.

† See illustrations in 'Early Hull Tobacco Pipes and their Makers.' (Hull Museum Publication No. 6).

many other pieces of information in the book, there would appear in the past to have been more connection between Hull and Halifax even than might be assumed from the well-known beggar's litany.

It is the third part of the volume, dealing with 'Prehistoric Halifax,' that will appeal more to the readers of this journal. In this the author justifiably deplores the careless way in which the tumuli in the district have, in early times, been ruthlessly rifled, resulting in more harm than good being done, particularly as in most cases the objects obtained have been for ever lost or destroyed. More recently, however, the Blackheath Barrow, near Todmorden, has been carefully opened, the valuable results of which are described in detail by Dr. J. L. Russell in the final chapter. The scientific results achieved by these more

Fig. 2.—Bronze axe found at Mixenden.

systematic excavations contrast with the few facts left to us as the result of the 'prospecting' of earlier workers. Mr. Roth describes in detail the various vases; flint, polished stone and bronze implements, etc., of British date which have been found in the Halifax neighbourhood. How much more interesting and valuable his story would have been if the Bankfield Museum had been in existence, say, during the last hundred years, and all the local objects found had been placed therein! Of a few of the implements figured, however, we are by no means sure of the authenticity, an opinion apparently shared by the author himself. The objects figured herewith (fig. 1), 'found in the outcrop of Hard Bed Clay in Shibden Park,' *if* an implement, is unique. In the other illustration which we are kindly permitted to reproduce, is a bronze axe head of the Palstave type, found at Mixenden. There are over 200 figures in the book, which greatly add to its value, the reproductions of Mr. Oddy's drawings being of exceptional worth. It may seem a little ungrateful to find fault with such a useful work ; but how much more valuable it would have been had there been a good index. The 'index of names' which has been given is only likely to prove of use very locally. The index is missed all the more when the somewhat peculiar arrangement of the articles in the book is taken into consideration.

ANCIENT BARTON.

IT is half a century since Mr. H. W. Ball published his ' History of Barton-on-Humber,' a work which is now out of print. Since then much more valuable information has come to light as the result of the continued researches of various workers. Prominent amongst them is Mr. Robert Brown, junr., whose recent work is before us.* This is a solid and scholarly contribution to the literature dealing with the early history of North Lincolnshire. It is divided into three sections, viz., 'Romano-British Times,' 'Anglo-Saxon Times,' and 'Norman Times.' Further sections are promised, dealing with later periods. The picturesque old town of Barton, with its glorious Saxon Church of St. Peter, its haven, and its old-world associations, lends itself peculiarly to a detailed description such as is sustained in Mr. Brown's volume. The British remains found and occurring in the neighbourhood unquestionably point to the occupation of the area in pre-Roman times. Of relics of the Roman rule there are scores—in fact, in evidences of all the more important historical periods the district abounds. These have been carefully gathered together by Mr. Brown, and we trust soon to see the completion of his work. On the vexed question of the site of the Battle of Brunanburh the author has an interesting chapter. For many years this theme has been a favourite one with antiquaries living in South-East Yorkshire and North Lincolnshire, and it is perhaps not remarkable that in each case the author of a paper on the subject has thought the site of this famous fight to have been in the district in which he lived. Mr. Brown writes, ' I think we shall have reason to conclude that the great

---

* 'Notes on the Earlier History of Barton-on-Humber.' By Robert Brown, junr., F.S.A. Vol. I., to the end of the Norman Period, A.D. 1154. London : Elliot Stock. 133 pages and plates.

D

B

E

The British Pipe Fishes.

(A) *Syngnathus ophidion.*   (B) *S. acus.*   (C) *S. typhle.*

(D) *S. lumbriciformis.*   (E) *Hippocampus brevirostris.*

fight was partly in the parishes of Wooton, Barton, and Barrow, but chiefly at Burnham, in the parish of Thornton Curtis.' We must at once admit, however, that Mr. Brown has brought forward much more evidence in favour of the opinion he holds than has been usually the case. The volume has several illustrations, not the least interesting being the restoration of the Church of St. Peter as it appeared in the time of Edward the Confessor. We presume and hope an index to the whole work will appear with the final part published.

—◆—

An Outline of the Natural History of our Shores. By **Joseph Sinel.** London : Swan Sonnenschein & Co., Ltd., 1906. 347 pages.

This is the best book on this subject that we have seen for some time, having regard to the reasonableness of the cost. Mr. Sinel's position at the

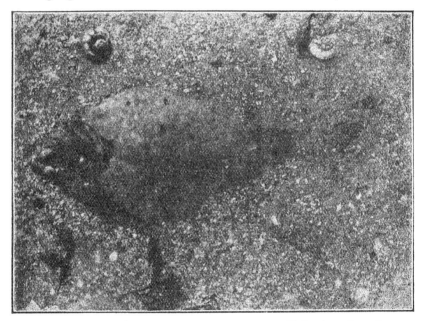

Plaice on Shell gravel, masking its outline, and modifying its colouration.

Zoological station, Jersey, has given him exceptional opportunities of collecting and observing the fauna of the shore, and of these he has taken full advantage. In a thoroughly scientific and up-to-date manner, and in plain language, the author describes almost all the forms of life likely to be met with by an ordinary worker on the coast, and the numerous illustrations, of which over 120 are from photographs, enable even a beginner to identify his specimens with ease. In addition to a concise review of the various divisions of the animal kingdom, there are chapters on collecting and preserving marine specimens, methods of microscopic mounting, etc., and on the Marine Aquarium. Two of the illustrations we are kindly permitted to reproduce (see plate XVI.).

In Memoriam.

SIR MICHAEL FOSTER,
K.C.B., F.R.S., D.C.L., D.Sc., LL.D., ETC.
1836-1907.

(PLATE XIX.)

It is with every sincere regret that we have to place on record the death of Sir Michael Foster, which took place as we were going to press with our last issue. In Sir Michael, Britain loses one of her leaders in scientific thought. He occupied a position in the nation's scientific welfare to which it will be exceedingly difficult to find a successor. As a physiologist he was the best known, but he was one of those who had so many interests, and took a leading part in such a variety of different channels, that he will be missed by very many indeed. Such was the esteem in which he was held by his fellow naturalists, that in 1899 he occupied the Presidential chair of the British Association. Another position of importance was that of the Secretary of the Royal Society, which he occupied for twenty-two years.

In both London and Cambridge, Sir Michael accomplished much as a teacher of Physiology, and in this way his influence has been far greater than can possibly be estimated.

In 1898 he was president of the Yorkshire Naturalists' Union, and many of our readers will remember the cheering Address he delivered at the Union's Annual Meeting at Scarborough, which was printed in this journal for July, 1899. He then took for his subject 'Integration in Science,' and after describing in detail the various ways in which science was specialised, so that even the Fellows of the Royal Society were 'no longer able to understand one another's speech,' he urged the Union to see that the old type of naturalist did not die out. 'It is for you,' he said, 'to gather and preserve the bits of knowledge which help to bind together diverging inquiries carried on in other places; it is for you to keep free from the rust of disuse the simpler way of asking questions from Nature without the complicated machinery which others use; the simpler way, which often brings answers of no little moment in their right places; the simpler way, which others may be apt to overlook.'

His contributions to literature are mostly relating to his favourite subject—physiology. He also helped in the editing of the well-known 'Scientific Memoirs of Thomas Henry Huxley.'

In his later years, he served his country as a Member of Parliament, and on such occasions as he spoke his utterances had always the greatest respect of the House. When his multifarious duties permitted him to spend a little time at his home in Great Shelford, Cambridgeshire, he devoted his attention to his fascinating hobby—that of gardening—and even here he was able to make most useful contributions to scientific botany.

Quite apart from his scientific attainments, however, will his death be deplored by a large circle of friends. To know him was to love him, and few had the greater respect of his fellow men than had the subject of these notes.—T.S.

Dr. W. M. BURMAN of Wath-on-Dearne.

On the last day of the old year, there passed away at Grange over-Sands, a veteran naturalist of the Gilbert White school. Dr. W. M. Burman was born at Wisbech, on the 23rd June, 1825. His father, two years later, removed to Wath, and there successfully built up an extensive practice in which, in later years, the deceased joined him, and continued after his death. It is interesting to note that this practice is now in the hands of the third generation; the eldest and only surviving son of the subject of this notice, after assisting his father for some years, having succeeded to it on the Doctor's retirement to Grange some ten years ago.

Naturalists, like poets, are said to be born, not made; be that as it may, we are sure the doctor was a born naturalist. There is little doubt that when very young he came under the influence of the clergyman, who, veiling his identity under the initial 'W,' contributed a list of the indigenous butterflies and moths of the district to the 'Wath Village Magazine,' in 1832. Later still, he enjoyed the friendship of another resident entomologist, the Rev. C. H. Middleton. In these congenial surroundings, notwithstanding the uncertain calls of a practice as extensive in area as in clientele, he gained a knowledge of the local fauna of an almost all-embracing character. He could remember when the glow-worm was a common insect in the lanes of 'The Queen of Villages,' where also the marbled white butterfly was not uncommonly met with. Botany, Entomology, and Astronomy were the chief branches of science in which he was at home, and he was ever ready to assist young beginners with advice; no question, however elementary, but would be answered in the most kindly manner, without a trace of impatience. As an entomologist he had studied most of the orders in a general way, while giving preference to the Lepidoptera. For this order he contributed many local records to Mr. W. E. Brady's 'List of the Macro-Lepidoptera of Barnsley,' which was published in the 'Transactions of the

Dr. W. M. Burman.

Barnsley Naturalists' Society,' 1883-1885. He devoted some considerable time to sericulture, and had reared most of the silk-producing Bombyces. For the last twenty years he had paid a good deal of attention to the Coleoptera, and at various times had spent a holiday at many of the famous collecting grounds, such as the New Forest (where he was successful in taking *Anthaxia nitidula*), Braunton Burrows, etc., etc.

Of the two orders mentioned, he had made fairly good collections, while Hymenoptera, Diptera, Arachnida, and

Acaridae contributed to its variety. The writer has pleasant memories of many happy hours spent with his old friend in the field and in his library surrounded by the many interesting natural objects with which it abounded. Though old when years are counted, his heart was ever young, his enthusiasm for nature study never waned. Although seventy-one when he retired from practice, and removed from Wath to Grange, he soon discovered kindred spirits there, with the result that the Grange Natural History Society was founded. Of this society he was the Secretary and prime mover. His delight on discovering *Anchomenus marginatus* in abundance at Grange was intense, his letters describing his captures abounding with careful details of first-hand observations. On December 3rd he lectured before the Grange Society on ' Incidents in an eventful life,' and a few hours later took a chill, which developed into pneumonia, and terminated fatally on the evening of the 31st. Thus at the ripe old age of eighty-one, came to an end a life of usefulness and many-sidedness. Here we are mainly concerned with the natural history side of his career, but it would be misleading to infer that this was the extent of his usefulness. He was all that we have said and more, indeed it is difficult to write with becoming restraint of the gap which the loss of our old friend has made in the lives of those who were privileged to enjoy his intimate friendship. Ever ready to take an intelligent and active part in any public or private work which commended itself to him, his place will be difficult to fill.

He was interred in the cemetery at Wath on January 4th.

E. G. B.

WILSON HEMINGWAY.

IT is with deep regret we have to announce the death of Mr. Wilson Hemingway of Dewsbury. Mr. Hemingway had been a member of the Yorkshire Naturalists' Union since 1893, and for long before that date as an Associate through his membership of the now defunct Dewsbury Naturalists' Society. Besides taking a great interest in the work of the Union at its meetings and excursions, which he often attended, he occupied a foremost place in connection with other Literary and Educational Institutions. He was one of the first members of the Brontë Society as well as the Dewsbury Naturalists' Society, and a good supporter of the present Technical School and of its forerunner the old Mechanics' Institution.

My last conversation with him was about three months ago, when he detailed to me some of his experiences at the Flamborough excursion and the Farnley Tyas Fungus Foray of the past year. Mr. Hemingway was a genial companion in the field, as the writer can testify on many an occasion.

In the Flamborough group ('The Naturalist,' 1906, p. 248), his life-like portrait * occupies a prominent place to the left of the picture.—P. F. L.

NORTHERN NEWS.

'Le Bambou, son etude, Sa Culture, son Emploi' is the title of a periodical published at Mons, Belgium, which is devoted exclusively to the culture and uses of Bamboo. The first volume has just been completed.

Mr. A. G. Tansley, of the University College, London, has been appointed Lecturer in Botany at Cambridge in succession to Mr. A. C. Seward, who has succeeded the late Prof. Marshall Ward in the Chair of Botany.

A White Blackbird and a Pied Sparrow are reported as living in the grounds of the Hancock Natural History Museum. When last we heard of them they were on the right side of the walls of the Museum.

At a recent meeting of the Doncaster Scientific Society the following gentlemen were elected honorary members of the Society :—Messrs. E. G. Bayford, C. Crossland, P. F. Kendall, G. T. Porritt, T. Sheppard, and J. W. Taylor.

Still further evidence of the spread of nature knowledge. The posters recently issued by the 'authorities' in reference to the new regulations relating to dogs, officially informs us that 'dogs *shall be animals* for the purposes of the following sections,' etc.

In the February Geological Magazine, Prof. E. J. Garwood has some 'Notes on the Faunal Succession in the Carboniferous Limestone of Westmoreland and neighbouring portions of Lancashire and Yorkshire.' The same magazine has an excellent portrait (with memoir) of Mr. W. Whitaker, B.A., F.R.S., etc.

In an article on 'Marble and Marble Working' in 'The Quarry' for February, we learn that '*Crinoidal*... applies to marbles made up of fossilised *shell fragments*. In some cases the shell formation is retained entire, in others it has been replaced by calcite crystals.' Before making any criticisms, geologists should read the article immediately following on 'The value of detonating caps in blasting!" A little further in the same journal a blasting accident is recorded at Penrhiwceiber, we hope an attempt to pronounce the name had nothing to do with it.

Another writer on crinoids, in 'The Country Side,' informs us that a pentacrinus consists of a cup-shaped body with a crown or arms 'attached by a stalk belonging to the Pentacrinidæ.' The stalk consists of 'ring-like or pentagonal joints,' and 'The whole group is *Palæozoic*, . . . these particular ones are Jurassic, ranging from the Trias to the present day,' '*In view of this scientific statement*,' the author adds 'it will be wiser to add nothing upon the "star-stone's" magical properties.' It will.

---

* With his hat upon his knee.

APRIL 1907.

No. 603
(No. 381 of current series).

THE NATURALIST.

A MONTHLY ILLUSTRATED JOURNAL OF
NATURAL HISTORY FOR THE NORTH OF ENGLAND.

EDITED BY

T. SHEPPARD, F.G.S.,

THE MUSEUM, HULL;

AND

T. W. WOODHEAD, Ph.D., F.L.S.,

TECHNICAL COLLEGE, HUDDERSFIELD.

WITH THE ASSISTANCE AS REFEREES IN SPECIAL DEPARTMENTS OF

J. GILBERT BAKER, F.R.S., F.L.S.,   GEO. T. PORRITT, F.L.S., F.E.S.,
Prof. P. F. KENDALL, M.Sc., F.G.S.,   JOHN W. TAYLOR,
T. H. NELSON, M.B.O.U.,   WILLIAM WEST, F.L.S.

Contents :—

LONDON :

A, BROWN & SONS, LIMITED, 5, FARRINGDON AVENUE, E.C.
And at HULL AND YORK.

Printers and Publishers to the Y.N.U.

PRICE 6d. NET. BY POST 7d. NET.

(*Continued on page 3 of Cover.*)

NOTES AND COMMENTS.

A NEW FALSE SCORPION.

In the Report of the Nottingham Naturalists' Society, which is reviewed elsewhere, Mr. Wallis Kew has a note on ' *Chernese cyrneus :* a recent addition to the known False-scorpions of Britain.' An illustration, which we are kindly allowed to reproduce herewith, and a very detailed description, accompany

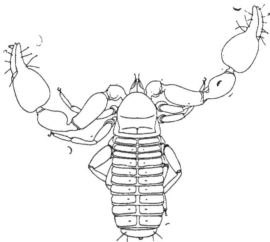

Mr. Wallis Kew's note. The specimen in question was obtained in Sherwood Forest, and of course is much magnified in the illustration. The specimen now added to the British list had previously been found in Corsica and other places abroad, and is of interest, as it is the largest of our false-scorpions.

THE MALTON MUSEUM

At the Annual Meeting of the Malton Naturalists' Society, recently held, it was announced that as the room in the Institute Buildings, in which the society's collections were housed, was required as a smoke and games room [!], the museum would have to be transferred elsewhere. It is sincerely to be hoped that a suitable suite of rooms will be secured for the exceedingly valuable specimens which the society possesses. Largely as the result of the labours of the late Samuel Chadwick, the Malton Museum contains a series of archæological and palæontological specimens of the greatest value, and most of the objects are of peculiar interest to Maltonians, as they have been obtained in, or close to, the town. British, Roman, and later relics of exceptional interest are included, whilst the fossils from the Corallian and other local deposits are known throughout the country for their excellence. It is sincerely to be hoped that Malton will be able to provide a suitable set of rooms—worthy of the collection—and thus enable the town to possess a local museum of great educational value.

AND ITS HISTORY.

At this meeting, Mr. M. B. Slater, who has done so much amongst the Yorkshire Mosses, gave an account of the history of the society. 'The commencement of the society,' said Mr. Slater, 'was immediately after a meeting of the Yorkshire Naturalists' Union, which was held at Malton in 1883. The late Prof. W. C. Williamson, then the President of the Union, was in the chair, and there was a good attendance of Malton people present.' Many of these were induced to take an interest in natural history, the local society was formed, and grew, and even at the present time has a balance in hand, although last year's subscriptions have not been collected !

MAMMAL *v.* ANIMAL.

Mr. E. Kay Robinson, having referred to 'Birds and Animals' when he should have said 'Birds and Mammals,' and having been corrected by a number of readers of *The Country Side*, seeks to justify his action in a recent issue of that journal. His substitution of the word 'animal' for mammal was the result of what he hopes will be *his* 'final decision' of a 'very troublesome question.' 'In ordinary conversation,' Mr. Robinson adds, 'we think of an animal as a hairy, hot-blooded creature with four limbs.' And in 'ordinary conversation,' too, a whale would be looked upon as a fish, we suppose. But this 'ordinary conversation' is incorrect, and just as inaccurate as it would be in ordinary conversation to refer to, say Mr. Robinson, as a raving maniac ; nothing could be further from the truth. Why, therefore, in a professedly popular journal, endeavour to perpetuate an error? Mr. Robinson admits that, *etymologically* speaking, 'mammals' is the correct word. Reference is made to the 'scientists' who strive to force upon us their strict distinction between mammal and animal, and we learn that 'there is no sufficient excuse for the *invention of such a word* as 'mammal' to replace the popular word animal. He has watched, with sympathy, the attempts of scientific writers to popularise the word mammal, but in his opinion they have completely failed. 'So *I propose* to drop the word mammal . . . and to use the word "animal" instead.' But, though Mr. Robinson may consider he has thus decided the future use of the word in scientific and popular literature, he condescends to ask to receive the opinions of 'naturalist friends' on the 'proposed *innovation* ;' hence these words. But we can assure Mr. Robinson that his 'proposed innovation' is merely a childish and unpardonable error, which, until recently, was made by most very small boys and girls, and also by illiterate 'grown-ups.' But we are thankful to say that in many of our schools to-day, the average intelligent child would be able to say that a whale was not a fish, and that animals were not necessarily all 'four-limbed, hot-blooded, and with hair.' Possibly, however, we are taking too seriously the nature of *Country Side*, but at present we have certainly the impression that it is intended to appeal to nature lovers, who, of all people, ought to be able to express themselves correctly.

NATURALIST ASSOCIATIONS.

In connection with the journal just referred to, is a 'British Empire Naturalist Association,' the aim of which would hardly seem to be to encourage an interest in nature study, so much as to encourage the sale of the journal. An application for particulars of the 'B.E.N.A.' means that one becomes a member of this association, willy-nilly, and a 'card of membership' is sent, which allows the holder 'and all members of the same household' to have 'all the benefits' of the association. One of these is that one may write 'B.E.N.A.' after his name, but not M.B.E.N.A., as the 'use of "M." to signify "Member," is only necessary in the case of societies which have separate grades, as "Fellows," "Members," and "Associates." In the B.E.N.A. *we are all equal.*' Lest the owner of this pennyworth of title should not be recognised, he can wear a badge, presumably to be obtained in due course, on receipt of stamp. One of these B.E.N.A. 'nature lovers' has just distributed 400 butterflies, and 430 birds' eggs, and wants further specimens. Strangely enough, one of the objects of the association is to secure protection for wild life ['animal,' bird, and insect presumably], wild plants, and interesting antiquities, *consistently with the legitimate interests of the sportsman and collector!* After perusing the 'objects' on the slip of paper sent, one gets an idea of another 'object' of the association, when turning the leaflet over, one is greeted with a recommendation in large type, to a certain firms ' dog-biscuits, dry.' (Similarly the prize for an ' Eyes and No Eyes' competition in the *Country Side* is divided amongst a number of successful competitors, who guessed the right tablet of Coal Tar Soap!). There is a list of the members of the B.E.N.A., but as we are informed that ' It must·be understood that this list is not accurate,' we need not seriously consider it. In one Yorkshire town at any rate, ' all the members' of one household are enjoying the benefits of the 'B.E.A.N.S.,' or whatever it is, and they represent six out of the seven for that town !

AND NATURALIST MAGAZINES.

Another advertising concern, which has already been referred to in these columns, the *Naturalists' Quarterly Review*, has reached its *fifth* number, notwithstanding the fact that the nature of its contents is as Davis-Westellian as ever. Under ' A Few Notes on Nature's Year,' Mr. Westell gives some characteristic notes. We are told ' to note the network of the bare trees, the robin sings again, rabbits become frolicsome, the song-thrush sings, February fill-dyke, earth-worms begin to move, beetles move, the mad March hare to be seen, and the woodpecker laughs.' And no wonder! The same issue contains a lengthy notice of ' Every Boy's Book of Natural History,' the writer of which apparently has a very different opinion of the merits of that work from that held by the writer of the review in our columns recently.

A DEFORMED BELEMNITE.

From time to time we have drawn attention to belemnite deformities, and have figured specimens from the chalk and from the Speeton Clay. Mr. C. G. Danford has favoured us

Deformed *Belemnites jaculum.*

with a further example, which is figured herewith. As will be seen from the tapering of the alveolar end, the species is *Belem-nites jaculum*, which is characteristic of bed C of the Neocomian clays at Speeton. It is nearly four inches long, and at the point of the guard is indented and twisted, obviously by damage or disease, during the life of the animal.

GRIFFITHIDES BARKEI.

In the 'Yorkshire Geological Society's Proceedings,' Dr. Henry Woodward describes a specimen of *Griffithides barkei* from Angram, in Nidderdale. This trilobite was found by Mr. E. Hawkesworth in the black carbonaceous shales occurring below the Millstone Grit, and a figure of the species, four times

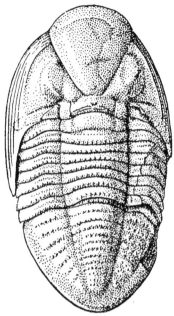

Griffithides barkei.

natural size, we are able, by the society's permission, to reproduce. Dr. Woodward is not justified in describing the trilobite as a new species, though he points out certain differences between Mr. Hawkesworth's specimen and that found by Mr. Barke on the same horizon in Glamorganshire.

———◆◆———

MAMMALS.

Remains of *Bos primigenius* near Doncaster.—Mr. Lloyd Roberts, the engineer for the South Yorkshire Joint Stock Railway, lately showed me a horn-core of Bos that had been found during excavations for the new line near Loversal. The core, of which the distal end is missing, measures 20 ins. along the outer side of the curve. Circumference at proximal end 14.5 ins., and at the broken distal end 6.25 ins.—H. H. CORBETT, Doncaster.

1997 April. 1.

A LARGE TROUT NEAR HARROGATE.

R. FORTUNE, F.Z.S.

A LARGE Trout was captured on September 28, 1906, at Beaver Dyke near Harrogate. Some little time after it was caught it weighed 8 lbs. Its length is 26 inches, and depth 6¼ inches. It was obtained when the reservoir, belonging to the Corporation, was being cleaned out. Beaver Dyke has, for some time, been renowned for its large trout, but never before has this weight and size been approached.

It was a matter of surprise to find that the water contained very few trout, only about fifty being observed, and yet it is

hardly surprising, for the amount of damage done by a fish of this size would be considerable, for, as a rule, these large fish are rabid cannibals, and worse than pike in their powers of destructiveness. The water is well rid of the monster.

A few Chub and Dace, and a quantity of Gudgeon were seen, probably it may be a puzzle how they got there, seeing that they are not found in the Oak Beck, or at any rate only a short distance from its joining the Nidd. There can be no doubt but that they have been introduced by anglers, who have had small specimens when spinning for trout, and at the end of the day have emptied their bait cans in the reservoir.

Cray fish are found here abundantly, many of them growing to a large size.

CORNUS SUECICA ON THE YORK MOORS.

HAROLD J. BURKILL, M.A., F.R.G.S.

THIS plant is recorded in Mr. J. G. Baker's 'North Yorkshire' as occurring in three localities, *viz.* Hole of Horcum, Cross Cliff Banks, and near Hackness. Mr. Gordon Home's 'Yorkshire Coast and Moorland Scenes,' published some two or three years ago, gave (erroneously) the Hole of Horcum as the only locality for the plant south of the Cheviots.

In may therefore be of interest to Yorkshire botanists to place on record the notes on this plant I have been able to make during the past fifteen years, partly from information supplied by others, and partly from my own observations.

(1) From information given me by the two authors mentioned above, I should say that the records for the Hole of Horcum possibly refer to the same patch of Cornus which is inside the Hole. In neither case, however, has the recorder seen the plant growing there.

(2) Cross Cliff Banks. This locality, mentioned by Mr. Baker, is probably identical with the one well-known to some members of the Scarborough Field Naturalists' Society, where there is a small bed of the plant, the precise position of which is carefully kept secret, and observations made every year by the society. In spite of these precautions, this patch may be in danger of extermination from ruthless collectors, as when I visited it in August 1904, a considerable amount of damage had been done to the plants by someone a few days previously. It was, however, fairly plentiful in 1905.

I have not heard the term 'Banks' applied to this part of Cross Cliff, but as this is by far the best known habitat of the plant, I have no doubt it is the one reported some years ago to Mr. Baker.

(3) Near the Derwent towards Hackness. This locality given by Mr. Baker is not one that I can identify. There are some likely places near Langdale End Village, and again lower down the valley on the hillside to the west of the stream, but I have not heard of the plant from this neighbourhood. Mr. Baker himself cannot give further information about this spot than he has in his book, and Mr. Wilson, the Hackness schoolmaster, who has succeeded in instilling a love of botany among his pupils, told me in 1905 he did not know of it in the district.

(4) The largest patch of the plant I know of is between the

Hole of Horcum and Goathland, and was pointed out to me by Miss Barker, of Scarborough, in 1904. Here it is strongly established, and extends for a distance of about one hundred yards in the heather. When re-visited in 1905 the plant seemed to have spread further down the hillside.

(5) Mr. R. H. Barker pointed out to me a small patch of the plant nearer Scarborough than the Cross Cliff locality known to the Scarborough Field Naturalists. As the plant here is nearly extinct, being apparently smothered by other plants, I prefer not to give the precise position, but it is not the place mentioned by Mr. Baker as being near Hackness.

(6) The plant was gathered some years ago near the head of Staindale, but I have not been on the hillside where it was found, and cannot say if it is still there.

(7) Last October I was told it had been gathered on one of the slopes of Troutsdale by a Scarborough botanist. This locality seems in every way a suitable one for the plant, and I should not be surprised to hear it occurs in more than one spot.

Thus the plant is well established on the moors between the coast and Newton Dale, and I should think there are probably several other localities besides the seven I have mentioned above. It may be grubbed up by collectors in some of the smaller patches, and so exterminated; but it seems to me that it does not get killed off when the heather is burnt, as it occurs in places where this is done at intervals, and near the largest patch I have mentioned the heather has evidently been burnt more than once during the last twenty or thirty years. I have not, however, come across the Cornus on a recently-burnt patch of moor, so my opinion is open to contradiction by observers who have more chances of studying the plant than I have in occasional holiday trips to the Scarborough district.

I was at Hawnby for three weeks this last summer, but did not meet with the plant on that part of the moors, though I was out on the hills nearly every day.

I understand Mr. Gordon Home has a second edition of his book in the press at the present time, and that the statement referred to above has been corrected.

━━◆◆━━

The first annual meeting of the Lancashire Amalgamation of Natural History Societies was held at the Accrington Mechanics Institute, on March 9th. Seventeen delegates attended. It was decided to commence an organ of the Union, to be called the 'Lancashire Naturalist.' This is to contain 16 pages a month, and is to be sold at one penny. If successful, greater things will be attempted.

THE DRIFTLESS AREA OF NORTH-EAST YORKSHIRE AND ITS RELATION TO THE GEOGRAPHICAL DISTRIBUTION OF CERTAIN PLANTS AND INSECTS.

FRANK ELGEE,
Middlesbrough.

ONE of the most remarkable features of the glacial geology of Yorkshire is the driftless area occupied by the moorlands of Cleveland, which also form one of the great botanical aspects of the county. In the words of Sir Archibald Geikie, 'these uplands appear to have formed an insular space round which the ice sheets swept, but which remained unsubmerged.' There is no question of any marine submergence whatever in this area, for Professor Kendall has conclusively proved to all who have examined the evidence fairly that a complicated system of extra-morainic lakes was held up in the valleys open to the ice sheets.* But the region occupied either by the ice or the lakes does not seem to have received the attention, from a natural history point of view, which it ought to have had. It has frequently been supposed that the driftless region was a barren desert, and only supporting a few Arctic plants, during the Ice Age, but no evidence can be adduced to substantiate this opinion. I hope to show that far from such being the case, the area in question probably maintained a fairly numerous fauna and flora during the Glacial Period. The object of this paper is, therefore, to briefly indicate what, in my opinion, was the botanical and zoological condition of the area at the climax of that period.

The driftless region measures, roughly, about thirty by eighteen miles, and embraces nearly the whole of the moorlands of north east Yorkshire. The dales on the north side of the watershed were, during the maximum extension of the ice, occupied by extra-morainic lakes, but the great valleys of Bilsdale, Bransdale, Farndale, and Rosedale were free of either ice or water. To the south the area was bounded by Lake Pickering.

With the oncoming of the glaciers, the animals and plants inhabiting the lowlands of England would either be driven southwards or exterminated. The views of Dr. Scharff and others are that parts of the British fauna and flora survived

---

* See 'Naturalist,' 1903, pp. 14-15.

in the south of England, and in sheltered places, throughout the Ice Age. With the retreat of the ice, the fauna and flora again spread into their former haunts, probably with the addition of further species from the Continent. But the south of the country need not have been the only place where the pre-glacial fauna and flora survived. It seems quite possible that many of its members escaped destruction on the extensive driftless area of north-east Yorkshire. The great dales south of the moorland anticlinal would afford ample shelter for numbers of animals and plants, whilst even the higher moors may not have been devoid of vegetation. Let us glance at some of the likely inhabitants of the region during the Ice Age.

The Flora.—When we consider the Arctic range of many of the moorland plants and the rigorous climates to which they are exposed, I think we need have little hesitation in concluding that they probably lived on the driftless area throughout the Glacial Period. Among them we may mention the Cotton Grass, Bilberry, Crowberry, Heather, Bracken, Birch, Sallow, the Lesser Twayblade, Potentilla, *Trientalis europaea*, and the famous northern species the May Lily, the Dwarf Cornel, *Carex pauciflora*, etc. In other words, the moors during the Ice Age must have been much the same as they are to-day so far as vegetation is concerned. It, however, necessarily follows that the moors must be of pre-glacial origin, though some of them are certainly of post-glacial age. Among these latter may be mentioned Eston Moor, the most northerly of all the Yorkshire heather-clad areas. As pointed out in my paper in this journal for August, 1906, p. 269, the whole of the Eston outlier was overridden by the Cheviot Glacier ; consequently, its moors are of post-glacial age. The drift here is very thin, and the hard sandstone crops out at the surface, forming the usual poor moorland soil. Some of the moors of the North Cleveland watershed are also of post-glacial origin, viz., Girrick, Wapley, and Easington Moors, which were heavily glaciated. It would seem as though the plants of the driftless area spread on to the lands deserted by the ice, and possibly in some cases across the lowlands of Cleveland. Round about Seamer, near Stokesley, Marton, etc., there are many place-names indicating that moors existed on the Cleveland plain since the period of the Scandinavian invasion of the district. This view receives confirmation when we learn that the superficial deposits at the above-named localities are chiefly of glacial sand and gravel, and Graebner has pointed out how heather moors develop on bare sand, *per se*

(Graebner, *Die Heide Norddeutschlands*). Cultivation has now reclaimed all these waste spaces, though whether they were always free from trees is doubtful.

The Fauna.—In considering the present moorland fauna we have to bear in mind that some of the species are of pre-glacial origin, and survived on this region, and that some have emigrated into the area after the Ice Age. It is not always easy to distinguish between the two, but a fairly safe guide is to be found in the present geographical distribution of the species. Those of northern origin, or that range into cold climates, are the most likely to have escaped destruction by taking refuge in the uplands; whilst those species of southern origin probably left the district, to reappear with the ameliora-tion of the climate. The former class is well represented by many species of Lepidoptera and Coleoptera among the insects, some of which are exclusively confined to the typical moorland plants enumerated above. Taking the Lepidoptera first, the following is a list of the species known to occur on the driftless region, with their geographical distribution and their food plants in the larval state. None of them, as a rule, are found else-where except on moors.

Agrotis strigula, North Europe (Calluna and Erica).

Agrotis agathina, West Central Europe (Calluna and Erica).

Celæna haworthii, North and North Central Europe (Cotton Grass).

Anarta myrtilli, North, Central, and South West Europe (Erica).

Hepialus velleda, North and Central Europe (Roots of Bracken).

Plusia interrogationis, North Europe and North Asia, in the British Islands becoming commoner northwards (Calluna and Erica).

Eupithecia nanata, Central Europe (Calluna and Erica).

Eupithecia minutata, Holland and Germany, probably a heath-frequenting form of *E. absinthiata* which is found in North Europe (Calluna and Erica).

Acidalia fumata, North, and the mountains of Central Europe, North Asia (Calluna, Erica, and Vaccinium).

Larentia caesiata, North, and the mountains of Central Europe, North West Asia, and America (Erica and Vaccinium).

Scodiona belgiaria Central Europe (Erica and Calluna).

Aspilates strigillaria, Central Europe (Erica and Calluna).

Cidaria populata, North Europe, Asia, and America (Vaccinium).

Cidaria testata, North Europe, Asia, and America (Calluna).

Ematurga atomaria, Europe, West-Central, and North Asia (Erica, Lotus, Trifolium).

Panagra petraria, Central Europe, West-Central, and North Asia, and Japan (Bracken).

Phycis fusca, North Europe and North America (Erica).

Polia solidaginis, North and Central Europe, North-West Asia (Vaccinium).

Venusia cambricaria, North and Central Europe, North Asia, Japan, North America (Mountain Ash).

In looking over this list we find that fifteen species are found in North Europe and North Asia, and therefore habituated to colder climates than that of Britain, and joining this with the further fact that species of the genera *Colias, Pachnobia, Plusia, Anarta, Cidaria* and *Eupithecia* are inhabitants of the Arctic parts of Europe and America, there seems little reason to doubt that many of the above species were enabled to live on the driftless region throughout the Ice Age. On the other hand there are some moor Lepidoptera which may have re-occupied the region in post-glacial times. Among them are *Saturnia carpini*, an insect of Asiatic origin; *Bombyx rubi, Lasiocampa quercus*, and *Spilosoma fuliginosa*, which, besides feeding on heather, live on many other kinds of plants.

The survival of insects on the driftless area receives remarkable verification in the following quotation from Heilprin's 'Geographical Distribution of Animals' (p. 280):—'The officers of the British North Pole Expedition, under the command of Sir George Nares, brought home a surprisingly rich fauna from the region [Grinnell Land] lying between the seventy-eighth and eighty-third parallels of latitude, comprising no less than forty-five species of true insects and sixteen arachnids, the former distributed as follows: Hymenoptera, five species (two humble-bees); Coleoptera, one; Lepidoptera, thirteen; Diptera, fifteen; Hemiptera, one; Mallophaga, seven; and Collembola, three. Among the Lepidoptera are a number of forms belonging to genera common in the temperate zones, such as *Colias, Argynnis, Lycaena*, etc., which appear the more remarkable, seeing that the species of this order are more limited in Greenland (with an insect fauna numbering eighty species), and that no forms are met with either in Iceland or Spitzbergen, although upwards of three hundred species of insects are represented in the former.'

Among the Coleoptera several distinctly northern and Alpine species now existing in North Yorkshire in all probability lived within the ice free region, e.g., *Pterostichus aethiops* and *P. vitreus*, both Alpine species ; and *Miscodera arctica*, recorded by Mr Lawson Thompson, from Stanghow Moor. Doubtless others have survived, especially those which live on the moorland plants, such as *Haltica ericeti*, *Ceuthorhynchus ericæ*, etc.

If these northern and Alpine species lived in sheltered places on the uplands during the Ice Age, it does not follow that no southern forms managed to struggle through thereon. If, as the Ice Age progressed, any species of southern origin contrived to escape on the driftless area, we ought still to find them either on the region itself or, allowing for dispersal since the close of the Glacial Period, just outside its bounds. The evidence, too, would be complete if the species also occurred in the south of England, but not in the intermediate area. Although such a case seems highly improbable, yet we actually have one in the Solitary Ant (*Mutilla europaea*), of which two specimens have been discovered in East Yorkshire, the first by Mr. Hey in 1903 ; and the second by myself in 1904.[*]

As this insect is a rare and interesting species, it will be worth our while to dwell a little longer on its geographical distribution, and see by what means it has arrived in East Yorkshire. Before the above records were made, the most northern locality for *Mutilla europaea*, according to Mr. Saunders, was Colchestsr in Essex ; nor since his great work on the Hymenoptera Aculeata of Britain appeared, have any except the Yorkshire examples been found north of that town (as he informs me in answer to a letter I wrote asking if such were the case).

We have here a very remarkable example of discontinuous distribution within our island, but before attempting an explanation of it we must glance at the further distribution of the species in England and Europe. In England it occurs principally in the sandy regions of Surrey, Dorset, Hampshire, and Berkshire, whilst on the Continent it is found in Sweden, Finland, Russia, Austria, Germany, and Italy.

If now we examine the distribution of the genus *Mutilla* in

---

[*] 'Naturalist,' 1903, p. 455 ; and 1905, p. 40.

Europe we find that its head-quarters are in France, and that the further north we go the rarer it becomes. Thus, in France there are thirty species, in Germany eleven, in Sweden two, and in Finland two. Russia has ten species (it is not stated whether all these are found in South Russia or not ; some occur in the neighbourhood of Elizabethgrad), Italy seven, and Greece three. Two species occur in Britain, *M. europæa* and *M. ephippium*, both practically limited to the south-east of England. From these facts it seems clear that the Mutillæ have spread over Europe from the south, and this inference is further confirmed by the fact that in North Africa twenty-one species live in Algeria, two in Tunis, two in Tangiers, twenty in Egypt, etc. Moreover, six species are common to Europe and Africa.

It is known from geological data that North Africa and the south of Spain were at one time connected by a land bridge, and across this !and must have come the European Mutillæ. The genus is certainly not of European origin, as over 1000 species are known from all parts of the world, chiefly the tropical parts of Africa, Australia, and South America. In the New World, the further north we go the rarer the genus becomes, as the following figures show :—

South America 133 species.
Central America 25 species.
North America 15 species.

The same decrease towards the north is shown here as in Europe.

It must, therefore, be concluded that ages ago the Mutillæ originated in tropical regions, whence they have spread over a greater part of the earth. *Mutilla europæa* and its congener entered England when our island was part of the Continent, and probably formed part of the oldest fauna of Britain, and belonged to the Lusitanian invasion of Dr. Scharff. It would then gradually spread over the country in pre-glacial times; with the oncoming of the Ice Age it would be exterminated in the glaciated area, but would manage to survive in the south and on the driftless area. This region is quite suited to its habits, possessing a sandy surface soil such as the insect loves. Of course, the bees on which it is parasitic must have survived as well. Moreover, the very fact that the ant lives at the present day in Sweden and Finland proves that it can withstand a cold climate. In this way would the curious distribution of this insect in England appear to be accounted for, and if we accept the opinion of Dr. Scharff 'that the climate of Europe during the Glacial Period was by no means so severe as we are often led to believe,'* the conclusions in this paper are strengthened. Many facts of geographical distribution can only be explained on the assumption of pre-glacial survivals.

Since the close of the Ice Age many of the upland survivors have no doubt spread on to the land formerly occupied by the glaciers, and in this respect it is worthy of note that the spot where I found *M. europæa* was the summit of the Brown Rigg oxbow, that grand memorial of the Ice Age in the neighbourhood of Robin Hood's Bay, described by Professor Kendall. The species has, therefore, apparently spread from the driftless region.

Further investigations are needed to firmly establish the views above set forth. More cases of the same nature as *Mutilla europæa* are required, and I hope will be forthcoming. The great valleys of Bilsdale, Rosedale, Farndale, etc., would well repay working, as they exhibit no traces of ice action ; besides, their natural history has not received that thorough investigation which it certainly deserves.

For the facts concerning the distribution of the Lepidoptera, I am indebted to Meyrick's ' Handbook ' ; for those concerning the Mutillæ to the catalogues of the British Museum, and to the 'Hymenoptera Aculeata of the British Islands' by Mr. Saunders.

———◆◆———

' One and All ' Gardening, 1907.—London : The Agricultural and Horticultural Association. Price twopence.

The Editor opens the work with an article on Country in Town, giving details of the movement for beautifying our towns and cities with garden features. The Hon. H. A. Stanhope writes on Some Useful Native Plants ; James Scott on Secrets of Garden Flowers and The Formation of Soil ; Horace J. Wright, F.R.H.S., on Onions ; Co-operative Gardens and Houses, by the late G. J. Holyoake ; Ivy Gardens ; Rhododendrons as Winter Flowers ; Shakespeare's Gardens ; The Colour of Flowers, etc. There is a Poet's Calendar of all the Months in the Year, by the late Nora Chesson.

Blackie's Nature Knowledge Diary, compiled by **W. P. Westell** (6d. net), is a pamphlet of ruled pages, resembling a school register, with an 'introduction.' It contains a ' Monthly Weather Chart,' with ruled squares, and marked ' bar ' and ' inches ' on the left. Then follows several pages ruled with headings, ' Date,' ' Barometer,' ' Thermometer,' ' Seaweed,' ' Rainfall,' etc. The book, however, is probably likely more to repel an interest in nature by the scholar than to encourage it. After justly urging the observer to make a ' note on the spot,' ' which is worth a *cart-load* of " recollections," it is pointed out that ' no record can be too trifling.' For a note book it may not, but we hope that when the scholar has filled his ' Diary ' he will not print it under the head of ' County Rambles,' or some such title !

———————

* ' European Fauna,' p. 68.

NOTES ON YORKSHIRE LEPIDOPTERA IN 1906.*

W. HEWETT.

THE past season has not been a good one for Lepidoptera, and Yorkshire Entomologists agree that 'sugar' has, with odd exceptions, been a failure all the season, and very few of even the commonest species have been reported as abundant.

The most noteworthy records are the following :—

Mr. Fieldhouse reports the capture of a hermaphrodite *Fidonia atomaria*, with one male, and one female antenna, one hind and two fore wings light like the female, and the other dark like the male.

Mrs. Lee, Huddersfield, took a specimen of the variety *Varleyata* of *Abraxas grossulariata* at large at Huddersfield, and Mr. Lee took one *Hydraecia petasitis*.

Mr. G. T. Porritt bred a brood of variety *Varleyata* of *Abraxas grossulariata*, the form breeding absolutely true. He also took a black form of *Fidonia atomaria* on Harden Clough Moors, Huddersfield, and *Tephrosia biundularia* at Huddersfield this year ; this last had not been seen in the district for probably forty years. *Selenia lunaria*, a species which is always a great rarity at Huddersfield, was taken by Mr. W. E. L. Wattam.

Mr. B. Morley, Skelmanthorpe, reports one long succession of failures from beginning to end of season, and says it has been very difficult to find anything to capture.

Cloantha solidaginis swarmed at Dunford Bridge at the end of August.

Mr. George Parkin, Wakefield, reports the capture of a specimen of *Chærocampa celerio*, the sharp-winged Hawk Moth at Wakefield, taken on the outside of a shop window on October 24th last, by Mr. H. Lumb. This specimen is now in the collection of Mr. Porritt. He also states that the dark variety of *Odontopera bidentata* is rapidly replacing the type form in the Wakefield district, just as the dark variety of *Amphydasis betularia* replaced the type of that species. Mr. Parkin also notes the sparrow as feeding on imagines of *Arctia lubricipeda* in his garden, and says sparrows are also particularly fond of 'Yellow Underwings' (*Triphœna pronuba*).

Mr. Beck, Bradford, records *Vanessa cardui* on Rombalds Moor, September 15th.

---

* Being the Report of the Entomological Section of the Yorkshire Naturalists' Union.

Mr. Stanger, Leeds, reports *Cilix spinula* plentiful at Beeston, June 6th. *Polia chi*, very dark, but not common, August 8th, and states that he bred a black variety of *Odontopera bidentata* at Leeds, and found *Vanessa io*, the Peacock Butterfly, flying in profusion at Bardsey, August 15th.

Mr. E. B. Tomkinson, Doncaster, reports *Argynnis paphia* and *Argynnis euphrosyne*, both at Edlington in August, *Vanessa cardui* at Edlington and Askern, and *Thecla W-album* at Edlington, *Sphinx ligustri* larva and imago from Hatfield, *Calligenia miniata* from Sandal Beat, *Acronycta alni* at Edlington, and three *Cerigo cytherea* at Edenthorpe. He obtained, last winter, twenty-six pupæ of *Amphydasis betularia* at Edlington, seventeen of these emerged black, and the others the ordinary form. He also captured a male *Orgyia gonostigma* flying in Greenhouse Park early in September.

Mr. L. S. Brady of Sheffield thinks melanism increasing in his district. The number of melanic specimens of *Venusia cambricaria* was certainly in greater proportion this year than formerly.

Mr. Hooper, Emby, near Wakefield, reports the capture of four male *Dasypolia templi*.

Mr. W. Hewett records the capture of a beautiful lilac coloured female variety of *Sphinx populi* at York, from which he obtained a large batch of eggs ; and also records *Hadena glauca* from Rombalds Moor ; *Cœnonympha davus* and *Hyria auroraria* from Thorne Moor, July.

Mr. J. Harrison, Barnsley, says the season seems to have been very bad ; he has come across *Prays curtisellus* and *Catopteryx cana* very freely (previously only odd specimens seen). He also bred and captured seven or eight varieties of *Pœdisca solandriana*, including the northern variety.

Mr. Porter, Hull, says *Cirrhœdia xerampelina* has been scarce this year, and *Agrotis ravida* gets scarcer year by year.

Mr. J. W. Boult, Hull, says he captured a few *Cœnonympha davus* on Thorne Moor, in June, and saw a swift fly after a male *Bombyx callunæ*, but it went down among the heather and the swift lost it. He has seen sparrows capture *Chelonia caja*, *Liparis dispar*, and other moths, as he released them, when rearing large numbers for varieties. He also saw a sparrow catch a cockchafer and fly away with it.

Mr. T. A. Lofthouse, Middlesbrough, says the season has been fairly good there. Took a single specimen of *Celæna haworthii* at sugar in his garden, also two or three *Anchocelis lunosa* and a few *Epunda lutulenta*, also the variety *capucina* of *Miselia oxyacanthæ*, in the Ayton district : although the type is generally common in his garden, has never seen the variety there, the place one would expect to find it if smoke has any-thing to do with melanism. *Noctua depuncta*, one or two specimens occurred ; a few *Vanessa cardui* also noticed. Autumn larvæ scarce.

Mr. Arthur Whitaker of Barnsley says 'sugar' a complete failure until October set in, when a few of the late species were attracted by it. A few *Cymatophora fluctuosa* were taken in its usual habitat in June. A specimen of *Sphinx convolvuli* was found at Hoyland, near Barnsley, early in September. He bred a black specimen of *Odontopera bidentata* from larvæ taken last autumn in Barnsley. The sudden appearance of a perfect swarm of *Callimorpha jacobeæ* on a waste uncul-tivated piece of land which he has known well for many years, afforded interesting ground for speculation as to how the insects reached, or were imported to the place, where he had never seen a specimen before. The nearest locality, he believes, where it previously occurred is distant somefive miles from the field in question ; some fourteen years since the field was used for growing grain and other crops, and *jacobeæ* certainly occurred nowhere near it. Owing to subsidence resulting from some shallow colliery workings, the surface of the field became too rough to cultivate, and was abandoned ; after this it gradually became over-run with ragwort, and for many years he had looked upon it as an ideal spot for *jacobeæ*, and wondered if the species would ever become established there. He was over the ground often last year, and saw no sign of it, and it came as a great surprise to him to find the perfect insects there in abundance, early in June this year.

Mr. Staniland, Doncaster, reports *Lobophora hexapterata*, and *Cidaria silaceata* at Wadworth in June, *Anaitis plagiata*, Edlington, August, *Chesias spartiata*, Edenthorpe, October, *Cucullia umbratica*, Edenthorpe, July, *Ino statices*, Wheatley Wood, *Asthena sylvata*, Edlington.

Mr. Thomson, Barnsley, took a dark form of *Agriopis aprilina* at sugar at Deffer Wood, Barnsley, in October ; base and centre of top wings almost black, sub-marginal band greenish bronze edged with white, outer band black, and hind wings perfectly black.

ON PECULIARITIES IN ATTIS SPIDERS.

W. W. STRICKLAND, B.A.,

THE Attis spiders are so extremely variable, that I doubt whether any sort of approximately exhaustive list has been made of their species. I have been several times to the Buitenzorg Gardens, and seated myself at the spot where I saw the first Attis spider that changed the colour of its eyes,* and many varieties have continually appeared. One of them was a small spider of a livid green colour, the same size as the green spider with yellowish markings previously described. To my great satisfaction I found that it too had the faculty of changing the colour of its eyes from black to green ; what was better still, I succeeded in catching it in a pocket-handkerchief, and taking it alive to the museum. Here Major Owens put it into a large glass bottle, and it speedily showed off its remarkable faculty, both to him and to the Curator, Mynheer Koningsbuyer. Finally, on my return to Batavia, I caught two specimens of a smaller kind of Attis, which possesed the same faculty. They were the colour of light whitey brown paper with a few darker markings. The change of colour of their eyes was from nearly black to nearly white (a very light whitey brown). Both the specimens were bottled and sent to our authority in spiders, Mr. R. I. Pocock, but whether they reached him alive I cannot say, as he has not yet acknowledged their receipt. At Gàroot, a mountain station on the Preanger (Battlefield), where the climate is relatively cool, a good many Attis spiders were caught ; two species of large brown ones, with nearly black eyes, and a medium vivid whitey brown kind with nearly black stripings and mottlings. There was also a fourth kind, intermediate in colour between the brown and the whitey brown ones. All I caught I bottled alive, and they lived together in tolerable harmony so long as they had nothing to eat; one day, however, in a bottle containing three of the large dark brown kind and one smallish whitey brown one I put a fly, thinking it would help to keep them alive and vigorous. The way one of the large brown Attides leapt upon the fly and killed it would have delighted those members of the sporting world to whom bull baiting and cock-fighting appeal irresistibly. Next morning, however, I found that the introduction of the fly had had tragic results. One of the large spiders was dead and another dying. Whether the fly had caused a surfeit, or been of a poisonous

---

* See 'Naturalist,' November, 1906, pages 401-402.

kind, or, what is most likely, the spiders had fought over the remains with fatal consequences, is uncertain. I now put a smaller specimen of the whitey brown spider into the bottle, also with fatal results. The larger specimen already there, flew upon the intruder with merciless ferocity and bit it. I withdrew the victim as quickly as possible, but it was already almost dead, and very soon breathed its last.

When left in the bottle without flies, these Attis spiders make the best of things, spin a small hammock and sit there, apparently perfectly happy, for an indefinite time. They seemed to keep awake, but may ultimately hybern- or astivate.

There are, then, at present known, four kinds of Attis spiders that have the faculty of changing the colour of their eyes :—

(1) A relatively large kind with exceptionally large pair of eyes, abdomen green : eyes change from black to green.

(2) A smaller kind : green with yellow stripes and markings : eyes change from black to green.

(3) Another, livid green : same size as (2), eyes change from black to green.

(4) A still smaller kind : very light whitey brown with darker markings : eyes change from nearly black to very light whitey brown.

Three of these kinds are insignificant in size, and it was only the mere chance of the observing of this faculty in the large and apparently very rare Attis spider that caused me to observe it in the smaller ones. It is not surprising, therefore, that the faculty has apparently never been observed before by naturalists, whose chief object is often to secure as great a variety as possible of species in spirits.

———◆●———

Birds Shown to the Children. By **M. K. C. Scott.** Described by J. A. Henderson. 112 pages, 48 coloured plates. Price 2/6. T. C. & E. C. Jack, Edinburgh.

Flowers Shown to the Children. By **Janet Harvey Kelman.** Described by C. E. Smith. 154 pages and 48 plates. Price 2/6. T. C. & E. C. Jack, Edinburgh.

These two volumes are of the 'Shown to the Children' Series, edited by Louey Chisholm, and Messrs. Jack are to be congratulated on producing two very tasteful little volumes, which will undoubtedly appeal to the young children for whom the books have been specially written. The coloured plates are usually very fine indeed, and are not misleading by being over-coloured. In the second book the plates are arranged according to the colours of the flowers, so that the child can readily find the picture and description of a specimen he may meet with whilst on a walk. In each case the commoner species are dealt with, and there is just sufficient letterpress—of the right kind—to interest and instruct the youthful reader.

NOTE ON A LIASSIC CONCRETION.

F. M. BURTON, F.L.S., F.G.S.,
Gainsborough.

A RATHER remarkable concretion has been met with in a well which has recently been dug near a cottage at Blyton, a village about four miles from Gainsborough, situated on the Lower Lias. It consists of indurated clay, of a round disc-like shape as big as a small cart wheel, thinning off at the outer rim. When broken—and it had to be broken in pieces before it could be got out—the interior showed, in places near the circumference, vertical, fusiform cavities, $2\frac{1}{2}$ inches in height and about an inch apart, crowded on their surfaces with small, almost pulverulent, dog-tooth crystals of calcite, much stained with impurities, giving the mass somewhat the appearance of a huge ammonite with its septa.* In the middle portion the nodule is $4\frac{1}{2}$ inches thick, and judging from the curve of the outer rim, when entire it must have measured quite 3 feet in diameter, and probably more. Besides these vertical openings near the outer rim, there were many slits and cavities in the middle portion of the nodule, all lined with similar crystals, but on the outside of it there was no appearance of anything, the surface was homogeneous and smooth throughout. There is, of course, nothing unusual in this inner shrinkage and crystallisation, the cause of it is well known. Some organic remains which had collected (perhaps in a slight hollow of the matrix assuming its shape) had become enveloped in a crust of clay, and the organisms, shrinking first before the outer integument, secreted the crystals and caused the fissures which are found in it.

Blyton lies on the fringe of the *Am. angulatus* zone of the Lower Lias, and on examining the heap thrown out in digging the well I found an ordinary septarian nodule, so frequent in these deposits, with some thin pieces of hardened limestone containing worn fragments of *Gryphœa* and other fossils. The well is 14 feet deep, and as the lowest zone of the Lower Lias, *Am. planorbis*, occurs in this neighbourhood at about the same horizon—being found near Corringham, Springthorpe, Lea, Gate Burton, and Marton—this concretion may belong to that zone which, in some places, shows signs of having been deposited

---

* The specimen sent to us is certainly remarkable for the regularity of shape, and distance apart, of the lenticular cracks, which are, of course, fusiform in section.—EDS.

1907 April 1.

in estuarine beds near the mouths of rivers, where organic accummulations of this nature might, one would think, be more likely to occur than in the *angulatus* zone above it.

I am sending characteristic specimens of this concretion to the Museums at Hull and Lincoln, which latter, to my great content, is now an accepted fact. May it, in course of time, imitate the former in its usefulness !

◆◆

Messrs. Charles Griffin and Co. have issued the twenty-third annual **Year Book of Scientific and Learned Societies of Great Britain and Ireland:** a record of the work done in Science, Literature, and Art during the Session 1905-6 by numerous Societies and Government Institutions. Not only does it contain useful information in reference to the various scientific societies—large and small—throughout the kingdom, but what is of more value to working naturalists—lists of the papers read at the meetings and published in the transactions of these societies.

Quarterly Record of Additions, No. XIX.—Notes on a collection of Roman Antiquities from South Ferriby in North Lincolnshire (two parts). **Guide to the Municipal Museum** (second edition). (Hull Museum Publications 37-40).

These latest issues sufficiently prove that the high standard of excellence set at the initiation of the series is being maintained. A cursory examination of them will show the value of making a museum a municipal institution. Most people would imagine that after nearly five years of such strenuous work as the Curator has done at this museum, the number of additions as well as their comparative interest would decline. The latest quarterly record would quickly disabuse any such notion. It opens with a highly interesting account of Leather Jacks, illustrated by figures of specimens now in the museum ; and is followed by a comprehensive sketch of relics of our grandfathers' days. As we look at the figures of moulds for ginger bread, we are carried back to our school-boy days when birds and beasts, such as could only exist in the Never-Never Land, ornamented with gold leaf, were not uncommon objects in old dames' windows. Whether this is a proof of a greater conservatism in the West Riding than in the East, or merely indicates that the writer is contemporaneous with the grandfather of the author of this pamphlet, we need not stay to consider. It is enough to know that in Hull, at anyrate, these vanishing land-marks of our progressive civilisation are being zealously collected and preserved. Our grandchildren will have cause to put a much higher value upon them than perhaps it is possible for us to do, for they will never have that familiarity which the proverb tells us breeds contempt.

The two pamphlets relating to Roman Antiquities from South Ferriby are very valuable indeed, and deserve careful perusal. Beyond stating that these antiquities are mainly brooches and the like, there is no need to say more. Without the pamphlets themselves, any notice would be inadequate. The second addition of the guide is a distinct improvement on the first, except in one particular, *viz.*, the absence of the index. The descriptive matter has increased from 35 to 40 pages, and a comparison of the plans in the different editions will show more eloquently than words how much the contents of the Museum have been added to. With this in hand, a visitor to the Museum will make a more intelligent survey of its contents. One interesting fact deserves special mention. We refer to the varied nationalities of the visitors. In a very limited period Denmark, The United States, Canada, Norway, and Australia were each represented, in addition to other visitors from widely different portions of our own country. In this way the excellent institution which Mr. Sheppard directs has a sphere of influence co-extensive with the globe.—E. G. B.

TWO NEW YORKSHIRE HEPATICS.

WM. INGHAM, B.A.

Lophozia atlantica (Kaal.) Schffn.

In the *Journal of Botany*, April 1902, Mr. Symers M. Macvicar published a short account of this plant as a new British Hepatic. It was gathered at Dirlot, Caithness, August 8th, 1901, by the Rev. David Lillie of Watten.

The plant is of about the size of *Lophozia gracilis*, but it may be separated from that by the absence of small-leaved attenuated stems, and by the leaves being very concave, also by the presence of Amphigastria or Stipules. The leaves are either two-lobed or three-lobed.

I found this plant at Hebden Bridge during the meeting of the Yorkshire Naturalists' Union there on 11th June, 1904. It grew in large patches on the blocks of Millstone Grit, and is no doubt a characteristic Hepatic of West Yorkshire on that rock. If any reader has gathered what is apparently *Lophozia gracilis* on this rock-formation, by re-examination he will probably find it is *L. atlantica*. The former seems confined to limestone; at least that is my experience.

The stipules (never found on *L. gracilis*) are well developed on the Hebden Bridge specimen. Kaalaas, who described the plant, allowed too little weight for this feature, but Dr. Arnell has pointed out their presence in Swedish specimens, as an important distinction from *L. gracilis*. During my examination a short time ago of the Hepatics collected by Mr. Waterfall of Bristol, *Lophozia atlantica* turned up, gathered by him on 10th July, 1886, on Mellbreak Fell, Cumberland, and last year Mr. Macvicar found it in v.c. 100, Clyde Isles.

In Norway it has been recorded near Stavanger, and on the Island of Stördo, and it has been found on the Faroë Isles by Herr C. Jensen.

Its distribution as far as known, is entirely western, hence the name *atlantica*.

Lophozia badensis (Gottsche) Schffn. (J. acuta a. Lindenb.; J. luridula Wils.).

This Hepatic, which has been much confused with *L. turbinata* (Raddi) Steph., and also with small forms of *L. Muelleri* (Nees). Dum, was published as a new British species in the *Journal of Botany* for February 1907.

The original description of this plant is given in ' *Musci*

Asiae *borealis*,' by S. O. Lindberg and H. W. Arnell, October 1888.

Mr. Macvicar and myself find the following points the most useful in distinguishing *L. badensis* from *L. turbinata* (the latter such a characteristic Hepatic on the magnesian limestone of West Yorkshire).

In *L. badensis* the leaves have a broad base (in *L. turbinata*, a narrow base, except generally in the male plant), the antical base is decurrent, and the cells smaller and slightly thickened at the angles (in *L. turbinata* the leaves are not decurrent at the antical base, and the leaf cells are larger without thickenings at the angles), and the stem has brownish copious rhizoids, with the postical side of stem brownish (in *L. turbinata* the stems are concolorous, with comparatively few rhizoids, and the leaves are generally more remote and lie flat).

Habitats.—Finmarken, Dovrefjeld (Jerkin), Foldalen, Sweden, Findland.

Scotland—Ayrshire, Jan. 17, 1883 (C. Scott); Edinburgh, Sept. 12, 1904 (J. McAndrew); Fife, Nov. 21, 1903 (W. Evans); E. Rosshire, Oct. 1903 (Miss K. B. Macvicar).

England—Sussex, Feb. 1907 (W. E. Nicholson). Yorkshire, Castle Howard (Spruce); Knottingley, in Magnesian Limestone quarry, Oct. 14, 1898 (W. Ingham); Marr, near Doncaster, Sept. 16, 1902, this plant very flaccid and mixed with mosses on wet floor of quarry (W. Ingham); Bowes, by the R. Greta, during the Y. N. U. Excursion, Aug. 3, 1903 (W. Ingham).

COLEOPTERA.

Rare Beetles from the Doncaster District.—Among some Beetles lately sent by me to Mr. Donisthorpe for identification, the following rare species are worthy of note. *Anistoma lucens* Fair. taken in Wheatley Wood, August, 1904. *Salpingus foveolatus* Ljun. taken in Wheatley Wood, March 1903. *Philonthus addendus* Shp. taken at Potteric Carr, March 1904. *Philonthus proximus* Kr. taken at Finningley, August 1903. *Ischnoglossa prolixa* Gr. taken in Sandal Beat, May 1905.

On Thursday, February 28th of the present year, I visited Wheatley Wood along with Mr. Bayford and my son, when we had the good fortune to take eight specimens of the very rare *Carpophilus sexpustulatus* F. They were obtained from dead Hoodie-crows on a 'keeper's tree!' This is a most unexpected habitat for the species.—H H. CORBETT, Doncaster, March 6th, 1907.

THE CHEMISTRY OF SOME COMMON PLANTS.

P. Q. KEEGAN, LL.D.;

Patterdale, Westmorland.

COMMON ORCHIS (*Orchis mascula*).—This plant varies considerably in stature, colouration and scent, but it is certainly a highly remarkable native or denizen of cold, elevated and moist localities. As might be anticipated, its chemistry presents some rather remarkable features. On 20th June the overground parts yielded only 1.7 per cent. in dry of wax, with some fat-oil and a mere trace of carotin ; the alcoholic extract contained a kind of tannin (or rather a resinous glucoside), which does not precipitate gelatine, and with iron salts gives a brownish-green colouration, it precipitates bromine water, and yields a decomposition product resembling a phlobophene when boiled with dilute HCl ; there was no tanoid, but some resin which dissolved in sulphuric acid with a brown colour passing to a splendid violet ; there was much sugar, mucilage, and oxalate of calcium, but no soluble proteid or starch ; the ash amounted to 6.5 per cent., and yielded 47.3 per cent. soluble salts, 5.1 silica, 20.5 lime, 3.8 magnesia, 7.8 P^2O^5, 3.8 SO^3, and 7.8 chlorine. It is evident from the foregoing analysis that the special feature of the plant is the great richness in carbohydrates and the organic acids produced thereby, the oxidation proceeding much further than in the case of the Parsley Fern, while the large amount of water in the tissues favours and makes the persist the phenomena of the dissolving power of the ferments (diastases) on the starch or other carbohydrates originally produced. Volatile oils and resin are the chief products of deassimilation. The root-knobs, which in this species are undivided, contain, according to De Dombasle, a volatile oil and a pungent bitter principle, and according to Robiquet have no starch. Another author asserts that in the autumn the old bulbs contain no starch and the young bulbs have very much starch, whereas during the flowering time it is absent in both. In July I found that the old bulbs had much starch, but there was none in the new organs, and there was no tannin, but much gummy matter (mannan) not precipitated by peracetate of mercury and not coloured by iodine.

BOG ASPHODEL (*Narthecium ossifragum*).—Boggy heaths and spongy bogs, the swampy and marshy areas of wild moorlands are the habitats of the elegant golden spires, stiffish stems, and

spear-shaped leaves of this attractive plant. The chemistry is
of special interest of itself, but as compared with that of the wild
hyacinth, orchis, and other Monocotyledons it becomes of very
eminent importance. About the year 1861 Waltz analysed the
plant and found therein a white bitter principle of very rancid
taste and a white non-volatile acid, both soluble in ether ; but it
is difficult to make out exactly what either of these constituents
is. On 3rd August the dried overground parts yielded 2 per
cent. wax with a little carotin, and a yellow substance soluble
in sulphuric acid with a deep brown colour ; the alcoholic
extract was bitter and acid, and contained a tannoid which
precipitated bromine water, and seemed to be like quercitrin, but
on boiling with dilute HCl it gave a granular crystalline deposit
soluble in ether, and with sulphuric acid a brown solution ; it was,
therefore, a resinous bitter principle (probably identical with the
scillin of Squill) ; there was also a considerable amount of free
resin, a little sugar, very little mucilage, a good deal of soluble
proteid, much starch, but no oxalate of calcium ; the ash
amounted to 4.3 per cent., and contained 66.3 per cent. soluble
salts, 4.5 silica, 8.7 lime, 5.9 magnesia, 5 P^2O^5, 6.2 SO^3, and
17.5 chlorine. The special features of this analysis are
extremely remarkable, viz., much starch in conjunction with
much chlorine, and much organic acids along with very little
mucilage or lime or its oxalate. In many of the allied monoco-
tyledonous plants there is no starch, although there may be
much mucilage and acids and much chlorine at the same time.
The tissues of the Bog Asphodel are comparatively dry, it
absorbs little water from the soil ; those of the other plants are
very moist, and, as we have seen, a large percentage of water
favours and prolongs the conversion of starch into sugar and
the production of mucilage, and of soluble carbohydrates
generally. On the other hand, the tendency in the Bog
Asphodel is towards the production of insoluble carbohydrates,
roughly the crude fibre amounts to 58 per cent. (that of the
Orchis is only about 40 per cent.). The slender root fibres
contain much starch internally. The aqueous extract of the
whole plant yielded when fused with potass protocatechuic acid
and phloroglucin.

BISTORT (*Polygonum bistorta*).—This is not a native plant in
Lakeland, but it has strongly asserted and secured itself in
certain special spots. It is what may be called a decisively
chemical plant. On 5th June the dried leaf blades yielded 2 per
cent. wax with a moderate quantity of carotin and a trace of

volatile oil ; the alcoholic extract contained quercetin and a phlobaphenic tannin which precipitates gelatine and bromine water, a trace of free phloroglucin, little or no sugar, and a resin which dissolves in sulphuric acid with brown to deep violet colour ; there is considerable pectosic mucilage extracted by dilute soda, some pararabin, no starch, and considerable oxalate of calcium ; the ash of leaves and petioles amounted to 8.5 per cent., and contained 41.4 per cent. soluble salts, 3.4 silica, 22.7 lime, 7.3 magnesia. 3.7 P^2O^5, 3 SO^3, and 5.4 chlorine. The analysis indicates the general features or condition of a normal plant, *i.e.* a plant whose vegetation is flourishing and wherein the albumenoids and soluble carbohydrates are fulfilling, so to speak, their chemical activities, oxidation proceeds far, and the products of deassimilation are well represented. The flowers contain a tannin same as in the leaves ; the pink pigment is not fully developed, but is shown off to best advantage by the configuration of the perianth ; the ash of the spike-like raceme amounts to 6.4 per cent. in dry, and contained 27.5 per cent. soluble salts, 20.3 lime, 9 magnesia, 14.7 P^2O^5, 5.4 SO^3, with considerable manganese and little carbonates. The twisted, much branched rhizome contains much starch which is edible and nutritious, also oxalate of potass, and a large quantity (about 20 per cent.) of soluble and insoluble tannin which is not a glucoside, and does not yield gallic or ellagic acid when fused with caustic alkali or heated with dilute sulphuric in a closed tube.

———◆◆———

A 'young man, aged 26,' advertising in a contemporary, 'desires a place as Assistant with Museum.' He is 'an excellent collector and setter of Lepidoptera,' and *can also make Bird-skins.* He ought to find a post readily.

We regret to record the death of Councillor J. E. Robson, of Hartlepool, at the age of 74. He was a member of the Yorkshire Naturalist's Union. Also of Mr. B. Hirst, of Oldham, who has been connected with the same society since 1890.

Sir Archibald Geikie has been elected President of the Geological Society of London for the third time. Though near seventy years of age, he is able to occupy many important offices, not the least of which is Secretary of the Royal Society, which post he has held since 1903.

Nos. 1 and 2 of Vol. 16 of the 'Irish Naturalist' have been issued together, and form perhaps the finest part of that journal yet issued. It is entirely devoted to 'Contributions to the Natural History of Lambay, County Dublin,' and is admirably illustrated.

At a recent meeting of the Lancashire and Cheshire Entomological Society, Mr. C. H. Forsyth exhibited about 90 species of North Lancashire tortrices, collected near Lancaster. These included *Sciaphila penziana* from Arnside, *Conchylis alternana, Aphelia ossiana, Grapolitha penkleriana,* and *Dicrorampha saturnana.*

In Memoriam.

THE RT. HON. LORD LIVERPOOL, F.S.A.

WE much regret to learn, on going to press, of the death of the Rt. Hon. the Earl of Liverpool, at his residence at Kirkham Abbey. Lord Liverpool, who was perhaps better known under his late title, Lord Hawkesbury, was much esteemed in the county. As president, for many years, of the East Riding Antiquarian Society, he has taken an active part in the further-ance of the study of the antiquities of East Yorkshire; he regularly attended the society's meetings, and for several years its Annual Volume of Transactions has contained a lengthy review from his pen. The last volume issued, which contained 168 pages and 80 plates, was entirely devoted to his work. His lordship was a life member of the Yorkshire Naturalists' Union, having joined that association in 1890.

One of the last occasions on which he made a public appearance, was on the occasion of the opening of the Wilber-force House Museum at Hull, in August last. It was then seen that he was suffering from a painful throat trouble, from which he does not appear to have ever recovered.

FIELD NOTES.

TRICHOPTERA.

Halesus digitatus at Huddersfield.—Among some Trichoptera, recently given to me by Mr. B. Morley, taken by himself during last year at Skelmanthorpe, was a fine specimen of *Halesus digitatus*. The species has not previously been recorded for south-west Yorkshire, although the closely allied *H. radiatus* seems to be abundant on all our larger ponds. Other recorded Yorkshire localities for *H. digitatus* are Bishop's Wood and Castle Howard, but then only as casual specimens. —GEO. T. PORRITT, Huddersfield, February 6th, 1907.

—:o:—

MOLLUSCA.

L. maximus vars. aldrovandi and bicolor in Lincoln-shire.—In the spring of 1906 I had occasion to remove my collection into more convenient quarters; when doing this I came across two bottles, labelled, and each containing a preserved slug, which had been laid aside and forgotten. Both slugs were found in my small garden, 8 Bridge Street, Louth, which is very cold and damp. Mr. Roebuck has identified one of them, found in 1903, as *Limax maximus* var. *aldrovandi* Moq-Tand. The other specimen will be noted in the forthcoming part of Mr. J. W. Taylor's Monograph.

On July 10th, 1906, 11 p.m., I had the good fortune to find, in the same place as the above, a very fine example of var. *bicolor*. It might almost have been the model for Mr. Taylor's excellent figure of that variety in the Monograph.—C. S. CARTER, Louth.

The example especially alluded to above by Mr. Carter, was a beautiful but somewhat aberrant example of my variety *bicolor* of *Limax maximus*, in which, although the ground tint remained of a perfectly pure white, and the spotting upon the shield, as well as the interrupted longitudinal banding upon the body, were of the usual intense black, yet the space or area occupied by the markings was greatly increased, and noticeably encroached upon, and in parts totally obscured the snow white ground colour.

In this connection, it may be advisable to avail myself of this opportunity to point out that the *Limax cinereus* var. *albus* of Am Stein, which has hitherto been regarded as a synonym of Lessona's var. *candida* of *Limax maximus*, is not referable to that form as has so long been thought, but really belongs to my variety *bicolor*, which is a variation of extremely rare occurrence, and Mr. Carter has been indeed fortunate to find two such beautiful examples within so restricted an area.—J. W. TAYLOR, February 20th, 1907.

—: o :—

BIRDS.

Malham Tarn Birds in Winter.—The following birds were seen on the Tarn this winter :—Tufted Duck, Pochards, Teal, Mallards (I saw a batch of each of these from 10 to 20 in each). Wigeon (a few). Little Grebe (this bird is there all the year round). Waterhens (plentiful). Grey Wagtails (a few birds). Wild Geese (I heard these at night, but did not see any during the day, although they are to be heard and seen every winter around the Tarn) ; and a couple of Golden Eye Ducks.

Pochard Duck.—I have a single record of this having nested there this last season (1906). A specimen of the *Great Crested*

Grebe and one of the *Goosander* have both been secured there in the past few years.

The Tufted Duck breeds there every season.

I might add that Mr. Walter Morrison, the owner of Mal-ham Tarn, protects all *Hawks* on his land around the Tarn and the neighbouring moors.—W. WILSON, Skipton, Jan. 20th, 1907.

Birds shot in Huddersfield District * during December 1906 to January 1907.—Jan. 12th—Two Kingfishers shot near 'Old Black House,' Dalton. Jan. 15th—Pair Bull-finches shot at Dalton Bank. Jan. 24th—Common Snipe near 'Old Black House,' Dalton. Dec.—Mallard.

These birds, of course, are simply stuffed for 'show cases.' —E. FISHER, Kirkheaton, Huddersfield.

———◆◆———

REVIEWS AND BOOK NOTICES.

YORKSHIRE GEOLOGY.

The 'Proceedings' of the Yorkshire Geological Society for 1906 † have been issued, though, judging from the very small proportion of papers it contains that have been read or discussed at the society's meetings, the word 'Proceedings' hardly seems the most appropriate. The publication, however, contains many contributions of interest, and some of distinct value. It commences with the papers on the Origin of the Trias, by Prof. Bonney and Mr. J. Lomas, respectively, which were read at the York meeting of the British Association. A summary of the discussion which also took place is likewise printed. On the Ingleton district the number is, as in past years, particularly strong. In fact, recently, few areas have been so frequently described as has this part of West Yorkshire. Prof. T. McKenny Hughes contributes Part IV. of his Notes on Ingle-borough; the present dealing with the Stratigraphy and Palæontology of the Silurian; it is well illustrated. Mr. R. H. Rastall has a paper on 'The Ingleton Series of West Yorkshire'; Mr. A. Wilmore describes the structure of some Craven Limestones; Dr. A. Vaughan contributes a 'Note on the Carboniferous Sequence in the Neighbourhood of Pateley Bridge'; and Dr. H. Woodward describes a Carboniferous Trilobite from Angram, in Nidderdale. These papers alone

---

* There *is* a Wild Birds Protection Act in force!—EDS.

† Vol. XVI., part I. 136 pp. and 14 plates. Price 5/-.

are evidence of the increased interest now being taken in the Carboniferous Series. Mr. E. E. Gregory has a brief record of a striated rock surface near Bingley, and the Rev. E. M. Cole gives a summary of our knowledge of the Roman relics found at Filey. The author demonstrates that far too much importance has been attached to the discovery made on Carr Naze in 1857.

A contribution of considerable interest is 'On the Speeton Ammonites,' by Mr. C. G. Danford. In this the Yorkshire Society is to be congratulated in obtaining a communication from this careful worker. It cannot, however, be commended on the way in which the blocks have been produced in illustration of his notes. The Rembrandtesque effect of the illustrations may, or may not, be artistic, but it certainly does not enable the species to be readily identified; in fact, some of the illustrations are quite worthless, and represent so much waste money. It is a pity the County Society did not exercise a little more tact in connection with these plates; had they done so they would have been of some practical use to workers amongst the Speeton fossils. When another society published Mr. Danford's paper on the Belemnites of the Speeton Series a year previously, it was illustrated in a way which was all that could be desired. The Ammonites Mr. Danford describes in the Yorkshire Geological Society's volume were exhibited and described at the Flamborough meeting of the Yorkshire Naturalists' Union in June last. (See 'Naturalist,' July 1906, p. 241).

—: o :—

One of the most recent productions from Browns' Savile Press, Hull, is **Essays upon the History of Meaux Abbey,** and some Principles of Mediæval Land Tenure. Based upon a Consideration of the Latin Chronicles of Meaux (A.D. 1150-1400). By the **Rev. A. Earle** of Wansford, Driffield. 192 pages, price 3/6 net. It deals with the Abbey of Meaux, at one time so prominent in the East Riding, but of which now not a trace remains above ground. Mr. Earle tells his story in an interesting way, and enables the student of local history to get much valuable information without having to peruse the lengthy "Chronicles,' which only the very enthusiastic would dare to undertake.

A Picture Book of Evolution. By Dennis Hird. Part I., containing lessons from Astronomy, Geology, Zoology. Watts & Co., 1906. 202 pages, price 2/6 net.

This little book is professedly for the beginner, and it can certainly be recommended to anyone wishing to obtain a concise statement of the doctrine of evolution. The work is apparently based upon six lantern lectures which have been prepared by the author, the illustrations being very largely those used with some of the lectures. Throughout, the author tells his story in simple language, and the wealth of illustration makes the book particularly suitable as a present.

NORTHERN NEWS.

We are pleased to note that our contributor, Mr. W. N. Cheesman, of Selby, has been elected a Justice of the Peace.

The greater part of the collection of rock specimens obtained during the voyage of the 'Beagle' in 1831-6, has recently been housed in the Sedgwick Museum at Cambridge, and Mr. A. Harker, M.A., F.R.S., gives a description of them in the March 'Geological Magazine.'

We regret to record the death of Miss Caroline Birley, formerly of Manchester. She was an exceptionally keen collector of fossils, and had a large collection. She regularly attended the meetings and excursions of the Geological Section of the British Association.

In the *Annals of Scottish Natural History*, for January, Dr. J. W. H. Trail appeals for a 'Natural History Society of Scotland.' The same journal contains a note ' On the occurrence of the Siberian Chiff-Chaff in Scotland : a new bird to the British fauna,' by Mr. W. Eagle Clarke.

Our contributor, Mr. T. Petch, who recently received an appointment in Ceylon, favours us with a reprint of an elaborate paper on ' The Fungi of Certain Termite Nests,' which appeared in the ' Annals of the Royal Botanic Gardens,' Peradeniya. It is illustrated by several very excellent plates.

The Grimsby Corporation is considering the question of the adoption of the Museums and Gymnasiums Act, under which a penny rate may be adopted for the purpose of maintaining a Museum, etc. In this way it is to be hoped that the useful collection belonging to the Grimsby Naturalists' Society, may become public property.

On February 24th the members of the Yorkshire Geological Society entertained their late Hon. Secretary, the Rev. W. Lower Carter, at the Hotel Metropole, Leeds. Prof. P. F. Kendall presided, and there was a representative body of Yorkshire Geologists present. Mr. Carter was the recipient of an illuminated address, a petrological microscope, a set of slides of typical rock sections, a library table, and a geological hammer. To Mrs. Carter the members had presented a set of furs.

On the ' Geological Section' at the back of the elaborate Menu prepared in connection with the dinner given to the Rev. W. L. Carter recently, the following names occured as typical ' fossils' in the beds shewn :—*Sircartera cammeri, Semicosmophyllum onlyjohnsii, Clevelandia nonsubmergis, Fornicula odorata, Cotsworthia cosmogonica, Muffia turritissima, Howarthrus scribibundus, Pastoria perpredatoria, Neupotatorella statheri, Monaspeetonia Bridlingtonensis (V. Zambesii), Dwerryhousia potoilensis, Cashia nummis (V. nobilissimus)*, and *Bingleya camerophora.* The ' fundamental rock ' was shewn as *Latina canina*, and when this is borne in mind it may be possible to identify most of the ' fossils' enumerated.

Invitations have already been issued in connection with the meeting of the British Association to be held at Leicester from July 31 to August 7th next. The Association has not previously met at Leicester. Sir David Gill, K.C.B., LL.D., is the President-elect, and the following presidents of sections have been appointed :—Section A. Mathematical and Physical Science, Prof. A. E. H. Love, F.R.S. ; Section B. Chemistry, Prof. A. Smithells, F.R.S. ; Section C. Geology, Prof. J. W. Gregory, F.R.S. ; Section D. Zoology, W. E. Hoyle, D.Sc. ; Section E. Geography, G. G. Chisholm, M.A., B.Sc. ; Section F. Economic Science and Statistics, Prof. W. J. Ashley, M.A. ; Section G. Engineering, Prof. S. P. Thompson, F.R.S.; Section H., Anthropology, D. G. Hogarth, M.A. ; Section I. Physiology, A. D. Waller, M.D., F.R.S., Section K. Botany, Prof. J. B. Farmer. F.R.S. ; Section L. Educational Science, Sir Philip Magnus, B.Sc., M.P.

MAY 1907.

No. 604
(No. 382 of current series).

THE NATURALIST

A MONTHLY ILLUSTRATED JOURNAL OF
NATURAL HISTORY FOR THE NORTH OF ENGLAND.

EDITED BY

T. SHEPPARD, F.G.S.,
THE MUSEUM, HULL;

AND

T. W. WOODHEAD, Ph.D., F.L.S.,
TECHNICAL COLLEGE, HUDDERSFIELD.

WITH THE ASSISTANCE AS REFEREES IN SPECIAL DEPARTMENTS OF

J. GILBERT BAKER, F.R.S. F.L.S.,
Prof. P. F. KENDALL, M.Sc., F.G.S.,
T. H. NELSON, M.B.O.U.,

GEO. T. PORRITT, F.L.S., F.E.S.,
JOHN W. TAYLOR,
WILLIAM WEST, F.L.S.

Contents :—

LONDON :

A. BROWN & SONS, LIMITED, 5, FARRINGDON AVENUE, E.
And at HULL AND YORK.

Printers and Publishers to the Y.N.U.

NOTES AND COMMENTS.

THE YORK REPORT OF THE BRITISH ASSOCIATION.

EARLY in April we received the York Report of the British Association, which includes a summary of the work accomplished at the seventy-sixth meeting of the Association. It contains not quite as many pages as usual, but is nevertheless a valuable record of the progress of science during the twelve months preceding the York Meeting. The volume contains the various presidential addresses, reports of committees, abstracts of papers read at York, etc., etc. Those of particular interest to northern readers were printed in this journal for September last.* A perusal of this report, it must be confessed, means that one meets with matter which was well published in various places seven months ago! Seeing that all the presidential addresses, reports of the various committees' investigations, abstracts of papers, etc., etc., were in type and circulated in August last, and that such papers as were not ready had to be in the secretary's hands within a few days after being read, it is difficult to understand why it should take eight months to arrange the pages in order and put a cloth cover on them. Surely some means could be devised for the more prompt appearance of these volumes.

LOCAL v. GENERAL MUSEUMS.

From the February issue of the 'Museum Gazette' it is evident that the writer in that journal is more in accord with our own opinion of the value of local museums than was apparent by a perusal of his first note on that subject, which we noticed in the February 'Naturalist.' We do not confine the word 'local' to those collections 'obtained within a radius of five miles from the Parish Church,' nor did we 'draw on our imagination' in citing certain local museums. And we hope we are able to distinguish between a local and a general museum. We should call the Driffield Museum a typical example of the former, as it is devoted exclusively to the geology, and antiquites, of a definite geographical area, the Yorkshire Wolds. The Hospitium at York—devoted to the antiquites of that ancient city—is another, though it is no fault of ours that the writer in the 'Museum Gazette' has never heard of it. The Selby Museum may be taken as a typical general museum, on a small scale.

* Several at greater length than they appear in the report just to hand.

THE SELBY MUSEUM.

In the same number of the 'Museum Gazette,' after quoting our statement that 'We were sorry to see from a report of an address recently delivered at Selby by Dr. Hutchinson, that this museum is not appreciated as it ought to be;' a writer, presumably Dr. Hutchinson, says ' *This statement is an error,* for nothing of the kind was suggested.' On this point, however, we can have no misunderstanding! Whether Dr. Hutchinson made this statement or not, it was certainly so reported in more than one of the dailies, and in another magazine, the 'Museums Journal' (Vol. VI., p. 144), an account of the same address appears, in which it states that Dr. Hutchinson 'very much regretted that more use was not made of Selby Museum by the inhabitants.' Our statement, therefore, was quite correct. However, as we now learn that the 'Museum Gazette' does not disparage local museums, but encourages them, and considers that every educational museum should have its local department, we have little to complain of, and we feel glad that the criticisms previously offered have enabled us to get a better idea of the views of the 'Museum Gazette' on the local museum question than was previously possible.*

THE FOURTH INTERNATIONAL ORNITHOLOGICAL CONGRESS.†

This handsome book, which forms Volume XIV. of 'Ornis,' has been edited by the Secretaries of the Congress, Dr. E. G. O. Hartert and Mr. J. L. Bonhote, under the direction of the President, Dr. R. Bowdler Sharpe. Its 'get up' is perfect, and reflects every credit upon the printers, Messrs. Witherby & Co. There were five sections to the Congress, dealing respectively with (1) Systematic Ornithology, Geological Distribution, Anatomy, and Palæontology; (2) Migration; (3) Biology, Nidification, Oology; (4) Economic Ornithology and Bird Protection; and (5) Aviculture. To each of these sections various communications were made by different

---

* Since writing the above we have seen the March issue of the 'Museum Gazette,' in which an article on 'Museums and Museums' appears. In this, much that appears in the February issue is repeated. We quite agree that it is a pity that misunderstandings should needlessly arise amongst those who are interested in the same pursuits, and we share with the 'Museum Gazette' the desire to further the study of Natural History. To put the whole matter in a nutshell, however, the 'Gazette' first expressed a preference of 'Educational Museums' to 'Local Museums.' We prefer 'Local Museums' (which should also be educational) to 'Educational Museums,' pure and simple.

† London, 1907. Dulau & Co., 696 pp. and plates.

authorities, and no fewer than forty of these were published in the volume. And when it is remembered that the work is printed with clear type on excellent paper, and is illustrated by numerous blocks in the text and several plates (some of the latter being coloured), it will be seen that not only is it a charming memento of this successful Congress, but it represents a sound and substantial contribution to ornithology.

AND ITS REPORT.

An idea of the variety of these papers may be obtained by the following titles, taken at hazard:—'What constitutes a Museum Collection of Birds?' 'The Migration of Birds,' 'The First Bird List of Eber and Peucer,' 'On Extinct and Vanishing Birds,' 'Monographie de la Sterne de Dougall,' 'Description of New Species of Neotropical Birds,' 'On the Origin of the Differences between Nestling Birds,' 'Sequence in Moults and Plumages,' 'Bird Protection,' 'On the Colour Variation in the Eggs of Palæarctic Birds,' 'The Food of Birds,' 'The Sparrow,' etc. Needless to say, amongst the authors are many of the leading naturalists of the day. As his presidential address Dr. Bowdler Sharp took for the subject, as might have been anticipated, the history of the ornithological collections in the British Museums and in this he refers to the great benefit to ornithological science that would accrue were the officers of the various important museums throughout the world to give a history of the collections under their charge.

FLAMINGOES.

Whilst many of the articles are of altogether exceptional worth, there is one which strikes the writer as of unusual interest and importance. This is by Mr. F. M. Chapman, and deals with the method of exhibiting birds in museums. The illustrations to these, indicating the way in which our American friends can show their specimens, are very fine indeed, and might be well taken as models to such other museums as have the space and funds necessary for such exhibits. The flamingo case, for instance, 'is twenty feet long and eight feet wide, and contains twenty-nine birds.' The background (birds) is painted by one artist, the landscape by another, and the birds are mounted by a third individual. The effect, however, is extraordinary, and the case appears to be nothing else than a huge flamingo colony (see Plate XX., which we are kindly permitted to reproduce). Methods of storing skins, etc., etc., are also dealt with. The only regret one has on closing this volume is that it will be four years before another such record is produced !

YORKSHIRE WILD BIRDS AND EGGS PROTECTION. COMMITTEE.

Mr. Riley Fortune reports that the response to the appeal by the above committee for funds to enable it to protect the nesting places of rare birds in the county has not been so generous as might naturally be expected. The aid of the Coastguards and Climbers has been promised for the protection of the Peregrines at Bempton ; arrangements have also been made for the protection of another pair of Peregrines nesting in the county. The paid watcher has commenced his duties at Spurn, and, as his whole time will be devoted to looking after the birds, we may hope for very good results. Other localities will be dealt with as funds allow, and we hope that members of the Union will do all they can to help us.

	£	s.	d.
Amount previously acknowledged	10	1	0
Bradford Naturalists' Club	1	1	0
W. D. Roebuck	1	1	0
E. W. Wade	0	10	6
York and Dist. F. Nat. Society	0	10	6
Scarborough Field Naturalists' Society ...	0	5	0
W. Wilson	0	5	0

DERBYSHIRE NATURALISTS.

The twenty-ninth volume of the Journal of the Derbyshire Archæological and Natural History Society, 1907, issued under the able editorship of Mr. C. E. B. Bowles, is one of the finest publications of a provincial society that we have seen for some time. It contains over 300 pages, and has a wealth of plates and plans that would do credit to any of the leading London societies. There is also an ample variety in the nature of its contents. Of the very many articles, the following are perhaps of principal interest to our readers :—'Some Notes on Arbor Low and other Lows,' by T. A. Matthews, 'Crich Ware,' by G. le Blanc Smith, 'Recent Cave Diggings in Derbyshire,' by W. Storrs Fox, 'Ornithological Notes for 1906,' by Rev. F. C. R. Jourdain, etc. Bound up with this Journal is the valuable and voluminous report upon the excavations on the Roman site at Melandra Castle, conducted by the Manchester and District branch of the Classical Association. This may be taken as a model which might be followed by others excavating sites of this character.

Flamingo Group in the American Museum of Natural History.

SPEETON AMMONITES.

By the kindness of the Yorkshire Geological Society we are able to give our readers a reproduction of one of the best of the plates (see Plate XXI) illustrating Mr. Danford's paper 'On the Speeton Ammonites,' which recently appeared in that Society's Proceedings. The originals of the figures are now in the Geological Museum, Jermyn Street, London, and are about twice the size shown on the plate. The specimens figured, are (1 & 1*a*.), *Olcostephanus* (*Polyptychites*) *bidichotomus* Leym., (2 & 2*a*.), *O. polyptychus* Keys, and (3 & 3*a*.), *O. keyserlingi* Neum. and Uhl. It should be added, however, that good examples of Ammonites are now-a-days exceedingly difficult to obtain at Speeton—at any rate, this is the experience of most collectors, Mr. Danford being a possible exception to this rule.

ANOTHER NEW MAGAZINE.

In April, the first part of a new monthly magazine, the 'Lancashire Naturalist,' made its appearance, under the editorship of Mr. W. H. Western, of Darwen, who is also the printer, a fact which probably accounts for this 16 pp. magazine being sold at the low price of one penny. The new venture is the official organ of the Lancashire Union of Natural History, Literary and Philosophical Societies, and it is to be hoped that the affiliated societies will liberally subscribe to this journal and thus ensure its success. We doubt, however, the advisability of printing the general reports of the various societies' general meetings ; particulars of important local exhibits or records might with advantage be extracted, but the space these full reports occupy might be put to better service by the insertion of original articles—it should surely be an easy matter to fill sixteen pages once a month in this way. The journal hopes to 'chronicle all interesting events, and to assist in recording the flora, fauna, etc., of our county.' If this local character is maintained, the new magazine may become an important and valuable addition to our monthly literature.

MARINE BIOLOGY.

The 'Twentieth Annual Report of the Liverpool Marine Biology Committee'* is to hand, and contains a record of a year's useful work accomplished at Port Erin. It is pleasing

---

* Liverpool University, 56 pp., and illustrations.

to find that there is a marked increase of visitors to the Aquarium ; and, as the Report points out, 'an institution where over fifteen thousand summer visitors are shown a number of the most interesting of our common sea-side animals and plants in a living condition and among natural surroundings, with labels, pictures, and other information, must surely be doing something to encourage nature study, and to foster an appreciation of biology.' Perhaps the most interesting exhibit was 'the Octopus of the Irish Sea, *Eledone cirrosa*' (which

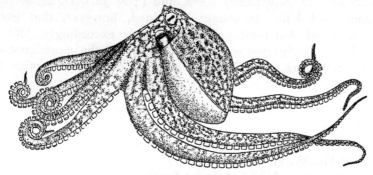

The Irish Sea Octopus.

we are enabled to figure herewith). A good supply of this Cephalopod was obtained in the early spring, and the tank in which they lived was a great attraction to the visitors. As an appendix to this Report is printed· an admirable address on ' Some Problems of the Sea,' by Prof. Herdman. Mr. Andrew Scott has also some useful ' Notes on Special Plankton Investigations,' which, like the other articles in the Report, is well illustrated.

LANCASHIRE AND CHESHIRE ENTOMOLOGISTS.

The vigorous Lancashire and Cheshire Entomological Society has just issued its thirtieth annual report,* and gives evidence of a successful year's work. It contains a record of all the important exhibits at the meetings; the Vice-President's address, by Prof. T. Hudson Beare; a portion of 'A Preliminary Catalogue of the Hemiptera-Heteroptera of Lancashire and Cheshire' (a valuable contribution), by Oscar Whittaker ; and a ' Note on the Remarkable Race of *Agrotis ashworthii*,' by Mr. W. Mansbridge. A portrait of Mr. F. N. Pierce appears as frontispiece.

---

* Fifty pages. Price 2/6. 'Visitor' Printing Works, Southport.

THEORIES OF EVOLUTION.

An Historical Outline.

AGNES ROBERTSON, D.Sc.

I.—INTRODUCTION.

ALL the animals and plants to-day inhabiting the world can be classified into small groups known as 'species.' This is recognised even by those who are not naturalists. We call an animal having a certain set of characters a horse, and a plant having a certain set of characters a dandelion, using these names without hesitation, although the individuals which we identify in this way differ considerably among themselves. We have in our minds a conception of a typical horse and of a typical dandelion, and an animal or a plant conforming to one of these types more closely than to any other, we regard as belonging to the horse or dandelion 'species.' The world, then, is peopled with a huge number of species which, on the whole, are sharply distinguishable, and further, these species are not altogether isolated, but can be classified into groups by means of the characters which they possess in common. For instance the Field Buttercup, the Lesser Celandine, and the Water Crowfoot, though no one would ever confuse them, yet resemble one another closely enough to be placed in the same group of similar species, or, as we say, in the same 'genus.' Again, it is found that these genera can themselves be arranged in larger groups called 'orders.' The species belonging to a genus resemble one another more closely than they do those of any other genus, and in the same way the genera belonging to the same order show more resemblance to one another than to the genera belonging to any other order. Similar orders can be grouped into larger classificatory units, and so on.

So far, we have been speaking of the organic world *descriptively*; the fact that organised beings naturally fall into a classification of the kind just outlined would be allowed by everyone, and is not a matter of theory, but merely of observation. But when we turn to the question of *how* this state of things has arisen, we plunge at once into the region of hypothesis. There are two main contrasting views; those of Special Creation and of Descent with Modification. According to the first view, the species of the animal and plant world were created just as we know them now, and have persisted un-changed to the present day.   The only way to test the value of a hypothesis is to ascertain how far it affords an explanation of observed facts, and it is on account of its failure to afford such an explanation that the doctrine of ' Special Creation ' has been rejected by biologists as a ' working hypothesis.'   On the other hand, the theory of Descent with Modification gives a luminous conception of the general scheme of organised life.   According to this theory species are not immutable, but all living beings had a common origin somewhere in the remote mists of antiquity.   The descendants of the original form, or forms (which are supposed to have been exceedingly simple in structure) have developed with progressive modification along innumerable lines, attaining at last to the state of differentiation and com-plication which we witness to-day.   The theory of Descent with Modification gives a real meaning to the existence of classificatory units, such as genera, orders, etc., for these are regarded as actual expressions of the degree of blood-relation-ship ; a genus, for instance, consists of a group of species which have a close affinity, being descended from a comparatively recent common ancestor.

The doctrine of Special Creation seems to have been accepted almost unquestioningly in ancient and mediæval times, though in certain classical writings a disbelief in it is vaguely foreshadowed. The idea that species are mutable was first definitely expressed by Sir Walter Raleigh in the seventeenth century.   In his great work on ' The History of the World,' speaking of the days of Noah's flood, he says, ' But it is manifest, and undoubtedly true, that many of the *Species*, which now seem differing, and of severall kindes, were not then *in rerum natura.*'

Buffon (1707-1788) was one of the first scientists to clearly suggest that species might have been gradually evolved.   It is not worth while to dwell at length upon his views.   He was by no means a great naturalist, but he was an amazingly brilliant writer ; some of his epigrams, such as ' Le style c'est l'homme même,' live to the present day.   He had occasional vivid flashes of scientific insight, and was the first to realise that fossils give evidence of the existence of extinct animals.   Buffon appears to have been much afraid of incurring the odium of his orthodox contemporaries, and though he sometimes dared to state the possibility of Descent with Modification quite clearly, at other times he definitely denied such views.   According to Professor Packard, ' The impression left on the mind, after reading Buffon, is that even if he threw out these suggestions and then retracted them for fear of annoyance, or even persecution from the bigots of his time, he did not himself always take them seriously, but rather jotted them down as passing thoughts.'

II.—ERASMUS DARWIN AND LAMARCK.

The germs of the doctrine of Descent with Modification seem to have been in the atmosphere in the last decade of the eighteenth century, for between 1790 and 1800 the conclusion that species were not immutable was reached independently in Germany by Goethe, in France by Geoffroy Saint-Hilaire, and in England by Erasmus Darwin. The last named, who was a physician, a poet, and grandfather of Charles Darwin, published in 1796 a remarkable book called ' Zoonomia,' which contained, amidst much speculation irrelevant to our purpose, a profession of faith in the doctrine of evolution. After discussing the metamorphoses which individual animals undergo, and the essential likeness of structure of all the higher animals, he goes on to say, 'would it be too bold to imagine, that in the great length of time, since the earth began to exist, perhaps millions of ages before the commencement of the history of mankind, would it be too bold to imagine, that all warm-blooded animals have arisen from one living filament, which THE GREAT FIRST CAUSE endued with animality.'

Erasmus Darwin not only believed in evolution, but tendered one or two suggestions as to *how* the process had been brought about. His main idea was that useful modifications, such as the hard beaks of certain birds, the elephant's trunk, and the rough tongue and palate of cattle, ' seem to have been gradually acquired during many generations by the perpetual endeavour of the creatures to supply the want of food, and to have been delivered to their posterity with constant improvement of them for the purposes required.' He also foreshadowed the theory of Sexual Selection, which we shall have occasion to speak of later in dealing with the work of Charles Darwin. The fact that these suggestions of Erasmus Darwin met with little recognition at the time is scarcely surprising when we remember that they were put forward more or less casually, and that no effort was made to establish them by proof.

The name of the French biologist Lamarck, who first published his views at the beginning of the nineteenth century, is better known in the history of evolutionary thought than that

of Erasmus Darwin. The opinions held by the two men were
very similar, but Lamarck, who was primarily a scientist, worked
out his theories much more completely and circumstantially, and
they received more attention from naturalists than the comparatively tentative ideas of a poet and dreamer such as Erasmus
Darwin. Lamarck was born in 1744. His father destined him
for the Church, but his own taste was for military things, and
as soon as his father died he joined the French army, then
campaigning against Germany. He distinguished himself so
much in an action, which took place the day he enlisted, that he
was made an officer on the spot. He was soon obliged to leave
the army owing to ill-health, with a pension of only about £20 a
year. He was obliged to work at a bank to make this pittance
up to a living wage, and at the same time went through some medical
studies. He worked at botany in his spare time, and produced
his ' Flore Française,' which brought him under the notice of
Buffon, who made him tutor to his son. For fifteen years he
lived precariously by his pen, but during the Reign of Terror he
obtained the appointment of Professor at the Musée d'Histoire
Naturelle. His colleague, Geoffroy St. Hilaire (then twenty-one),
was responsible for the Vertebrate Zoology, while Lamarck
cheerfully undertook everything else ! In 1809 his great work,
the ' Philosophie Zoologique,' saw the light.

Lamarck had, in many ways, an unfortunate life. For many
years he had a great struggle to make both ends meet, and for
the last ten years of his life he was blind. His daughter Cornélie
devoted herself to him absolutely, and became his scientific
secretary. His poverty and his blindness do not complete the
full tale of his sorrows, for, in the words of Professor Packard,
his biographer, ' Lamarck's life was saddened and embittered
by the loss of four wives.'

To try and state Lamarck's philosophy in a few words is
extremely difficult. He believed that, in the main, two great
factors had coöperated in producing organic evolution. The
first of these was a tendency inherent in all organisms to progress from the simple and undifferentiated to the complex and
highly differentiated. The second was the power of organisms
to adapt themselves to their environment through habit and the
use and disuse of organs, and to transmit to their offspring the
adaptations so produced. For instance, Lamarck would suppose
that the giraffe was descended from a short-necked ancestor,
whose perpetual efforts to browse on higher and higher branches
of trees produced a certain elongation of its neck. This

elongation, he believed, it would transmit to its offspring, which would continue to make efforts to reach higher and higher, so that in each generation some progress would be made towards the immensely elongated neck of the modern giraffe. We know, of course, of many instances in which habit and use profoundly modify an organ during the life of the individual : the size and strength of the blacksmith's arm would be a case in point. But Lamarck demands more than this, for such modifications as a giraffe's neck could never be produced if each generation had to start afresh from the beginning. Lamarck's theory stipulates that characters acquired during the lifetime of an organism shall be transmitted to the next generation, and it is here that the main difficulty comes in. For no one has at present been able to offer indisputable proof that such transmission takes place.

(To be continued).

------◆◆------

At the recent annual meeting of the Craven Naturalists' Society Mr. J. T. Davison was elected President.

At the annual meeting of the Hull Geological Society, held on April 11th, Mr. T. Sheppard was elected President.

We are pleased to notice that Miss Mary Johnstone, B.Sc., LL.A., the head mistress of the Grange Road Secondary School, Bradford, has recently been elected a Fellow of the Linnean Society.

Mr. S. B. Steelman points out that on the Lincolnshire Wolds the var. *doubledayaria* of *Amphidasys betularia* appears to be the dominent form there, as in the South-west Riding of Yorkshire (April 'Entomologist').

The ' Journal of Conchology' caters well for its Lancashire readers; the April issue contains two lenghty contributions, viz., 'The Land and Fresh-water Shells of Morecambe, Lancaster and district,' by J. Davy Dean; and 'Bibliography of the Non-Marine Mollusca of Lancashire,' by J. W. Jackson.

On a plate illustrating an article on 'A permanent record of British Moths in their natural attitudes of rest,' by Mr. A. H. Hamm (Trans. Ent. Soc., issued Jan. 23rd, 1907), are some of the best examples of 'protective colouration' that we have seen for some time.

Since our last impression was published, 'Hull Museum Publications,' Nos. 41 and 42, have appeared. The first is an admirably illustrated guide to the new Wilberforce House Museum, and the second is the twentieth Quarterly Record of Additions. The latter has illustrated accounts of Querns, Mantraps, Coins, etc., etc. The same institution has recently acquired the fine collection of British and Roman Querns formed by the late Dr. H. B. Hewetson, at Easington.

Part IV. of 'The Birds of the British Islands,' by Charles Stonham (E. Grant Richards, 7/6), has appeared, and completes the first volume of this work, upon which no expense appears to have been spared to make it a really attractive publication. With the present instalment are two excellent coloured maps—the first being an orographical map of the British Islands, and the second 'The Zoological Regions of the World.'

WHITE LESSER BLACK=BACKED GULL, AND THE REPORTED NESTING OF THE IV RY GULL, ON THE FARNE ISLANDS IN 1906.

HARRY B. BOOTH, M.B.O.U.

MR. H. A. PAYNTER (of Alnwick), the Honorary Secretary and Treasurer of the Farne Islands Association, in his interesting report for 1906, states :—'I heard that a pair of Ivory Gulls nested on the Wamses, but I did not see them.' In a recent letter to me on this subject, Mr. Paynter says that when the occurrence was reported to him, it was too late for him to verify it.

It is exceedingly unlikely that this species should breed at any spot in the British Isles, and in company with Mr. Fortune I visited the Farne Islands twice during the breeding season of 1906. On the North Wamses there was a white gull, which I spent a considerable time in stalking and watching with my field glasses, whilst my friend was engaged photographing. Our boatmen said they had been informed that it was a rare Arctic species, and they considered it to be half as large again as the Lesser Black-backed Gulls. In this they were deceived by its conspicuousness—it was exactly the same size—and I made it out to my entire satisfaction to be a partial albino of the Lesser Black-backed Gull, and paired to a normal one of the same species. It was of a slightly dirty white shade, with the exception of a few faint streaks of grey, chiefly on the secondaries, which showed from above, but more particularly from below, when the bird was just overhead. Its beak and legs showed the faintest tinge of yellow, its eye appeared to be of the normal colour, and its note and cry were similar to those of the other Lesser Black-backs. In early August I attempted to single out its offspring, but could not do so as they ran away with the other young gulls of various sizes, and in a small herd like chickens. I was prompted to examine them out of curiosity, for even if they should develop into partial albinoes, they would not necessarily show any difference whilst in the downy stage. However, I could not detect the slightest variation in any of the young gulls in the vicinity.

I should like to testify to the admiration and esteem in which I regard Mr. Paynter for the great share he has played in the protection of the breeding birds of the Farne Islands— an illustration probably unequalled in the history of Bird Protection.

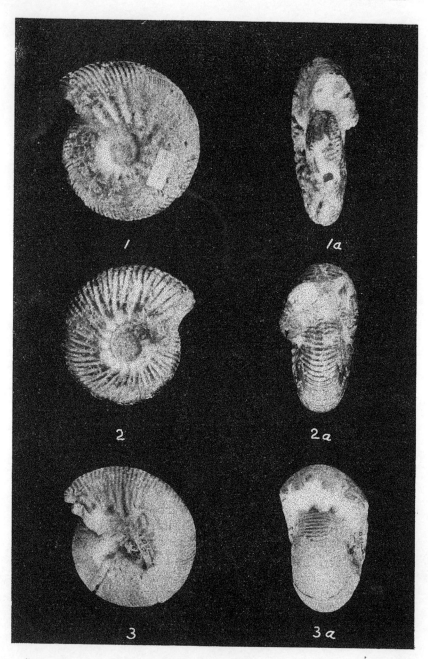

Speeton Ammonites.

NOTES ON *SUCCINEA OBLONGA*, DRAP., AND OTHER SPECIES AT GRANGE-OVER-SANDS, LANCS.

J. WILFRID JACKSON,
Manchester.

ON Good Friday, March 29th, a small party spent the day in hunting for shells in this favourite locality. The weather was magnificent, the temperature being more like July than March. Our main efforts were devoted to ascertaining the distribution of *Succinea oblonga,* taken for the first time, in August last year, by Mr. H. Beeston of Havant ('Naturalist,' Jan., 1907, p. 31). Our first efforts were, however, disappointing, and we decided to try fresh ground some little distance away from the ditch where Mr. Beeston discovered his specimens. In this we were more successful, and were soon rewarded by finding a dead adult specimen. This was shortly followed by others, among which were a few full-grown living ones. We then became aware that the sides and bottom of the damp ditches in which we were working contained numbers of juvenile examples, all on the crawl, along with a number of other species.

We were surprised to keep coming across numbers of dead slugs, chiefly *Agriolimax agrestis* and *A. lævis,* which had the appearance of having been drowned—no doubt during the recent floods, when many of the marshes in this district were covered by the heavy tide. The animals were mostly in an extended condition, and some were covered with a whitish mould.

The other species of mollusca noticed living in the ditches and their vicinity were *Hygromia hispida,* fine specimens and very hispid ; *Succinea elegans,* a few juvenile examples ; *Carychium minimum,* abundant ; *Punctum pygmæum ; Cochlicopa lubrica ; Agriolimax agrestis,* A. *lævis ; Arion circumscriptus ; Limnæa pereger* var. *maritima,* L. *truncatula ; Aplecta hypnorum ; Pisidium pusillum,* and *P. obtusale.*

Crossing over out of Westmorland to the Lancashire side of the river Winster, we made our way back to Grange, examining other likely habitats for *S. oblonga* on the way. Near to a triangular piece of brackish water and marshy bit of ground we were surprised to find the species again in evidence (all young specimens) at the roots of grass on the top of a low wall—a most unusual habitat for *Succineæ.* Flood work was again noticeable, and numbers of dead land shells were strewn about among the debris, and it is quite possible that the *Succineæ* had been driven out of their usual marshy habitat by the inroads of the sea.

Among the dead shells nothing striking was noticed, the species being mostly of the commoner kinds, such as *Hygromia rufescens, H. hispida, Vitrea cellaria,* etc., though more time spent at this place might have resulted in the acquisition of some of the rarer kinds, as is often the case in the north of Ireland after floods. The finding of this species in the above two widely separated places, coupled with the fact that Mr. Beeston also obtained some immature specimens of *Succineæ,* which he believes to be *S. oblonga,* in another situation on the same marsh last August, leads us to think that it will prove abundant all over the marsh.

It is interesting to compare this habitat with that of Braunton Burrows, in Devonshire, both of which, whilst dissimilar in physical and general aspects—the one being among the sand dunes, the other (near Grange) being on ' salt-marsh '—agree in the fact of the species inhabiting places near the sea and where the water is brackish.

We proceeded to Eggerslack Wood, where we were soon amongst an abundance of *Acanthinula lamellata.* This species was first found here in October, 1905, by the writer,[*] when only a single specimen was taken. In August last Messrs. Booth and Rhodes were more fortunate, obtaining thirty or more specimens in a short time.[†]

On the occasion of our joint visit the shell was very much in evidence, almost every dead beech leaf having one or more examples adhering to it. The shells were mostly of a somewhat depressed form.

Another object of our visit—*Acicula lineata*—we were not so fortunate with, only about eight examples being noticed, three of which were white. A number of the usually more prolific species was observed during our collecting.

It would be interesting here, also, to note that Mr W. H. Heathcote, of Preston, found *Arion ater* var. *alba* sub-var. *marginata* in abundance at Woodhead, near Grange, when collecting on Easter Monday, as well as a fine example of *Helix aspersa* var. *exalbida,* the first record for Lancashire.

---

[*] See ' Journ. Conch.,' Vol. 11, 1906, p. 361.

[†] *Op. cit.,* Vol. 12, 1907, p. 19.

NOTES ON LEPIDOPTERA IN THE WILSDEN DISTRICT IN 1906.

ROSSE BUTTERFIELD.

FROM the end of June to the end of the season I worked with the object of ascertaining what changes had taken place in the local lepidopterous fauna during recent years.

The latter part of June and the whole of July and August were very favourable months. September was characterised by a copious amount of sunshine, but a prevailing north-east wind militated against successful collecting. October, on the whole, was fairly good, while November was a very wet month.

The predominating entomological features of the year were the abnormal abundance of *Acronycta psi*, *Neuronia popularis*, *Amphipyra tragopogonis*, and *Dasypolia templi*. *N. popularis* swarmed around the gas lamps on suitable nights from August 19th to the first week in September. Hitherto only one or two isolated examples of this species have been recorded for this district. *D. templi* appeared as early as the second week in September. On the night of the 18th October, the lamps in the immediate suburbs of Bradford attracted an unprecedented number of males of this species. This moth is, I believe, one of the commonest Noctuæ within the city boundary. It is, however, a very capricious insect, and although the night in question was dark, damp, and overcast, such conditions were accompanied by a cold northerly wind.

In addition to the foregoing, *Eugonia alniaria*, *Venusia cambrica,* and *Melanippe galiata* deserve special mention as being unusually common, and in a lesser degree *Noctua glareosa.*

Putting out of the question certain moths which formerly occurred, but which appear to have disappeared, *Apamea basilinea* and *Hadena oleracea*, which are usually common, were characterised rather by their absence, as also was the first brood of *Melanippe fluctuata.*

In June, July, and September sugar failed as an attraction to insects. *Xylophasia monoglypha* on some nights appeared in numbers, the type greatly preponderating. A few *Orthosia suspecta* and a single *Miana literosa* were all worthy of mention which came to sugar in those months. The flowers of the Ragwort and Ivy proved an absolute failure, although both flowered freely. Ragwort is the chief attraction to *Noctua dahlii*, yet in spite of this flower failing, I observed a few. In October insects came freely to sugar. *Orthosia macilenta* was extremely abundant; other common ones were *Miselia oxyacanthæ* (mostly var. *capucina*), *Agriopis aprilina, Calocampa exoleta, Agrotis suffusa*, a few *Scopelosoma satellitia*, also two *C. vetusta*. I believe only one example of the latter species has heretofore been recorded in the Bradford district.

Melanic forms do not appear to have been particularly conspicuous. It is noteworthy that dark specimens of *Scoparia cembrae* and *S. ambigualis* were observed, and I secured several nearly black specimens of the ubiquitous *Noctua xanthographa*. Mr. Porritt, in his interesting paper on ' Melanism in Yorkshire Lepidoptera,' read before Section D. at the British Association at York (1906), does not include in the list therein given the two latter species. It is singular that the only specimen of *Amphidasys betularia*, captured by my father, Mr. E. P. Butterfield, was a normal type. I secured a few black *Boarmia repandata* from Blackhills. Among the black forms captured were a few *Cidaria immanata* and *Miana strigilis*. My father and I caught at sugar one night two striking dark forms of *Agriopis aprilina*. The most interesting capture of the year was a rather dark example of *Agrotis agathina* in perfect condition. It was caught whilst it was circling around a lamp at the base of Harden Moor.

Among moths which are associated with the moor, or rather moor edges, *Acidalia fumata, A. inornata, Eupithecia nanata*, and *E. minutata* were fairly common, as were also the Tortrices *Penthina sauciana*, and *Amphysa gerningana*, and the pretty Tineæ *Exapate congelatella*. On Harden Moor I caught three *Cloantha solidaginis*. Whilst searching with a lamp for this species, I noticed an abundance of the larvæ of *Hadena pisi*. *Scodiona belgiaria* appears to have been rather scarce, for I only saw two males.

Among the Geometers caught in September were several *Oporabia autumnaria*. This has not previously been recorded for the district. Mr. Porritt, in his ' Yorkshire Lepidoptera,' does not give it specific rank, though it is included as a species in ' The Entomologist Synonymic Reference List.'

Butterflies, as usual, have been conspicuous by their absence —that is to say, in the immediate neighbourhood of Bradford— with the exception of Cabbage Whites; a Small Tortoise Shell, caught on December 4th, is the only record I have !

GEOLOGICAL NOTES ON THE ROBIN HOOD'S BAY DISTRICT.

Professor PERCY F. KENDALL, M.Sc., F.G.S.

THE Whitsuntide excursion of the Y.N.U. to Robin Hood's Bay will afford the members of the Geological Section an opportunity, of which they will probably not be slow to avail themselves, of studying a series of rocks unsurpassed in Britain for the completeness of their development and the excellence of the exposures, both on the coast and inland. The physical features are of equal interest to the stratigraphy, and the palæontology to both. But the geological fare is even more generous still, and students of geological tectonics and of glacial geology will find *entrées* as subtly compounded as the most fastidious palate could demand. The *piéce de resistance* is, of course, the Jurassic series, and with Spring Tides (the moon is new on Whit Sunday) the magnificent exposures on the scars and in the cliffs will be easy of access.

The whole Liassic succession can be made out in the Bay with the sole exception of the lowest Zone, that of *Ammonites* (*Psiloceras*) *planorbis*, and of that evidence is occasionally obtainable in blocks thrown up from submerged reefs. At Blea Wyke can be seen the only certain occurrence in Yorkshire of the Zone of *Lytoceras jurense* and of the overlying Blea Wyke Beds which form a complete passage from the Lias into the Inferior Oolite. These are succeeded by the most important development of the Dogger to be found in the country, and that is followed by the great Estuarine Series with its occasional marine beds, which have played an important part in the correlation of the Lower Oolites of Yorkshire with their south-country equivalents.

All these Jurassic beds are well furnished with the fossils characteristic of their age and of the conditions under which they were deposited. The principal problem which these rocks present is connected with the movements of a large fault which is exposed in the upper part of the cliff at Peak, and bifurcating seaward produced Peak Steel, a triangular reef of Middle Lias let in between the Lower Lias of Robin Hood's Bay and the Upper Lias of the foreshore to the south of the point.

1907 May 1.

M

Peak Fault seems to belong to a system of dislocation completely surrounding the folded, but not faulted, mass of the Cleveland Hills in much the same way that the Craven and related faults surround the unfaulted mass of Carboniferous rocks of the Yorkshire Dales.   The analogy may be traced even further, and just as it has been argued that contemporaneous movements of the Craven system of faults affected the deposition of the Carboniferous and Permian rocks, so, and with even greater clearness and certainly, it can be inferred that the Cleveland system of faults affected the deposition of Jurassic and Cretaceous sediments.   It is not improbable that the Cleveland phenomena are the actual complement of those of Craven, and that if either the secondary rocks had been partially preserved in the western area, or the palæozoic rocks exposed in the eastern one, we might see that repetition of movements of the dislocations had continued intermittently from Carboniferous times, or earlier, down to Cretaceous times, or later. It can be readily demonstrated that some of the folds were lines of persistent movement.

The Peak Fault, which belongs to the same series as the Speeton Fault and the dislocation that runs across the peninsular of Scarborough Castle, has long been a subject of speculation among geologists, and within the last two or three years Mr. Rastall and Mr. Herries have both written on the subject.   The feature to which most attention has been directed is the effect of the fault upon the contrasted development of the Upper Lias and Lower Oolites upon opposite sides of the fault.

On the downthrow (seaward) side the Dogger shows a very large development, and includes important fossiliferous marine beds, and the Blea Wyke beds and the *Lytoceras jurense* beds are of great magnitude.   While on the upthrow (landward) side the succession, which is splendidly exposed in the great range of old Alum Works extending from Peak Tunnel to Stoupe Brow, shows a greatly attenuated Dogger, consisting of only a few feet of sandstone, with some inches of basal conglomerate resting on the Alum Shales of the Upper Lias, the *jurense* zone and the Blea Wyke beds being absent.   Various explanations of the discrepancy have been offered, and they will form excellent provocation to those dialectic encounters for which the geologists have long been renowned.

Glacial Geology will, no doubt, have its day, and if there should chance to be a geologist present whose jaded palate can still respond to the stimulus of glacier lakes and overflow channels, he can indulge his taste with the original and authentic brand of those delicacies. The coast sections of the glacial deposits are, or at least were last autumn, in a very favourable condition for observation. One noteworthy feature then visible was a basement layer of coarse rubble consisting of, local rocks, chiefly Jurassic Sandstone, with a few fragments of Lias. In this deposit no foreign boulders have as yet been detected.

———◆◆———

LINCOLNSHIRE MITES.

RHYNCHOLOPHIDÆ—(continued).

C. F. GEORGE, M.R.C.S.
Kirton-in-Lindsey.

ERYTHRÆUS appears to be a fairly common Lincolnshire mite, as I have found several specimens at different times. Mr. Soar has succeeded in making a very characteristic likeness of the creatures (fig. 7), as well as a capital, and almost diagramatic sketch of the palpi and mandibles (fig. 8). This genus differs from the preceeding ones in having the mandibles only retractile, and their lancet-like shape and barbed extremities are well shown in the figure. The palpi are seen to be of five joints ; the last and smallest joint is bag-shaped, and attached rather near to the base of the fourth, altogether the organ is very similar to that found characteristic of the Trombididæ. The colour of the mite varies considerably ; sometimes it is rust coloured and at others a rather deep red, the hairs being considerably darker than the body of the mite. They are rather short, stout, and thickly pectinated. Those on the body are rather blunt at the free end and sometimes slightly curved, whilst those on the legs and palpi are sharply pointed. The eyes are simple, very convex, bright, like red sealing-wax, and give the mite a rather fierce appearance under the microscope ; they are two in number, embedded in the skin, one on either side of the cephalothorax, and in the middle between them is a dorsal groove containing a chitinous rod, looped at either end, having two very fine, rather long curved hairs in each loop, best seen in a recently dissected mite. The legs are long, the tarsus of each leg compressed sideways, and that of the first pair larger than the others. The Epimera are in four groups, rather small and far apart. There are, no doubt, a good number of species in this group requiring examination and description. I should be glad to receive specimens, living or

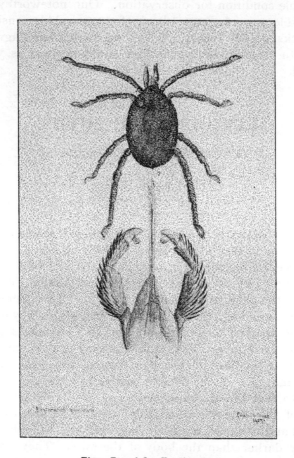

Figs. 7 and 8. Erythræus.

dead, for examination and dissection; they would come by post in a corked bottle or a closely fitting box containing a little slightly moist moss.

The Board of Agriculture and Fisheries desire to announce that a new edition of their leaflet on the Black Currant Mite has been published, in which information on the treatment of this pest with lime and sulphur has been incorporated. Fruit growers whose bushes have been attacked with the mite are advised to experiment with this process. Copies of the leaflet may be obtained gratis and post free on application to the Secretary of the Board of Argriculture and Fisheries, 4, Whitehall Place, London, S.W. Letters so addressed need not be stamped.

THE VICTORIA HISTORY OF YORKSHIRE.*

THE first Yorkshire volume of this magnificent work has at last appeared, and is certainly well up to the admirable standard already attained by this monumental series. Yorkshire is to have nine volumes devoted to it—three dealing with the county generally and two having reference to each of the three Ridings. The volume just issued is perhaps that which will interest readers of this journal more than will any of the subsequent volumes, as it contains an account of the geology, botany, zoology, and pre-historic remains of the county. Yorkshire, and the Yorkshire Naturalists' Union particularly, can be congratulated upon the prominent part which workers in the county have played in the production of the volume, though certainly in scanning its pages names of contributors which one might have expected to have seen are missing, whilst others, equally unexpected, appear. The valuable monographs issued by the County Society have probably had much to do with the selection of many of the authors of the chapters under notice.

After the usual preliminary matter, we find the first article is on the Geology of Yorkshire, by Professor Kendall. For this we have nothing but praise, and it can be safely stated that it represents one of the finest productions from the pen of that author, and unquestionably he has given a summary of the geology of the county in a way which only a thorough master of every branch of the subject could have done. In this contribution, which extends to just upon one hundred pages, it is evident that the writer is thoroughly acquainted not only with the principal sections in the various strata referred to, but, what is of more importance, with the literature relating thereto. Professor Kendall has always been a keen worker at bibliography, and this contribution to Yorkshire geology is an admirable instance of the value of such work. In carefully perusing the pages it cannot be seen that he has omitted references to anybody's work ; in fact, he has, if at all, erred on the side of giving too much prominence to the work of certain individuals. One great, though pardonable omission is the adequate recognition of the valuable contributions to Yorkshire geology which have been made by Professor Kendall himself. As was truly stated at a recent gathering of geologists, the study of the features of our broad-acred shire and the causes to which they are due,

---

* Vol. I., 524 pp., 12 plates and 4 maps. Constable & Co.; price (together with Vols. II. and III.) 4½ guineas.

1907 May 1.

received an impetus a few years ago when Professor Kendall.
first came to the county, and from that moment to the present
time might aptly be termed the Kendallian era in our knowledge
of the very early History of Yorkshire. Probably no one has
contributed more in a variety of ways to the elucidation of
the many intricate problems in our county as he has, and
whilst anyone familiar with his work might detect its influence
upon the contribution under reveiw, it is more than possible
that an outsider might study the article and have a very imperfect idea of the proportion of it which is really the result of
Professor Kendall's own researches. This contribution to
Yorkshire geology forms a fitting and firm foundation to the
series of monographs which are to follow. There is also a really
beautiful geological map of the county, in four sections.

The next article is on Palæontology, by Mr. R. Lydekker,
who is unquestionably well qualified to deal with the subject ;
but in perusing his very brief contribution, one is struck with
the obvious disadvantage under which he has laboured by not
being familiar with the district upon which he writes, nor with
the various collections of organic remains preserved in different
parts of the county. We must congratulate this author, notwithstanding, upon the thoroughness with which he has examined
the recent literature on the subject. The result of this, however,
is that too much prominence appears to be given to the work of
certain investigators of recent years, whilst that of the pioneers in
palæontological research, to whom we owe so much, seems to
almost be taking a secondary position. For example, we should
be the last, in any way, to deprecate the excellent work now being
done in the south eastern corner of the county ; and, largely
as the result of the frequent contributions during the last few
years by the Curator of the Hull Museum, there is no doubt
that the specimens under his charge have received a publicity
which, though deserving, has not been followed in the case of perhaps even more interesting examples in other parts of the county.
Mr. Lydekker has had very largely to depend upon *published* records and as a consequence one finds that the Hull Museum, or the
specimens it contains, or Mr. Sheppard, are referred to on almost
every page of this contribution, and on some pages several times.

Speaking, at any rate for the vertebrate remains found in
recent deposits, it can be safely stated that the records given in
this volume might be very largely increased.

The botanical section occupies 160 pages, and has been
prepared under the guidance of Mr. J. G. Baker, the veteran

Yorkshire botanist. Not to have secured the services of Mr. Baker would, indeed, have been a calamity. His familiarity with the botanical features of the county and his various contributions on the subject (including his recently issued second edition of 'North Yorkshire') are a sufficient guarantee of the excellence of the work, and in this section every acknowledgement is made to various Yorkshire workers and to the publications of the Yorkshire Naturalists' Union. The botanical section is divided into 'Introduction'; 'Botanical Districts' (1. the North Riding : 2. the East Riding : 3. the West Riding); 'Mosses'; 'Liverworts'; 'Marine Algæ'; 'Lichens'; and 'Fungi.' Mr. Baker gives the following approximate estimate of the number of plants known in the county :—

Flowering plants and ferns	1050.
Mosses	400.
Hepaticæ	150.
Lichens	500.
Algæ	1200.
Fungi	2626.
Total species ...	5926.

Mr. Baker's contribution is accompanied by an admirable coloured map of the county, divided into twenty-two botanical districts, based on the river basins.

We are agreeably surprised to find that a fairly comprehensive account of the marine zoology of the county appears, from the pen of Mr. John Oliver Borley. In this the author gives a description of the nature of the coast-line and its bearing upon the fauna. He supplies lists of the species according to their northern, southern, etc., types, and follows with apparently carefully compiled lists of the various forms of marine life occurring on the coast, for which he is admittedly indebted very largely to the 'Naturalist' and to the 'Transactions of the Hull Scientific and Field Naturalists' Club.' At the present moment, when there seems to be a desire on the part of Yorkshire naturalists to renew the study of the marine zoology of the coast, the appearance of this list seems particularly appropriate, and forms a suitable basis for future work. It is quite possible that a few records already published may have been overlooked by Mr. Borley, but it is more than probable that his lists can be enlarged by even a few carefully planned collecting excursions to Filey Brig, Flamborough Head, or other suitable points. Probably something in this direction will be accomplished during the coming summer.

The account of the non-marine mollusca of the county seems very brief indeed, occupying four-and-a-half pages only. This is probably due to the fact that it has been-prepared by Mr. B. B. Woodward, of the British Museum, who naturally cannot be familiar with the district he is describing, as are at least half-a-dozen prominent conchologists in the county. Perhaps the omission from the list of contributors to the present volume of the names of undoubtedly the best qualified conchologists in Yorkshire, we had almost said Britain, is the greatest surprise we have had in perusing it. Mr. Woodward points out that of the 146 species of non-marine mollusca now known to inhabit the British Isles no fewer than 122 have been recorded from Yorkshire. He regrets that there is, as yet, no complete memoir dealing with the land and fresh-water shells of the county as a whole, and this regret we certainly share with him. He points out that such a list was begun by Taylor and Nelson, in 1877, in the 'Transactions of the Yorkshire Naturalists' Union,' but it remains a fragment. We believe the conchological section of the Union is now considering the question of the completion of this list; and the sooner this appears the better. Possibly, had it existed, Mr. Woodward's contribution to the Victoria History would have been more substantial. It is gratifying to notice that 'the compilation by Mr. T. Petch of a published record of the land and fresh-water mollusca of the East Riding, with additions (Trans. Hull Sci. and F. Nat, Club, iii., 121-181), is the best planned local list it has been his good fortune to meet with, especially in the matter of the maps, and it is a great pity it has not been extended to the whole county.'

Dealing with the insects there is a very satisfactory contribution of eighty pages, under the general editorship of Mr. G. T. Porritt, who has supplied the lists of Orthoptera, Neuroptera, Trichoptera, and Lepidoptera.* Mr. Denison Roebuck is responsible for the Hymenoptera, Messrs. E. G. Bayford and M. L. Thompson for the Coleoptera, and Mr. P. H. Grimshaw for the Diptera. There does not appear to be a

---

* The list of Lepidoptera contains the following records, which have been made since the second edition of Mr. Porritt's 'List of Yorkshire Lepidoptera' appeared:—*Plusia moneta, Euchromia mygindana, Orthotænia antiquana, Catoptria fulvana, Acidalia emutaria, Ephippiphora grandævana, Gelechia atriplicella, Bedellia somnulentella.*

list of Hemiptera, though surely there must be some printed records, although possibly these are not numerous. It will be for the new Yorkshire Hymenoptera, Diptera, and Hemiptera Committee to remedy this defect. As might be expected, as a result of the exceptionally complete and excellent monographs already issued in the county, much of the material occurring in the contribution under 'insects' has already appeared elsewhere, though, of course, it is necessary in a work such as the Victoria History that it should occur again.

The Arachnida (Spiders, Harvest-men, and False-Scorpions) are dealt with by the Rev. O. Pickard-Cambridge, and include about 221 species for Yorkshire, out of about 540 known for Great Britain and Ireland. Mr. Pickard-Cambridge gives a useful bibliography.

The Rev. T. R. R. Stebbing has written a contribution dealing with Crustaceans, and gives an exceedingly readable and useful account of them ; in fact, he seems to have taken every possible care to gather together the various scattered records and to have presented the information in an attractive form. We cannot find that he has omitted anything of importance, and he has certainly examined the literature on the subject in a way which merits praise.

For the Fishes, Reptiles and Batrachians, Birds, and Mammals, Mr. Oxley Grabham is responsible, and his contribution to these subjects occupies, for fishes 8 pages ; reptiles and batrachians, $1\frac{1}{2}$; birds, 28 ; mammals, 6. With this limited allocation of space of course it is not possible for Mr. Grabham to have done every justice to these important departments. With regard to the fishes, the author has added his own notes, together with others supplied by Messrs. W. J. Clarke and T. Newbitt, to the list, 'now, however, considerably out of date,' in ' The Vertebrate Fauna of Yorkshire.' This chapter, however, had to be written at very short notice. Of the birds, we observe Mr. Grabham includes 326 species in his list. Amongst these is a red-throated pipit which the late Mr. John Cordeaux reported as seen near Kilnsea, though it was not obtained. We also observe that reference is made to the fact that quite recently ten eggs and a stuffed specimen of the Great Auk located at Scarborough have, with the exception of one egg, now been lost to the town, and that there are two stuffed specimens in the York Museum. These, of course, have no connection with the county, as Mr. Grabham recognises, as the information is given in square brackets. He refers also to the forthcoming ' Birds of

Yorkshire,' and whilst, of course, the somewhat brief account in the Victoria History is necessarily of a different type altogether from that about to appear in the two thick volumes dealing with the birds of the county, it is much to be regretted that the Yorkshire Naturalists' Union was not able to get its monograph before the public first. In examining Mr. Grabham's list we notice a preference for the records of specimens which have been shot near York or are in the York Museum, but this is quite pardonable. 'Mr. G. Hewett of York' seems unfamiliar, and we take this opportunity of informing Mr. Grabham that Mr. Hewitt's christian name is William. In the account of the Cetacea it might have been worth while to have added that the specimen of Sibbald's Rorqual caught at Spurn some years ago is the type specimen of that species, as was described in this journal for August, 1901.*

(*To be continued*).

———◆◆———

We have received the Annual Report for 1905 of the Public Museums and Meteorological Observatory of Bolton. It has been prepared by the Curator, Mr. T. Midgley, and contains a record of a useful year's work.

Under the title 'Nature's Night-Watchman,' Mr. Frank Finn figures and describes several species of owls in the April 'Animal World.' His photographs of the different species are very useful.

At a recent meeting of the Lancashire and Cheshire Entomological Society Mr. Sopp exhibited the cockroach *Phoraspis leucogramme*, taken in the Liverpool docks, this being a Brazilian species not previously recorded as having occurred in Europe.

An interesting illustrated pamphlet on 'Allotments,' by T. W. Sanders, F.L.S., has been issued by the Agricultural and Horticultural Association (one penny). It is the eighth issue of the 'One and All' practical gardening handbooks, edited by Edward Owen Greening. The author gives practical illustrations of the obtaining and working of allotments, and the editor adds some detailed advice 'how to proceed.'

From the 'Keighley Museum Report, 1906' we learn that it has been decided that Airedale shall be the District represented by its Museum, and towards securing specimens from this area, the Curator, Mr. S. L. Mosley, proposes to devote a very large portion of his leisure time, and all 'holidays.' The labels etc., printed at the Keighley Museum are reprinted on 8vo. sheets of paper, and issued as 'Keighley Museum Notes.' Of these 58 have already appeared. They principally deal with botanical and entomological subjects.

Miss M. V. Lebour favours us with a reprint of her paper on 'Larval Trematodes of the Northumberland Coast' (Trans. Nat. Hist. Soc. of Northumberland, Durham, and Newcastle-on-Tyne, New Series, Vol I., pt. 3). The authoress deals with this very neglected order in detail, and gives particulars of the various species of mollusca which are infested with Trematodes. Illustrating the paper are plates showing *Monostomum flavum, Cercaria ubiquita, C. pirum, Monostomum (Cercaria [lophocerca), C. oocysta, Distomum (Echinostomum) leptosomum,* and *Bucephalus haimeanus.*

In Memoriam.

JOHN EMMERSON ROBSON.

NORTHERN LEPIDOPTERISTS have sustained a great loss in the death of John E. Robson of Hartlepool, which event took place on February 28th last, after an illness of some weeks' duration. Mr. Robson was seventy-four years of age. For a very long period he was known in the north of England as an ardent and successful lepidopterist, and since his connection with the 'Young Naturalist' (afterwards the 'British Naturalist'), equally so throughout the country. Mr. Robson edited the journal just alluded to for the fourteen years from 1879 to 1893, the first several years in conjunction with Mr. S. L. Mosley. The journal was very popular and did much good, and will long be remembered on account of the lively but thoroughly good-natured discussions between prominent lepidopterists of the time on various entomological problems. Mr. Robson also issued ' A List of British Lepidoptera, and their named Varieties,' but his greatest literary work was probably 'The Lepidoptera of Northumberland, Durham and Newcastle-on-Tyne,' the concluding part of which he was engaged upon at the time of his death. He had been busy on this work for some years, and three parts had already been issued, bringing it to the end of the Tortrices, and leaving only the Tineæ and Pterophori to be dealt with. We are glad to know that this part will not suffer through the author's death, as Mr. E. R. Bankes has kindly undertaken to see it through the press, and it would have been impossible to have placed it in better or more suitable hands. Mr. Robson was an enthusiastic and genial companion as we know from experience, and a charming correspondent. He had been a Fellow of the Entomological Society of London since 1890.

Besides his business and entomological pursuits, Mr. Robson took great interest in public work, especially educational, and was formerly on the old School Board, and more recently on the Education Committee at Hartlepool. He was, too, until his death, a member of the Borough Council, of which body his father was Mayor so long ago as 1855. The funeral took place at Hartlepool Cemetery on March 4th, and was of a public character, being attended by the Mayor and Corporation and very many of the leading inhabitants of the town, whilst the streets en route to the cemetery were lined with people assembled to show their respect.—G. T. P.

1907 May 1.

FIELD NOTES.

LEPIDOPTERA.

Crambus falsellus : **An addition to the Yorkshire List of Lepidoptera.**—When on a visit to the Rev. C. D. Ash, in early July, last year, I netted a fine specimen of this Crambus at dusk in his garden at Saxton, near Tadcaster.—T. ASHTON LOFTHOUSE, Middlesborough.

Tortrices, etc., at Guisborough.—The following were taken on the occasion of the Y. N. U. meeting held at Guisborough in August, 1906 :—

Peronea sponsana.
Penthina variegana.
Hedya dealbana.
Grapholitha ramella.
Grapholitha penkleriana.
Pædisca corticana.
Pædisca solandriana.

Aphelia osseana.
Prays curtisellus.
Cerostoma radiatella.
Plutella cruciferarum.
Argyresthia nitidella.
Argyresthia gædartella.
Argyresthia spiniella.

T. ASHTON LOFTHOUSE, Linthorpe, Middlesborough, April 1907.

Lepidoptera taken in the Cleveland district during 1906.—In most cases the species named below are additions to the local list :—

Eudorea cratægella. At sugar. Kildale.
Tortrix unifasciana. Redcar.
Peronea caledoniana. Battersby.
Argyrotoza conwayana. Great Ayton.
Ptycholoma lecheana. Guisborough.
Hedya lariciana. Kildale, only two previous Yorks. records for this species.
Hedya neglectana. Middlesborough.
Sciaphila subjectana. Bred from near Middlesborough and Redcar.
Sciaphila hybridana. Redcar.
Pædisca sordidana. Among alders at Great Ayton in September.
Pamplusia mercuriana. Battersby.
Retinia pinivorana. Great Ayton.
Stigmonota internana. Eston in June.
Stigmonota regiana. Bred from sycamore, Kildale.

Trycheris aurana. Great Ayton.
Xanthosetia zoegana. Marske and Redcar.
Eupæcilia atricapitana. Redcar.
Epigraphia steinkellneriana. Saltburn.
Tinea semifulvella. Great Ayton.
Incurvaria masculella. Middlesborough in April.
Depressaria costosella. Marske and Kildale.
Depressaria angelicella. Kildale.
Gelechia ericetella. Swainby in Cleveland.
Gelechia dodecella. Kildale.
Gelechia ligulella. Kildale.
Gelechia rufescens. Redcar.
Ornix anglicella. Swainby in Cleveland.
Coleophora albicosta. Eston.
Coleophora fabriciella. Great Ayton.

T. ASHTON LOFTHOUSE, Linthorpe, Middlesborough.

BIRDS.

A Spotted Crake (*Porzana maruetta*) ♂ was picked up dead at Corwen on Sunday, April 14th. It had evidently met its fate by colliding against the telegraph wires, under which it was found. Its lower mandible was completely broken. The body of the bird was in a very wasted condition and the stomach contained only a few grains of sand. As this species is most secretive and loth to fly unless positively obliged, there is very little doubt the bird was migrating.—A. NEWSTEAD, Grosvenor Museum, Chester.

Rare Birds in Craven.—On 20th March, 1907, I saw a fine Pink Footed Goose feeding on the bank of the River Aire near Skipton ; it is not often we hear of this bird being so far inland.

Last October a Great Crested Grebe, an adult bird in winter plumage, was shot at Malham.

In February this year a fine specimen of an adult Oyster Catcher was shot near Skipton, probably one which had lagged behind from a batch of these birds which were seen here a few days earlier.—W. WILSON, Skipton-in-Craven, April 22nd, 1907.

An Albino Carrion Crow was reported in the last October issue of this journal by Mr. H. B. Booth, as seen on Burnsall Fell, Skipton-in-Craven.

One of the Duke of Devonshire's keepers has informed me that he saw the bird many times during July and August, and distinguished it as a Carrion Crow by its croak, which he heard on several occasions. The bird was also seen and identified by Mr. Alf. Downs, the Duke's agent, on August 3rd, 1906. It appears to have frequented that part of the moor around Crookrise, near Skipton, and was always accompanied by three other Carrion Crows, probably its nest-mates. The crow was last noticed in this district on September 3rd, 1906, near Embsay Crag, Barden Moor. I have since had reported to me that a White Carrion Crow was seen above Penyghent Gill, during the month of November last, in company with three other crows. If these are the same birds, it is interesting to note that this supports the suggestion that broods in this species keep together until they attain maturity. Many other persons saw the bird during July and August, 1906, and identified it as a Carrion Crow. The latest report I have is that it was seen on April 3rd, 1907, near Barden.—W. WILSON, Skipton-in-Craven, April 20th, 1907.

YORKSHIRE MOSSES.

Leskea catenulata (Brid.) Mitt.—In some way this moss has been omitted from Mr. Slater's list in the new edition of North Yorkshire. It is given on page 130 under the name of *Pseudoleskea catenulata* in the list of plants on the summit and higher slopes of Micklefell, and I gathered it there in 1906.

Orthothecium rufescens (Dicks.) B. and S.—This distinct moss I got at Easter this year at the bottom of Park Gill, Buckden. It is not mentioned in the account of the Langstrothdale and Buckden Mosses by Mr. W. Ingham, B.A., in the Sept., 1904, 'Naturalist,' and does not seem to have been reported in Wharfedale proper, though known from Malham to Arncliffe.—CHRIS. A. CHEETHAM.

———◆◆———

REVIEWS AND BOOK NOTICES.

The Rev. E. A. Woodruffe-Peacock, F.G.S., of Cadney, Brigg, continues to publish his useful 'Rural Studies Series,' the last issued, No. 9, dealing with 'Pasture and Meadow Analysis' (J. W. Goulding & Son, Louth, 20 pp., 1/-). In this the author gives an account of his method of carefully analysing the flora of a meadow by dividing an area into squares and carefully investigating the contents of each square.

The Common Wild Birds of Great Britain, by David T. Price, M.D. Gurney & Jackson, 1907, 62 pp., 1/- net. We have tested this book, and it appears to be thoroughly reliable and to admirably answer its purpose as 'A ready aid to distinguish the Common Wild Birds of Britain.' The birds are arranged according to their habitats and sizes, and by a series of cross-references it is quite an easy matter to distinguish most of the birds likely to be met with in a ramble in any part of the country. It is just the book for a beginner who does not want to be troubled with details of the birds he is very unlikely to see.

List of British Seed-Plants and Ferns, by **James Britten** and **A. B. Rendle**, Department of Botany, British Museum. 44 pp. 4d.

British Botanists will feel indebted to the compilers of this List for placing at their disposal the results of the International Rules of Botanical Nomenclature, adopted by the Botanical Congress at Vienna, 1905, in so far as they affect British plants. For the convenience of Botanists they have correlated the names adopted in the three principal handbooks, viz., Bentham's 'Handbook,' Hooker's 'Students' Flora,' and Babington's 'Manual,' so that one sees at a glance the changes that have been made in cases where the name is not adopted in the List. The List has been shortened by the omission of Channel Island plants and critical forms of genera, such as *Rubus*, *Hieracium*, etc. Much time and money was spent by the Congress in framing these rules, and it is hoped their general adoption will lead to something like stability in Botanical Nomenclature.

Text Book of Fungi, including Morphology, Physiology, Pathology, Classification, etc. By **George Massee.** Duckworth & Co., London, 1906. 427 pages. Price 6/- net.

Mr. Massee is well known in the north. His attendance at the annual Fungus Forays of the Yorkshire Naturalists' Union has enabled many mycologists to benefit by his personal acquaintance in the field; and as joint author of the Fungus Flora of Yorkshire, and as Chairman of the Yorkshire Mycological Committee, he has greatly helped forward the study of the lower forms of plant life with which he is so familiar. In the work under notice he has placed all students under a debt of gratitude for the careful

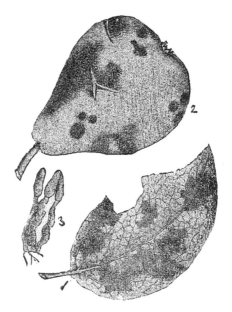

Fusicladium pirinum, a destructive Parasite on Pears.

1. Fungus forming minute, velvety, dark-coloured patches on a pear leaf. 2. Fungus forming scurfy patches, and causing cracking of the flesh of a pear. 3. Conidiophores bearing conidia. Figs. 1 and 2 reduced; fig. 3 magnified.

way in which he has put in a presentable form all the modern views on the subjects referred to in the sub-title, as well as by giving an introduction to the comparatively new lines of research. A useful chapter on 'Biologic Forms' is contributed by Mr. E. S. Salmon. Not the least valuable part of the book is to be found in the numerous illustrations and the descriptions thereof, which from an economic point of view are invaluable. We believe they are all from Mr. Massee's own pencil, and prove him to be a most capable artist as well as a learned author. By the courtesy of the publishers, we are able to reproduce one of the 141 figures.

The March 'Naturalists' Quarterly Review' has secured one or two new contributors. Mr. Westell (W. Percival, F.R.H.S., M.B.O.U.) informs us that in April the ' Lambs become *stronger*,' in this way, presumably, differing from the young of other animals. In the usual praises for new books we once again learn ' that these are quite the finest photographs we have ever seen,' and that 'this is one of the very best volumes we have ever seen,' etc., etc. The same 'reviewer' apologises for describing 'The Evolution of Man' in a previous issue as 'The Evolution of *Mars!*' An advertiser (whose name we might perhaps guess), wants 'list of duplicates with lowest prices' of 'finely marked *recent* clutches of British Birds' Eggs.' Is it possible that the same individual writes upon the advantages of the Acts for the Protection of Birds and Eggs?

NORTHERN NEWS.

The recent leaflets issued by the Board of Agriculture and Fisheries deal with Crimson Clover, Dodder, Poultry Fattening, The Dogs Act, 1906, Insurance of Farming Stock against Fire, and Cleansing of Water Courses. A significant post-script appears to the last named :—'This leaflet *does not apply to Scotland!*' Stands Scotland where it did ?

Hints on self-advertisement. 'As a member of the School Nature-Study Union, and as a writer of books and lecturer on Natural History subjects designed to interest, elevate, and amuse young and growing children, etc., etc.'... 'Listen to what Sir Herbert Maxwell says in his Introduction to one of my books.' (Extracted from an article on 'Nature Teaching, through the senses to the mind,' by a F.R.H.S., M.B.O.U.).

The Report of the Corresponding Societies' Committee of the British Association, and of the Conference of Delegates (York Meeting) has been issued, and may be obtained at the offices of the Association for one shilling. In addition to the papers read, and the discussions which took place at the Conference of Delegates at York, it contains particulars relating to the various societies affiliated with the Association, as well as a useful bibliography of papers which have appeared in the various societies' proceedings.

The B.E.N.A.'s have been having a 'beano.' Forty-nine of the readers of 'The Country Side' have guessed correctly a 'What is it' problem (a thimble); the prize has been divided, and each 'successful competitor' has received the sum of *fivepence*. May we suggest that these forty-nine show their gratitude by purchasing a B.E.N.A. badge with the proceeds, and be thus easily identified. In the same journal we learn that 'country people love to keep a jay in a wicker cage!'.

At a recent meeting of the York Naturalists' Society a discussion took place on 'the Young Naturalist.' From the press report we learn that 'Mr. William Hewett, a born entomologist of fame, drew timely attention to the fact that it would be a sad day if all the children, youths, and maidens in the land were made mere collectors of rare plants, insects, etc. The county of York could not afford to have a single species exterminated by too ardent collecting, of which there was always a lurking danger present.' To this we say Amen !

Perhaps the principal article in the April 'Reliquary' deals with 'Damme.' To this city of the Netherlands a good American once paid a visit. 'There was no hotel, the door of the one estaminet was too narrow to admit his trunks, and, sitting down upon them in the deserted Grand Place, he softly whispered the word which is at the head of this chapter.' In the same journal Mr. T. P. Cooper, of York, has a paper on 'The Story of the Tobacco Pipe.' In this he figures, as early York-made pipes, two or three examples which were most probably made in Hull.

As an illustration of the way in which even common objects can be examined with fruitful results, Mr. C. Gordon Hewitt has recently issued 'A Preliminary Account of the Life History of the Common House Fly' (Memoirs and Proceedings Manchester Lit. & Phil. Soc., Vol. 51, part 1). In the the same proceedings, Mr. R. L. Taylor draws attention to the remarkable luminosity produced by rubbing or knocking together various forms of silica. A correspondent in a local paper some time ago pointed this out as a property peculiar to the white pebbles found on the coast at Whitby.

The taste for natural history is growing, and 'Punch' is the latest journal to have a column devoted to 'Nature Studies.' The subject in the issue before us is 'The Motor Bus,' though the Latin name of the species is not given, which seems unfortunate. Judging from the description, however, we presume it is a variety of a *Slitheranda damdûm* (L). In another part of the same issue, under 'Zoological Sequels,' are given some advertisements for patent medicines for animals. Amongst these we find 'Leopards try Pumacea. It touches every spot.' Another is a cure for lobsters blushing, which reminds us of a story. [Thanks, but this is a natural history journal.—ED.].

JUNE 1907.

No. 605
(No. 383 of current series).

THE NATURALIST.

A MONTHLY ILLUSTRATED JOURNAL OF
NATURAL HISTORY FOR THE NORTH OF ENGLAND.

EDITED BY

T. SHEPPARD, F.G.S.,

THE MUSEUM, HULL;

AND

T. W. WOODHEAD, Ph.D., F.L.S.,

TECHNICAL COLLEGE, HUDDERSFIELD.

WITH THE ASSISTANCE AS REFEREES IN SPECIAL DEPARTMENTS OF

J. GILBERT BAKER, F.R.S. F.L.S.,   GEO. T. PORRITT, F.L.S., F.E.S.,
Prof. P. F. KENDALL, M.Sc., F.G.S.,   JOHN W. TAYLOR,
T. H. NELSON, M.B.O.U.,   WILLIAM WEST, F.L.S.

Contents :—

LONDON:

A. BROWN & SONS, LIMITED, 5, FARRINGDON AVENUE, E.C
And at HULL AND YORK.

Printers and Publishers to the Y.N.U.

PRICE 6d. NET. BY POST 7d. NET.

BOTANICAL TRANSACTIONS OF THE YORKSHIRE NATURALISTS' UNION, Volume I.

8vo, Cloth, 292 pp. (a few copies only left), price **5/-** *net.*

ontains various reports, papers, and addresses on the Flowering Plants, Mosses, and Fungi of the county

Complete, 8vo, Cloth, with Coloured Map, published at **One Guinea.** *Only a few copies left,* **10/6** *net.*

THE FLORA OF WEST YORKSHIRE. By FREDERIC ARNOLD LEES, M.R.C.S., &c.

This, which forms the 2nd Volume of the Botanical Series of the Transactions, is perhaps the most omplete work of the kind ever issued for any district, including detailed and full records of 1044 Phanero. ams and Vascular Cryptogams, 11 Characeæ, 348 Mosses, 108 Hepatics, 258 Lichens, 1009 Fungi; and 382 reshwater Algæ, making a total of 3160 species.

680 pp., Coloured Geological, Lithological, &c. Maps, suitably Bound in Cloth. Price **15/-** *net.*

NORTH YORKSHIRE: Studies of its Botany, Geology, Climate, and Physical Geography.

By JOHN GILBERT BAKER, F.R.S., F.L.S., M.R.I.A., V.M.H.

And a Chapter on the Mosses and Hepatics of the Riding, by MATTHEW B. SLATER, F.L.S. This Volume orms the 3rd of the Botanical Series.

396 pp., Complete, 8vo., Cloth. Price **10/6** *net.*

HE FUNGUS FLORA OF YORKSHIRE. By G. MASSEE, F.L.S., F.R.H.S., & C. CROSSLAND, F.L.S.

This is the 4th Volume of the Botanical Series of the Transactions, and contains a complete annotated list f all the known Fungi of the county, comprising 2626 species.

Complete, 8vo, Cloth. Price **6/-** *post free.*

HE ALGA-FLORA OF YORKSHIRE. By W. WEST, F.L.S., & GEO. S. WEST, B.A., A.R.C.S., F.L.S.

This work, which forms the 5th Volume of the Botanical Series of the Transactions, enumerates 1044 species, with full details of localities and numerous critical remarks on their affinities and distribution.

Complete, 8vo, Cloth. Second Edition. Price **6/6** *net.*

LIST OF YORKSHIRE LEPIDOPTERA. By G. T. PORRITT, F.L.S., F.E.S.

The First Edition of this work was published in 1883, and contained particulars of 1840 species of Macro- and Micro-Lepidoptera known to inhabit the county of York. The Second Edition, with Supplement, contains much new information which has been accumulated by the author, including over 50 additional species, together with copious notes on variation (particularly melanism), &c.

In progress, issued in Annual Parts, 8vo.

TRANSACTIONS OF THE YORKSHIRE NATURALISTS' UNION.

The Transactions include papers in all departments of the Yorkshire Fauna and Flora, and are issued in separately-paged series, devoted each to a special subject. The Parts already published are sold to the public as follows (Members are entitled to 25 per cent. discount): Part 1 (1877), 2/3; 2 (1878), 1/9; 3 (1878), 1/6; 4 (1879) 2/-; 5 (1880), 2/-; 6 (1881), 2/-; 7 (1882), 2/6; 8 (1883), 2/6; 9 (1884), 2/9; 10 (1885), 1/6; 11 (1885), 2/6; 12 (1886), 2/6 13 (1887), 2/6; 14 (1888), 1/9; 15 (1889), 2/6; 16 (1890), 2/6; 17 (1891), 2/6; 18 (1892), 1/9; 19 (1893), 9d.; 20 (1894), 5/- 21 (1895), 1/-; 22 (1896), 1/3; 23 (1897), 1/3; 24 (1898), 1/-; 25 (1899), 1/9; 26 (1900), 5/-; 27 (1901), 2/-; 28 (1902), 1/3 9 (1902), 1/-; 30 (1903), 2/6; 31 (1904), 1/-; 32 (1905), 7/6; 33 (1906), 5/-.

THE BIRDS OF YORKSHIRE. By T. H. NELSON, M.B.O.U., WILLIAM EAGLE CLARKE, F.L.S., M.B.O.U., and F. BOYES. Also being published (by Subscription, **One Guinea**).

Annotated List of the LAND and FRESHWATER MOLLUSCA KNOWN TO INHABIT YORK-SHIRE. By JOHN W. TAYLOR, F.L.S., and others. Also in course of publication in the Transactions.

THE YORKSHIRE CARBONIFEROUS FLORA. By ROBERT KIDSTON, F.R.S.E., F.G.S. Parts 14, 18, 19, 21, &c., of Transactions.

LIST OF YORKSHIRE COLEOPTERA. By REV. W. C. HEY, M.A.

THE NATURALIST. A Monthly Illustrated Journal of Natural History for the North of England. Edited by T. SHEPPARD, F.G.S., Museum, Hull; and T. W. WOODHEAD, F.L.S., Technical College Huddersfield; with the assistance as referees in Special Departments of J. GILBERT BAKER, F.R.S. F.L.S., PROF. PERCY F. KENDALL, M.Sc., F.G.S., T. H. NELSON, M.B.O.U., GEO. T. PORRITT F.L.S., F.E.S., JOHN W. TAYLOR, and WILLIAM WEST, F.L.S. (Annual Subscription, payable in advance, **6/6** post free).

MEMBERSHIP in the Yorkshire Naturalists' Union, **10/6** per annum, includes subscription to *The Naturalist* and entitles the member to receive the current Transactions, and all the other privileges of the Union A donation of **Seven Guineas** constitutes a life-membership, and entitles the member to a set of the completed Volumes issued by the Union.
Members are entitled to buy all back numbers and other publications of the Union at a **discount of 2 per cent.** off the prices quoted above.
All communications should be addressed to the Hon. Secretary,

T. SHEPPARD, F.G.S., The Museum, Hull.

NOTES AND COMMENTS.

THE BRADFORD SCIENTIFIC JOURNAL.

The Bradford Scientific Journal for April completes the first volume of twelve parts, and contains an index. The journal is to be congratulated on its success, and particularly from the fact that a very fair proportion of the articles have been of local interest. The present part has a useful and well illustrated paper on 'The Vegetation of some Disused Quarries,' by S. Margerison. Amongst the other items are 'Our Birds in Winter,' by C. A. E. Rodgers, and 'The Diversion of a Wharfedale Stream into Airedale,' by E. E. Gregory.

THE BRADFORD MUSEUM.

In an editorial we notice that our Bradford friends have 'one regret. Our scheme for the furtherance of the Bradford Museum still languishes in abeyance. Not for want of enthusiasm amongst local Scientists, nor for want of appreciation by the public, who have evinced a lively interest in the little that has been done. The obstacle *is official indifference, if not actual opposition on the part of our municipal authorities.* North, south, east, and west of the borough we find museums. Leeds, Keighley, Halifax, Huddersfield, and now Spen Valley, all show us how it is done, and still Bradford occupies its position of 'splendid' isolation.' We share the hope of the Bradford naturalists that before long the Corporation will exhibit a larger public spirit and provide a museum worthy of the city. A good deal of the material which has been put on exhibition, we understand, is utter rubbish, valueless for any educational purpose, and in some cases actually the laughing-stock of the informed naturalist. It is surprising that Bradford having so excellent a building for the purpose of a small museum should be content to allow it to be wasted. We are ashamed of Bradford.

THE INCLOSURES ACTS IN LINCOLNSHIRE AND EAST YORKS.

The Footpaths Preservation Society has reprinted Dr. Gilbert Slater's most valuable paper on 'The Inclosure of Common Fields considered Geographically,' which appeared in the 'Geographical Journal.' This contains several plans and diagrams, and the subject is dealt with in a way which is of great value to a naturalist. Some peculiar cases of the effect of the Inclosures Acts on the village communities of Lincolnshire and the East Riding of Yorkshire are referred to. ' In the Isle of Axholme every cottager possessed a right of common over the vast swampy pastures which separated the Isle from Yorkshire on one side and Lincolnshire on the other ; the owners of land, as such, had no rights of common. In consequence, the cottagers were able, when the marshes were divided, drained, and inclosed, to defeat the proposal to also inclose the arable fields, and these remain to the present day, to a very great extent, open and intermixed.'

LAKELAND RAVENS.

From a recent issue of the *Yorkshire Post* we regret to learn that the few ravens that continue to nest in Lakeland are exposed to continual persecution by well-to-do egg collectors. Recently a pair that had nested on a crag on Melbreak was robbed of five eggs. Two parties went to secure the nest. Both took 20ft. ladders for the purpose, but finding them too short, agreed to join them together, and divide the spoil. A spin of a coin decided the fate of the odd egg. The eggs of the raven are protected in Cumberland, but it is a mere paper protection. The nests are regularly harried by collectors, and probably in a short time the Lake District will know the noble bird no more.

We suppose there are police in the Lake District? though possibly they are of the variety of some of those in south-east Yorkshire, where a police inspector was recently noticed to be one of a party enjoying 'sport' amongst protected birds, though as they were on a boat on a river they may possibly not have been in a *place* within the meaning of the Act !

PLANT ASSOCIATIONS AND GOLF.

A recent writer in one of the dailies has found an original excuse for losing a golf match. He was so absorbed in the fact that the Fescue grasses occupied the hillocks and the higher and drier slopes, whilst others of softer texture and harder names flourished in the ' damp hollows.' * He continues : ' In

---

* This kind of hollow is frequently referred to by golfers.

a particular little hollow, at which I had a good look because my ball came to rest in it, I could not detect a single stalk of Fescue grass. The verdant carpet seemed to me to be composed chiefly of the poa grasses. Thirty feet higher up, at the same hole, there were nothing but the Fescues.' Who says there is nothing in the study of plant associations after that?

A DARLINGTON 'FIND.'

Some little stir has been caused in local circles by the discovery, at Darlington, of a bone two feet in diameter and three feet long, found at a depth of six feet, 'immediately below a bed of glacial clay and above the gravel.' . . . 'Theories innumerable have been propounded by geologists, scientists, anatomists, and others who have inspected the find, but nothing at all probable has as yet been forthcoming. It will be produced at a meeting of the Darlington Naturalists' Field Club, when anyone interested may attend and take part in the inquest which is to be held. What the verdict may be it is impossible to say, but for the guidance of the jury we append the jaw-breaking designation of some of the largest known extinct animals, and they may choose any they wish :—Anthracotherium, anapoltherium, dinotherium giganterum, ichthyosaurus, lopiodon, mastodon giganteus, megatherium, mosasaurus, palæosaurus, palæotherium, plesiosburus, teleosaurus, and tetracaulodon.' We have printed the names as they appear in the local press! Mr. Edward Wooler has kindly favoured us with a photograph of the 'discovery,' and as was suspected, it turns out to be a part of the lower jaw of a whale—probably part of a gate-post —as fifty years ago hundreds of these jaw-bones were sold for this purpose and distributed over the country. In East Yorkshire dozens such still exist.

THE SCARBOROUGH MUSEUM.

We have received the Annual Report of the Scarborough Philosophical Society for 1906, in which is included the report of the Scarborough Naturalists' Society. We are glad to notice that, through the energy of its members, its debt has been wiped off, the much needed renovations of the property have been partly carried out, and there is a good balance in hand. The floor of the geological room, which had collapsed, has been concreted. Substantial alterations have also been carried out in other parts of the Museum, and a generous gift of £100 enables the Society's library to have proper attention.

SCARBOROUGH NATURALISTS.

·It is also evident·that the Field 'Naturalists' Society at Scarborough is doing·useful work, judging from the Recorders' reports. We learn the membership ' is probably about equal ·to that of last year.' Amongst·the more interesting items· we· notice· that a small·colony of Black-headed :Gulls nested near Scarborough. 'On May 16th there were about forty nests, but all had been plundered, either by human ·or avian depredators, and not an egg was to be found.' Reference is made to the " self-denial " of the climbers at Speeton in allowing the young· peregrines last year to fly, and to the fact that this is the fourth ·occasion ·upon which peregrines have· nested within the Scar- borough district since 1900·

RECORDERS' REPORTS.

Mr. W. J. Clarke is responsible for the report on the· Vertebrata, Mr. W. Pearson for Coleoptera, Mr. A. S. Tetley for Lepidoptera, in which report it is recorded, somewhat un-- expectedly, that 'the season of 1906 has been a good one for Lepidoptera,' Mr. E. R. Cross for Flowering Plants, Mr. W. Gyngell for Conchology, Mr. E. B. Lotherington for Micro- Botany and [Micro-] Zoology, and Mr. R. Gilchrist for Geology and Arachnida. All these gentlemen have carefully recorded the season's work, and in several departments additions to our knowledge have been made. If we might make a suggestion for the benefit of future reports, it is that the scientific names of the species referred to be printed in italics (this is· done, partly, in the Botanical Report), and that greater care be taken in the matter of proof-reading. Amongst some of the more: glaring misprints are ' Fuses,' ' Cornbrach,' and ' Slathwait.' The commas also appear to have been dropped in anyhow in parts of the report. Perhaps the most unexpected item was that Prof. Kendall had sent for inspection ' a special model of the Glaciatioro, in North Yorkshire.' Something·Italian, surely !

MODEL OF EBBING AND FLOWING WELL.

In Evan's '·How to Study Geology,' which we are noticing· elsewhere, an interesting illustration is given. of the model of the ebbing and flowing well at Austwick Hall, the property of T. R. Clapham; Esq., which was constructed by the late Richard Clapham ·about the year 1851. By the courtesy of the· publishers we are able to reproduce this diagram. for the benefit of our readers. A. is the side of a large slate cistern, in which is a fountain. B. is a cylindrical cistern of lead, from the bottom of which, at E., projects a syphon C., the other extremity of which is carried down the garden for a distance of fifteen yards to a stone well D., the outlet of which is then at F. B. is supplied with water from the tap shown. This tap has to be regulated to give a certain supply. As the cistern B. gets filled with water to the level of the dotted line it is filling the right

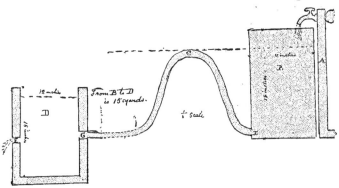

limb of the syphon also. When the water has risen to the dotted line, all air being driven from the tube, the action of the syphon begins, and as the base of the syphon pipe is greater than that of the tap, the water in B. is drawn off until it falls down to E., when air gets into the syphon and stops its action until the cistern B. is again filled from the tap. When the syphon is drawing all the water from B. it is flowing into the well D., and is, of course, emptying itself during the time that B. is again being filled.

---●◆---

The Kingdom of Man, by E. Ray Lankester, D.Sc., F.R.S., etc. Constable & Co., 1907. 192 pp., plates, 3/6 net. This volume contains three essays by Professor Lankester, which, under other titles, have previously appeared elsewhere. The first, 'Nature's Insurgent Son,' was the Romanes lecture for 1905; the second was the presidential address of the British Association, delivered at the York meeting last year, and contains an account 'of the progress made in the last quarter of a century towards the assumption of his kingship by slowly-moving man.' This address was dealt with in these columns at the time it was delivered. The third paper, reprinted from the 'Quarterly Review,' is an account of the recent attempt to deal with the Sleeping Sickness of South Africa, The volume takes its title from the first of the essays named. All three will be familiar to our readers, though many may desire to possess them in this convient form, particularly as numerous additional illustrations are inserted. We notice, in a description of a photograph of six Eoliths of the 'shoulder-of-mutton' type, 'the descriptive term "trinacrial" is suggested by me for these flints in allusion to the form of the island of Sicily which they resemble (Original).' Future writers please note.

1907 June 1.

YORKSHIRE NATURALISTS AT ROBIN HOOD'S BAY.

MAY 18th to 20th, 1907.

DURING Whit week-end (May 18-20) over forty members of the Yorkshire Naturalists' Union stayed at Robin Hood's Bay. This number was augmented on Bank Holiday, and a few remained longer. As has been the practice in recent years, the evenings were devoted to the reading of papers and discussions thereon, and in this way additional value and interest was added to the work accomplished during the field excursions.

As might be expected from the nature of the district, and its geological traditions, the hammermen were predominent, and had an exceptional opportunity of examining the sections in the cliffs and in the quarries adjacent. These places have also long been known as haunts of interesting birds, which, on this excursion, were the means of attracting several camera and field-glass ornithologists ; the old-fashioned type—the egg collector pure and simple—is now almost as rare as are some of the birds he has all but exterminated.

The numerous steep-sided, well-wooded ravines around 'Bay Town' proved irresistible to many besides the botanists, and the shore yielded shells and sea-weeds to those interested in marine fauna and flora, the last being an attraction not always present on our rambles.

The geologists had the advantage of the leadership of Prof. P. F. Kendall, who was as familiar with the problems of 'solid' geology as with those of the glacial series, in connection with which latter his work in north-east Yorkshire is now so well known.

On Saturday morning an early start was made, and the beach, scars, and cliffs between ' Bay Town ' and Ravenscar were examined ; sections in the Lias, ' Dogger,' and Estuarine Series being available. Particular attention was paid to the fault in the strata at the Peak. This yielded evidence that a movement in the earth's crust was taking place during the deposition of the beds, the thickness of the layers of sandstone, etc., varying on each side of the fault. The continuance of the fault seawards and the way it bifurcates was easily seen on the scars. The area traversed was a favourite hunting-ground for geological specimens, and several were secured. A fitting finish to the day's ramble was the ascent of the cliff at Raven-scar by means of a 'path,' which would be an excellent one for rabbits.

On Saturday evening, under the chairmanship of the President of the Union, Mr. C. Crossland, F.L.S., a well-attended meeting was held at the Union's headquarters, the Grosvenor Hotel. Prof. Kendall gave an address on some of the geological problems of the district, paying particular attention to 'Persistent faulting,' which caused a good discussion.

On the following day the botanists investigated the woods and fields, the ornithologists hied to the cliffs, and the geologists were conducted over the old Alum Works at the Peak. The Liassic shales, which had been exposed during the working of this by-gone industry, yielded a large number of Ammonites and other fossils. The relative positions of the 'Dogger' and Estuarines were also demonstrated, though on account of the varied applications of the word 'Dogger' it was seen that misunderstandings might easily arise in regard to its precise place in the geological sequence in the area.

After dinner there was even a larger attendance than on the preceding evening. Mr. F. Elgee read a paper on 'Glacial Survivors,' and introduced a subject of which, doubtless, more will be heard at future meetings of the Union. The secretary exhibited and described the bronze bridle-bit and other objects from a British Chariot burial which he had recently excavated.

On Monday most were early astir, and, as on previous days, owing to the not unfavourable weather, many paid visits to the fields and shore before breakfast. None could say that the excursions of the Yorkshire Naturalists' Union are idle holidays! The geologists walked to Hawsker, where they descended the cliff by another 'path,' which fortunately was dry. From there they walked round on the beach to Robin Hood's Bay, and had some excellent collecting en route. In addition to the Liassic shales and the capping sandstone, the party had the opportunity of examining a six-inch bed, the diminutive representative of the Cleveland Ironstone, which further north is so well developed.

At the general meeting, held later in the day, the representatives of the various sections made their reports, which are referred to below. The President congratuled the Society on its present position. At this meeting no fewer than fourteen new members were elected, and two societies became affiliated with the Union. Votes of thanks were passed to the land-

owners for permission to visit their estates, and particularly to
J. W. Barry, Esq., who personally looked after the party visiting
Fylingdales and gave every assistance in his power. To Prof.
Kendall for leading the geologists, to the readers of the papers,
and to the Divisional Secretary (Mr. J. J. Burton), for the
excellence of the local arrangements, similar compliments were
paid.

Later in the evening Mr. Burton read some valuable notes
on the sylvan vegetation of Fylingdales, which had been kindly
prepared by Mr. Barry for the benefit of the Union.

In the matter of the weather the party were favoured—one
short shower being all that was experienced—and top-coats
were not needed on either of the three days. Judging from
the reports in the press, elsewhere in the county the weather
had been anything but pleasant ; *verb. sap.*

BOTANY.—Mr. J. HARTSHORN writes :—Most of the Botanists
did not reach Robin Hood's Bay until Saturday at noon, or
later, hence the principal ramble was the official one on the
Monday through Ramsdale, returning by the moorland edge to
Fyling Hall. This was in every way a success, and very largely
due to the kindness of J. W. Barry, Esq., who personally conducted the party over his grounds, and acted as guide in
the Dale.

Near the Hall there was much of interest, and some fine
Araucarias were especially admired. Throughout the valley
such was the richness and profusion of bloom that the compiling
of a complete list of the species present was impossible in the
time available. The reluctance of the members to leave the
stream was forgotten when, almost immediately, a mass of
Chickweed Winter-green (*Trientalis europæa*) was discovered.
In characteristic situations were seen the Butterwort, Sundew
(*Drosera rotundifolia*), Sweet Gale, Needle Green Weed or
Petty Whin, and Bog Bean.

During the week-end over 130 species of flowering plants
were noted, mostly in bloom. Especially beautiful was a hedge
bordered uniformly by an unbroken band of Stitchwort (*Stellaria
holostea*). In the adjoining meadow the Green-winged Meadow
Orchis was growing in plenty. And on the cliffs there was
found in fruit the Spurge Laurel. Other interesting plants seen
were the Wood Vetch (*Vicia sylvatica*), Slender Vetch (*V.
gemella*), Golden Saxifrage (*Chrysosplenium oppositifolium*),
and a white specimen of the Early Purple Orchis (*Orchis mascula*).

MARINE MOLLUSA.—The Rev. F. H. Woods, B.D., writes :—
The Mollusc Fauna of Robin Hood's Bay does not appear to be
very extensive, but is not without interest, being determined by
the rocky character of the coast.   No examples were found of
those genera which live in sand and mud, while those which feed
on the algæ and live in or upon the rocks were very fairly represented.   For this reason there was a large preponderance of
univalves over bivalves.   Of the two borers, *Pholas crispata* is very
characteristic of the Yorkshire Coast, being exceedingly abundant
at Redcar and frequent at Bridlington.   The louse-like mail shell,
Chiton cinereus, was very abundant, and one was found very near
high-water mark.   One remarkably fine specimen, over an
inch long, of the lesser hairy mussel, *Modiola phaseolina*, was
found.   The blue-striped limpet, *Helcion pellucidus*, was very
abundant, and the beautiful little pink limpet, *Tectura virginea*,
was fairly frequent.   The most interesting were the smaller,
and often minute shells.   About two table-spoons of fine shingle,
taken from the surface of the drift, were exhaustively examined
under a lense.   There were three distinct species of *Rissoa*.
Curiously enough, while *R. striata* was very abundant, *R. parva*,
which can be gathered by the thousand on a favourable day at
Bridlington, was comparatively scarce.   Of the third species, *R.
punctura*, there were two specimens.   This is found rather
sparsely about the Yorkshire coast, and appears to be a northern
species.   All three are remarkably beautiful seen under a lense.
Of the two *Pleurotomas*, *P. turritella* I found at Bridlington occasionally, the other I have not found nor seen before.   Except for the
channelled mouth it is remarkably like a *Rissoa parva*.   It is probably, however, an immature specimen.   There was also a very
good example of *Tornatinus truncatulus*.   A single valve of a shell
of the *Venus* type appears to be an immature *Circe minima*.   It is
certainly not *Venus Gallina*, which in its young state is easily
recognisable.   Among the drift were also many examples of
several common shells in a very young state.   Well worth
mentioning is the *Odostomia spiralis*, a shell found fairly
frequently along the Yorkshire coast, but I have not seen it
elsewhere.   The specimens here are always very small, about
$\frac{1}{12}$ of an inch or less, but I have carefully compared them with
the smallest in the Barlee type collection at Oxford, and have
no doubt about their identity.   The other *Odostomia*, is very
common about the coast, but is evidently immature ; it may be
conoidea or possibly *acuta*, but it agrees with none in the Barlee
collection.

The following is a complete list of those found :—

Pholas crispata.
Saxicava rugosa.
Circe minima ?
Cardium fasciatum ?
Mytilus edulis.
　,,　modiolus (young).
Modiola (Mytilus) phaseolina.
Pecten opercularis (young).
Anomia ephippium.
　,,　patelliformis.
Chiton cinereus.
Patella vulgata.
Helcion pellucidus.
Tectura (Acmæa) virginea.
Trochus cinereus.
Littorina littorea.
　,,　obtusata.

Littorina rudis.
Lacuna divaricata.
　,,　puteolus (fragment).
Rissoa parva.
　,,　　,,　var. interrupta.
　,,　striata.
　,,　punctura.
Odostomia spiralis.
　,,　conoidea ?
Natica catena (small fragment)..
Purpura apillus.
Nana incrassata.
Buccinium undatum.
Pleurotoma turricula.
　,,　? species.
Cypræa europæa
Tornatinus truncatulus.

COLEOPTERA.—MR M. L. Thompson reports that the following beetles were met with along the route through Ramsdale :—

Notiophilus biguttatus, F.
Bembidium obtusum, Stm.
Patrobus excavatus, Payk.
Tachyporus chrysomelinus, L.
　,,　hypnorum, F.
Conosoma pubescens, Gr.
Philonthus decorus, Gr.
Adalia bipunctata, L.
Coccinella 10-punctata, L.
　,,　7-punctata, L.
Brachypterus urticæ, F.
Micrambe vini, Panz.

Adrastus limbatus, F.
Rhagonycha pallida, F.
Gastroidea polygoni, L.
Longitarsus suturellus, Duft.
Polydrusus pterygomalis, Sch.
Phyllobius pyri, L.
　,,　argentatus, L.
Sitones tibialis, Hbst.
Liosoma ovatulum, Clair.
Cæliodes quercus, F.
　,,　quadrimaculatus, L.
Ceuthorrhynchus pollinarius, Först.

Proceeding across the moor above the dale through the peat bogs to Robin Hood's Bay, additional species were found, with the help of Mr. J. T. Sewell. These were :—

Cicindela campestris, L.
Carabus monilis, F.
Bradycellus cognatus, Gyll.
Harpalus æneus, F.
　,,　latus, L.
Pterostichus madidus, T.
　,,　vulgaris, L.
Calathus flavipes, Fruic.
　,,　melanocephalus, L.
Anchomenus albipes, F.
Bembidium nigricorne, Gyll.
　,,　atrocæruleum, Steph.

Agabus bipustulatus, L.
Hydroporus pubescens, Gyll.
Gyrinus natator, Scop.
Drusilla canaliculata, F.
Homalota analis, Gr.
Xantholinus linearis, Ol.
Geotrupes stercorarius, L.
　,,　sylvaticus, Pz.
Dolopius marginatus, L.
Haltica ericeti, Al.
Strophosomus lateralis, Pk.
Ceuthorrhynchus ericæ, Gyll.

(*To be continued*).

THE FALSE-SCORPIONS OF CUMBERLAND.

G. A. and R. B. WHYTE.
Edinburgh.

THE only species of Chernetidea recorded, to our knowledge, from Cumberland previous to our visit were *Chernes nodosus* Schrank, and *Obisium muscorum*, Leach. These were sent by the late Rev. F. O. Pickard-Cambridge to his uncle, Rev. O. Pickard-Cambridge, before the year 1892. The former (*C. nodosus*) was sent from Carlisle, and the latter (*O. muscorum*) was merely recorded as from Cumberland.

We here record five species of false-scorpions, four of which are new to the county list—the fifth being *O. muscorum*.

Cheiridium museorum Leach.

This very small species was obtained in five different barns near the head of Derwent Water. One hay-loft is worthy of special note, for in it we discovered a colony, the numbers of which exceeded our counting powers. We took over sixty specimens from one stone, on which there were innumerable nests. A few young, never more than three at a time, were found in them. The actual habitat of *C. museorum* was under loose stones, which lay on the top of the walls close under the roof. Some moults and one or two living specimens were obtained by sifting dust and hay-seed from the dark, undisturbed corners of the loft. We watched with interest an adult *Cheiridium museorum* feeding on a mite. The mite was held firmly in the chelicerae, and the life-juices were sucked out.

Chernes rufeolus Simon.

We found a single specimen of *Chernes rufeolus* on April 11th, and a week later obtained many more. This rare false-scorpion was found under the stones set fast in the earth floors of three old barns. The nests were also obtained, and proved to be of this species by the moults inside them.

C. rufeolus has only been obtained in a few of the southern counties in England, and never in Scotland, so that this is the most northerly record for Britain.

Chernes dubius Cambridge.

This species, which is smaller than its congener, *Chernes rufeolus*, was discovered in some numbers under stones firmly embedded in the soil, both in the woods and in the open country, south of Derwent Water. It is a chernetid which closely resembles the stones on which it lives, and consequently a careful search is required to find it, especially as it is so small.

We obtained about fifty specimens, many of which were immature. Empty nests were also found, and these probably belonged to *Chernes dubius,* as they were on the same stones, and no other species were present

Obisium muscorum Leach.

This, the commonest false-scorpion, was found in abundance wherever it was searched for, from the Styhead Pass to the slopes of Helvellyn. Most of these were in nests, and the majority had the egg-mass attached. Those free were always smaller and of a darker colour, probably all males. The difference in size is accounted for, as the females were swollen with eggs.

Chthonius tetrachelatus Preyss.

On April 14th, six of this species were found under flower-pots in the hot-houses of Mr. James Moorsom, on his grounds near Keswick. *C. tetrachelatus* is the commonest false-scorpion found in hot-houses ; but it is also found under stones in the open country.

[By the courtesy of the Editor, the writer of this note has been permitted to read the above paper, and is able thus early to congratulate the authors on the substantial additions to the known fauna of Cumberland, and on the interesting observations with which the records are accompanied. Through the kindness of a mutual friend, the Rev. Robert Godfrey, M.A., specimens of the five Pseudoscorpions referred to have, moreover, been seen and examined. In 1903, when the writer published a preliminary paper on ' North of England Pseudoscorpions' (3), it was only possible to quote for Cumberland Mr. Cambridge's records of *Chernes nodosus* and *Obisium muscorum* (1), and the repetition of these records in ' The Victoria History of the County of Cumberland' (2), where the latter species is noted for Carlisle, Armathwaite, and Wreay : common in woods amongst dead leaves. More recently, through the kindness of Dr. A. R. Jackson, this species (*Obisium muscorum*) has been seen from Scafell ; and *Cheiridium museorum* from Penrith, three specimens' found in a starling's nest, as already recorded by Dr. Jackson (5). Perhaps the most interesting of the new records is that of *Chernes rufeolus,* a species recently added to the British list by Mr. Cambridge (4). It appears to occur chiefly in stables and farm-buildings, the writer having collected or received it from such places in Kent. Essex, Wiltshire, Derbyshire, Lincolnshire—and now from Cumberland. (1) Cambridge, O. P.—'On the British Species of False-Scorpions.' ' Proceedings of the Dorset Natural History and Antiquarian Field Club,' XIII. (1892), pp. 199-231. (2) Cambridge, F. O. P.—'The Victoria History of the County of Cumberland,' I. (1901), p. 157. (3) Kew, H. W.—'North of England Pseudoscorpions.' ' The Naturalist,' 1903, pp. 293-300. (4) Cambridge, O. P.—' On some New and Rare British Arachnida.' ' Proceedings of the Dorset Natural History and Antiquarian Field Club,' XXVII. (1906), pp. 72-92. (5) Jackson, A. R.—'Rare Arachnida captured during 1906.' ' Proceedings of the Chester Society of Natural Science,' Pt. VI. (1907), pp. 1-7.—H. Wallis Kew.]

THE BIRCH TREE.

Betula alba, L.

P. Q. KEEGAN, LL.D.,
Patterdale, Westmorland.

THE sylvan bank, the pastoral hollow, the steeply acclivitous ledge, and the fell rising into mountain solitude are invested by the silvery bark and feathery evolution of foliage of this the 'queen of the woods.' The tree belongs more to lowlands and uplands than to mountains, but it serves frequently to charmingly hide or relieve the savageness of the mountain wastes, whereto it imparts an aspect of picturesqueness and witchery. It shuns the dingle and the coomb, imperatively demands light, and is highly accommodating as regards power of resistance to great extremes of heat and cold. For all these and other reasons, this tree intrudes itself very forcibly on the attention of the naturalist and the nature-lover ; and a brief description of its chief anatomical, chemical, and physiological characteristics will therefore, it is hoped, prove acceptable.

Stem.—The wood is of medium hardness and weight (specific gravity 0.5 to 0.76), entirely white, homogeneous, with no distinct alburnum, *i.e.* it is a sap-wood tree, no heart-wood. The medullary rays are mostly in one row (three or four rows in old wood), are 6 to 8 per millimetre of arc, rather irregularly spaced, and about 1 mm. high ; the vessels are rather large, being about 85 μ average width, are isolated or disposed in radial rows of two to four through the whole width of the annual ring, their lateral walls are very thickly beset with minute pores, and their transverse wall is scalariform ; the fibres are about 12 μ broad, have thickish walls beset with a few simple pores ; the parenchyma is sparse and occurs in very straight transverse bands, or else isolated. In the bark, the fiber parenchyma forms tangential bands of one to four rows of cells, which are narrow and high, and have large simple punctures on their radial walls ; the sieve-tubes are elongated, have wide lumen and thin walls provided with sieve plates ; the bast fibres appear as isolated bundles in the first year of growth, and later on sclerenchyma is formed between these groups so that ultimately a completely closed sclerous ring (peri-cycle) is developed ; there are no secondary bast-bundles, they being replaced by groups of scleroblasts (stone-cells) representing highly lignified parenchyma. After the fall of the epidermis in the three year old branches, a suberous envelope of flat brown cells is developed; in about three or five years later, a cubical white thin-walled tissue is interposed in thin layers between the zones of the still forming brown cells; and so it proceeds up till the age of fifteen to twenty years, at which time the periderm presents the appearance of numerous alternate layers (each of about ten rows of cells), of which the two or three external ones have thin walls, and the internal ones have thick walls; by the decortication of one of these layers situated on the outside, the cells of an internal layer are torn or broken, and the white resinous matter (betulin) which they contain, escapes and plasters the entire surface of the periderm with a chalky incrustation. The Birch is a fat tree, *i.e.* the starch completely disappears from bark and wood during the winter, and reappears a month or six weeks before the swelling of the buds in spring. The wood shows traces of tannin and phloroglucin in the pith and medullary rays, much glucose in summer, much oily matter in winter, about 30 per cent pentosan, 0.7 nitrogenous substance, and 0.3 ash, which has 26 per cent soluble salts, 35.4 lime, 8.8 magnesia, 10.5 P^2O^5, etc. The trunk bark contains 34 per cent. of a white resin or camphor (betulin), 4.8 soluble and insoluble tannin, about 4 gummy matter, 60 impure suberin, and 2.3 ash in fresh which has 6.4 per cent soluble salts, 2.6 silica, 52 lime, with traces of magnesia, phosphoric acid, and considerable manganese. The spring bleeding period of the Birch, commences at the end of March, first from the root, and then step by step towards the crown, and lasts some six or seven weeks. The sap contains about 1.8 per cent. dry matter, of which from 0.8 to 1.4 consists of glucose, and the remainder of nitrogenous matters, malic acid, and ash mostly of potass. The young spring shoots on dry ground are covered with a white resinous secretion in the form of vesicular papillæ, a circumstance that seems referable to the vigorous rapidity of their early growth.

Leaves.—The mesophyll is composed of one layer of palisade cells, narrow and close joined, and a lacunar tissue of equal thickness, made up of cells of varying size with air-spaces between each; the cuticle is feeble; the cells of the upper epidermis are much larger than those of the lower epidermis, their lower portion is blocked with mucilage, and there are scarcely any hairs on either surface of the adult organ; the stomata are small, and have no accessory cells, their number per square mm. is about 237; the leaf is only 200 μ thick; at the base of the petiole are three vascular bundles united into a U, higher up, the ends of these branches emit two bundles, and at the blade form a V, the system is therefore ' open,' whereas in Alder and Hazel it is ' closed.' On 4th August, the leaves had 58 per cent. of water, and the dried substance contained 3 per cent. wax and resin with a little carotin, some palmitic acid, and traces of volatile oil, 11 albumenoids, 4.3 tannin and quercitrin, considerable glucose and starch, a large quantity of pectosic mucilage with tartrate and oxalate of calcium (mostly in the lacunar tissue), and 3.5 ash which had 34 per cent. soluble salts, 4 silica, 25.3 lime, 8.4 magnesia, 11.3 P^2O^5 and 4.4 SO^3. The ash of the yellow and red autumn leaves rose to 5.8 per cent., and contained 6.9 per cent. silica, 39.8 lime, and 1.4 P^2O^5. The chlorophyll of the leaf is disorganised about mid October ; but the nerves and parenchyma still retain considerable starch much later on. The resin glands of the young leaf become less active in September, and completely dry up in October.

Flower and Fruit.—The inflorescence is in the form of catkins, the male flower being terminal, entirely naked, and born from a bud of the previous year, while the female flower is axillary, and springs from a bud of the current year, formed of five or six membranous scales, and borne on a lateral peduncle. At the time of fertilization in the female catkin there is visible only a single rounded scale, bearing at its base six pistils ; there is no sign of a placenta or of ovules. Later on these pistils unite, and at maturity of the organ there is seen three bilocular carpels monosperm by abortion. The fruit is a flattened samara with a membranous and transparent wing at each side. The seed is anatrope (the nucellus straight, and the chalaza distant from the hilum), without raphe, and pendulous ; the cotyledons are flat, and there is no endosperm ; it ripens in October, but the germinative capacity is lost in about six months. According to Jahne, the fruit (shell and seed) contains 10.5 per cent. water, 18.2 fatty matter, 14.4 albumenoids, 11.4 sugar and dextrin, 27.2 fibre, and 4.2 ash which (pure) has 27 per cent. potass, 9 silica, 23.7 lime, 9.2 magnesia, 10.9 P^2O^5 and 4.8 SO^3. The reserve materials are aleurone and oil only, no starch. The Birch is a pretty prolific seed-producer, but few of our native trees generate more seed that is sterile, only about 20 per cent. thereof being fit to germinate.

Physiological Summary.—A study of the chemical changes occurring in the Birch leaf during its life, teaches that the vegetation of this tree so far approaches perfection. There is no serious decrease in the production of starch, cellulose, or lignin, the albumenoids and sugars diminish considerably in autumn, while there is no special or heavy fixation of insoluble matters (silica, lime, or magnesia salts) in the old organ. With respect to the wood there are indications of a rapid growth of all the tissues simultaneously with defective differentiation. For instance, in May and June, *i.e.* the period of great vegetative activity, the outer wood-rings of the stem and branches are almost devoid of starch, *i.e.* there is only a feeble reserve of starch in the wood, the vessels are still free and open for the circulation of sap, and the amount of tannin produced is very small, so that everything forbids the differentiation of the wood elements into alburnum and duramen. In the region of the bark, moreover, the rapid growth and so far perfect vegetation lead the way to a subsequent inanition, whereby a powerful periderm is constructed. The pressure of the actively growing internal tissues rends the epidermis, imparts a potent stimulus to the phellogen (cork-cambium), and the attractive phenomenon of the silver-barked Birch is the result. The production of tannin being, however, comparatively poor, the outer bark is white, and not brown, as in most other instances.

But whereas the vegetation so far as the foliar organs are concerned, is comparatively perfect and unexhausted, the chemical development, so to speak, of the tree is decidedly backward. Thus the young leaves produce a not insignificant quantity of volatile oil, resin, and tannoid, while, nevertheless, in the old organs the amount of tannin and phloroglucin is distinctly insignificant. The mature bark contains only some 5 per cent. of tannin and only a trifling indication of free phloroglucin. Thus it appears that the process of deassimilation is very incomplete, and stands in this respect in marked contrast to that of the nearest ally, *viz.* the Alder. In fact, this is the chief reason of the comparatively short life of the Birch tree, *viz.* some 110 years. The wood never becomes perfect, no true duramen is formed ; indeed, when cut and exposed to the air, it putrifies very rapidly and completely owing to a serious poverty in tannin and resin. The copious ' bleeding ' in spring is dependent on a portion of the crude sap, taken up by the roots, being pressed into the mature air-containing vessels, which (as aforesaid), although mature, are nevertheless so free from the thickening and incrusting substances ordinarily characteristic of ' perfect ' wood, that the circulating sap (the spring maximum of starch production in the wood having been attained), readily finds a place of rest, so to speak, on its passage upwards, and from whence it may be readily withdrawn by the simple process of tapping.....

THEORIES OF EVOLUTION.

An Historical Outline.

AGNES ROBERTSON, D.Sc.

(Continued from page 171.)

III.—NATURAL SELECTION.

We must now turn to the most famous theory ever brought forward to account for the origin of species—that of *Natural Selection* or the *Survival of the Fittest*, which was propounded independently by Darwin and Wallace. Charles Darwin (1809-1882) was the son of a doctor, and the grandson of Erasmus Darwin, whose evolutionary views we have already mentioned. His intellectual development was not rapid ; at Cambridge he merely took the poll degree, and he says of himself at the age of twenty-two, 'I should have thought myself mad to give up the first days of partridge-shooting for geology or any other science.' Charles Darwin is, in fact, a most striking instance of the truth of Keats' saying—' Nothing is finer for the purposes of great productions than a very gradual ripening of the intellectual powers.' He always had a taste for collecting, and at Cambridge he came under the influence of Professor Henslow, with whom he went for long walks and discussed natural history. Through Henslow, soon after he went down from Cambridge, he received the offer of the position of naturalist on *H.M.S. Beagle*, then just starting for Tierra del Fuego and the East Indies with a view to surveying the southern extremity of America. He closed with the offer, and these travel years had a most profound effect on his after-life. On the voyage he began to reflect on the problem of the origin of species, and after his return home, in the summer of 1837, he opened his first note-book on the subject, a subject at which he worked continuously for the next twenty years. Curiously enough, in 1858 Alfred Russel Wallace, who was studying the natural history of the Malay Archipelago, sent him an essay he had just written embodying the identical idea which he had himself reached, namely, that of Natural Selection ! Darwin, by the advice of his friends, published some extracts from his existing manuscripts with Wallace's paper, and the next year saw the appearance of the first edition of the ' Origin of Species.' It is a lasting honour to English science that, instead of the embittered struggle for priority which has sometimes occurred when the same discovery has been made independently and almost simultaneously by two workers, Darwin and Wallace each showed the greatest anxiety to give each other all possible credit.   Wallace arrived at the idea of Natural Selection much more rapidly than Darwin.   He tells us that after three years pondering on the subject it came to him in a sudden inspiration while he was suffering from a severe attack of ague in the Moluccas ; he sketched out the whole theory in a day and sent it straight to Darwin.

The 'Origin of Species' is a long book, which from cover to cover consists of one closely-wrought argument of which no shortened and condensed account can be anything but unsatisfactory ; however, I will try in the briefest possible way to summarise the main contentions.   In the 'Origin of Species' Darwin attempted to answer two distinct questions :—

1. Were the animals and plants of the present day created just as we now know them, or have they been gradually evolved ?

In answer to this question Darwin brings forward much evidence to show that the Evolution hypothesis is the only tenable one.

2. If it be granted that Evolution has occurred, *how* has it been brought about ?

Darwin's answer to this was that it had been mainly brought about by the action of Natural Selection on small spontaneous variations, occurring in all directions, good, bad, and indifferent.

In discussing the 'Origin of Species' considerable confusion often arises because these two questions are not kept distinct. People talk as if Darwinism or Natural Selection were the same thing as Evolution, and as if the idea of Evolution would stand and fall with the idea of Natural Selection.   But in reality, if Darwin's hypothesis of Natural Selection were disproved, this would leave the Evolution theory quite untouched, since the hypothesis of Natural Selection is simply put forward to explain *how* Evolution occurs.   A concrete illustration may perhaps make this clearer.   Suppose a passenger on a ship takes observations of the stars at intervals; and to explain his observations puts forward the theory that the vessel is moving, and that the movement is taking place in a certain direction.   Then suppose that other evidence, such as the presence of sails, leads the passenger to form a second theory— namely, that the agency by which the ship is moved is wind. Now imagine that he presently discovers that the ship is really worked by steam, and that the sails are comparatively useless —imagine, that is, that he proves to be wrong in his second theory, which he propounded to answer the question, ' *how* does the ship move?' This error does not in the least shake the truth of his first theory, namely, that *the ship is moving in a certain direction.* The question of *how* this is brought about is secondary, and has nothing whatever to do with the primary question as to whether movement in a certain direction is or is not occurring.

We will now shortly consider Darwin's Natural Selection hypothesis. There is much evidence that since man first began to domesticate animals and plants these have changed enormously under his hands. For instance, there are now more than a score of distinct breeds of pigeons, which an ornithologist, if he met them in the wild state, would probably name as distinct species, but which, according to the common opinion of naturalists, are all descended from the common rock pigeon. According to Darwin, who made a special study of these birds, all the different races have been produced by ' artificial selection' carried on through many generations. The expression 'artifical selection' almost explains itself, but it may perhaps be well to say a few words about it. No two individuals belonging to a species are exactly alike, and domesticated animals are particularly apt to vary. A man possessing domesticated animals or plants showing considerable variation among themselves will naturally choose those to breed from in each generation which possess in the highest degree the characters which he particularly wants ; that is, he will pick out the most favourable variations. For instance, if he is breeding race-horses, he will select his fleetest animals in the expectation that their offspring will resemble them. Or if he has a fancy for pigeons with exaggerated tails, he will always by preference breed from the individuals with the longest tail feathers. The idea of improving domesticated animals and plants by selection is by no means a modern one, though it has only been reduced to a definite method in comparatively recent times. The effect of artificial selection in producing differing races of domesticated animals and plants is quite extraordinary. This is brought home to us when we think of the difference between the race-horse and the dray-horse, the greyhound and the bloodhound, the crab-apple and Cox's orange pippin, and realise that it is more than probable that these differences have been produced by selecting and breeding from those individuals in each

generation which possessed in the highest degree some special
quality at which the breeder was aiming. The material upon
which the breeder has to work is the variations which
spontaneously arise. He cannot compel a variation in the
direction which he desires, but he can make use of it if it
happens to arise. The variability of domesticated plants and
animals is particularly marked, but species in the wild condition
vary much more than is generally supposed. Misapprehension
on this point arises from the fact that systematists are inclined
to take no account of variations from the type form of a species
unless the variation is sufficiently well marked to be named and
classified as a sub-species or variety. Darwin held that between
species and variety there is no hard and fast line. If this is
true, and species are only strongly marked and well defined
varieties, it follows that species belonging to a large genus
will be more likely to present varieties than species belonging
to a small genus. For a large genus is a group within which
the manufactory of species in past times went on with special
vigour, the numerous species of the genus existing at the
present day being a legacy from this former activity. If this
activity is not only a matter of the past, but continues into
modern times, we shall find a number of species now actually
in the making—that is, many of the species in the genus will
show sub-species and varieties. Darwin tested this sup-
position by tabulating the plants and beetles of twelve
countries according to the size of the genera and the number of
recorded varieties, and found that it was borne out by the facts.

If we grant that variations exist, and that no sharp line
can be drawn between species and varieties, the question next
arises—how is it that varieties become so far differentiated from
their parent species as to be converted into fixed and stable
species, and how is it that these species show such wonderful
adaptations to their surroundings? In other words—what
factor in the natural evolution of species takes the place of that
' artifical selection' by which man produces his domestic races?
The answer is that this factor is the ' natural selection' brought
about by the struggle for existence between living things.
Since many more animals and plants are brought into the world
than can possibly survive, variations, however slight, which
happen to be in any degree profitable to the individuals
exhibiting them, will tend to the preservation of such individuals,
and will very likely be inherited by their offspring. This process is conveniently known by Herbert Spencer's name of the

Survival of the Fittest. Animals and plants are subject to immensely severe competition, on account of the high rate at which they all tend to increase if they are unchecked. The excess of individuals produced over the number which have any chance of surviving is quite startling. Linnæus calculated that if an annual plant produced only two seeds, and its seedlings two each, and so on, in twenty years the descendants of the original plant would number one million! So it is clear that stringent checks to increase must be perpetually operating to keep the numbers of plants and animals down to those which we meet at the present day. Darwin dug and cleared a piece of ground three feet by two feet so that there could be no choking from other plants, and marked all the seedlings of native weeds that came up. Of 357, 295 were destroyed, chiefly by slugs and other insects, and yet he had eliminated what is probably the chief cause of seedling mortality, namely, germination in ground already stocked with other plants. We are too apt to think of the struggle for existence among plants and animals as being a struggle against the elements only. But since so many more animals and plants are produced than can possibly survive, the struggle is really in the first place a competition with other organisms, notably individuals of the same species, since they frequent the same districts, require the same food, and are exposed to the same dangers. In the struggle for existence Natural Selection may be compared to a sieve which is perpetually sifting out all but the more favourable varieties. 'It may metaphorically be said that natural selection is daily and hourly scrutinising, throughout the world, the slightest variations; rejecting those that are bad, preserving and adding up all that are good; silently and insensibly working, *whenever and wherever opportunity offers,* at the improvement of each organic being in relation to its organic and inorganic conditions of life.'

As a corollary to the theory of Natural Selection, Darwin brought forward a subsidiary hypothesis, that of Sexual Selection, which had been originally suggested by his grand father, Erasmus Darwin. According to this theory the secondary sexual characters in the male arise through the struggle between the individuals of the male sex for the possession of the female. In those animals in which the 'law of battle' prevails, weapons such as antlers are gradually evolved, because the individual possessing these in the most highly developed form will be most successful in the struggle with his compeers, and so will be most likely to leave offspring. Further, the brilliant plumage and beautiful song of some male birds is attributed to a more peaceful rivalry in which the birds compete for the favour of the female, displaying before her their plumage and their voice. The female is said in such cases to actually select the most attractive male. Wallace is in entire disagreement with Darwin as to the possibility of the plumage and song of the cock bird being evolved in response to the taste of the hen. He thinks that where the male has gayer plumage than the female this simply means that the need for protection *represses* in the female the bright colours which are normally produced in both sexes by general laws. He also believes that brightness of colour is correlated with vigour, and hence the male exceeds the female in brilliancy owing to his higher vigour and vitality. It further follows from this correlation that the more vigorous males which succeed best in the struggle for the female will be the most brilliant, and hence vividness of colour will be selected incidentally by the ' law of battle.'

Numerous criticisms of the theory of Natural Selection have been brought forward from time to time. The most serious collection of difficulties in the way of theory is probably that brought together by Darwin himself in the ' Origin of Species,' for his intense intellectual honesty made him state all possible drawbacks to this theory in a more forcible and telling way than his opponents could! Perhaps there is no more cogent difficulty than that of accounting for the *incipient stages* of useful modifications. Many adaptations, which are obviously useful when highly developed, do not seem likely to have been of any advantage to the creature in their rudimentary stages. A second difficulty of a slightly different type is that of the *utility of specific differences*. On Darwin's theory utility is the sole test by which a variation stands or falls, but in many cases the very constant differences distinguishing two allied species have, so far as we can judge, no utility whatever. In Mr. Bateson's words :—' as to the particular benefit which one dull moth enjoys as the result of his own particular pattern of dulness as compared with the closely similar pattern of the next species, no suggestion is made.'

These difficulties are very real, and they show us that the ' Origin of Species,' magnificent as it is, ought not to be accepted as the last word on the subject. Darwin himself (unlike some modern ' Darwinians ! ') never looked upon Natural Selection as the one and only key to evolution, but regarded it as 'the most important, but not the exclusive, means of modification.' Further, we must remember that Darwin never attempted to explain *how* the variations arise upon which Natural and Sexual Selection work. He simply assumed that organisms *do* vary, and that these variations take place in all directions. As has been tersely said, 'Natural Selection may explain the Survival of the Fittest, but it cannot explain the arrival of the fittest.'

(*To be continued*).

————◆◆————

Animal Artizans and other studies of Birds and Beasts, by C. J. Cornish, M.A., F.Z.S. Longmans Green & Co., 1907, 274 pp., 7/6 net. This volume, on account of the thickness of the paper on which it is printed, is very massive. It contains a collection of the late C. J. Cornish's contributions to the 'Spectator,' 'Country Life,' etc., and in this form will be very welcome to the admirers of that writer. In her preface his widow has reason to believe 'that some account of his life and work might be welcome to those of his readers who never knew him personally and to others who admired his often unsigned writings without being aware of the identity of their author. And it seemed appropriate that such an account should be embodied in the preface to this book.' In her memoir Mrs. Cornish describes the life and doings of her late husband, with details of trivialities which are perhaps pardonable in the circumstances. There are thirty-five essays gathered together in the book, on a great variety of subjects, and some are very suggestive. The illustrations, however, are not quite as successful as one might wish. The artist has not grasped the full value of light and shade, and the pictures have consequently a 'thin' appearance; the flight of swans on the plate facing page 82 seems to represent their 'ghosts' rather than the birds themselves.

The Stone Implements of South Africa, by J. P. Johnson. Longmans, Green & Co., 1907. 53 pp., price 7/6 net. In this volume, which is illustrated by 258 line drawings of implements, the author has given an account of his researches amongst the stone implements of South Africa. Some idea of the nature of the work may be gathered from the following extract from the preface :—'The object of this little volume is to co-ordinate the various discoveries of stone implements I have made during the last four years—discoveries that I venture to think mark a new era in our knowledge of the Stone Age of South Africa. No attempt will be made to review the abundant, but unsatisfactory, literature already in existence.' Notwithstanding this, there is evidence that the author is fairly familiar with the recent contributions to our knowledge of man's early weapons, not only in Africa, but much nearer home. He uses typical Palæolithic (Acheulian) implements as his datum line ; earlier types are 'primitive' and later are 'advanced.' He recognises Eolithic, Palæolithic, and Neolithic types, but takes them as stages in the general progressive evolution from very primitive to advanced forms. He describes 'pigmy' implements, and also the ostrich-egg shell beads for the manufacture of which these diminutive forms were used. We were relieved to find that he did not wish to invent a special pigmy race of men to account for their presence. His study of the stone implements of South Africa shows that 'the sequence is the same as in Britain and other parts of the Old World, though it is not so complete, having lagged behind somewhat.' There is no index, and the price is quite sufficient.

THE PEREGRINES AT BEMPTON.

E. W. WADE, M.B.O.U.

WE may congratulate ourselves that the efforts of the Y.N.U. in protecting the Bempton Peregrines are once more rewarded. It had been rumoured that the birds were not at their old haunts, but, like most birds, they are very quiet when occupied in family cases, and as the situation of the eyrie, halfway between Old Dor and the Pig-Trough, is so chosen as to be invisible except from the bottom of the cliffs, where no one had been previous to the commencement of the climbing season, they had entirely escaped notice. At Whitsuntide, however, when egg-gathering began and their security was disturbed by the climbers, the birds at once showed their disapproval of the intruders. The cock bird did not remain long on the scene, but so long as any human beings appeared on the cliff-top the hen soared round in wide circles, uttering harsh cries more in anger than distress, for who had the right to dispute her hitherto un-challenged supremacy on the face of the cliff? Now and again she closed her wings, and, swift as lightning, descended vertically down the face of the cliff, scattering to right and left the swarms of sea-birds. Then, by simply extending her broad wings, she rose to the level of the cliff-top again. Apparently the wings were never flapped except to steady her flight, for, without visible motion of these, she repeated the manoeuvre a dozen times, sometimes twisting completely round in the air or turning over with the wings extended in order to alter the direction of her flight, but always with the same perfect ease, which made Gulls, Guillemots, Razorbills, and Puffins appear clumsy and slow in comparison. Numerous feathers scattered about on the various grassy nabs along the cliffs attest the voracity of this prince of robbers, but sea-birds and cliff-pigeons, which apparently form the whole diet, are numerous, and one need not grudge the few required for his larder. A descent to last year's eyrie showed that the falcons had moved a little to the west, where they are better hidden from observation, and unfortunately in a position where a photograph cannot be secured. The climber entertains a prejudice against this bird, which they say frightens the Guillemots off their eggs, which thus become spilled and broken; but as the same Guillemots sat on the ledges all round the eyrie all the while the birds were under observation, and apparently showing not the slightest trace of fear, we may dismiss this objection as prejudice only.

THE VICTORIA HISTORY OF YORKSHIRE.

(Continued from page 186).

PRE-HISTORIC MAN.

THE section of this volume dealing with pre-historic remains (other than earth-works, which are being described later) has been written by Mr. George Clinch, F.G.S., who must be congratulated upon the thoroughness with which he has accomplished his task. In perusing his notes, it is obvious that not only has the writer made himself conversant with the voluminous literature on the subject, but he has also visited the principal collections in the county. Mr. Clinch gives a useful summary of the relics (pottery, dwellings, etc.) of the Neolithic Age, the Bronze Age, the Iron Age, etc. His notes are illustrated by a large number of drawings and photographs of typical vases, implements, weapons, ornaments, carved stones, etc. He has obviously and admittedly been greatly assisted in his work by the fact that two such excellent memoirs as Canon Greenwell's 'British Barrows' and Mortimer's 'Forty Years' Researches' have appeared, dealing very largely with the county ; and, in addition, the museums in this broad-acred shire are also exceptionally well furnished with relics of early man. As might be expected, the spelling of both personal and place names indicates a want of familiarity with the county, and this might have been prevented by allowing any of the numerous Yorkshire antiquaries to have read the proofs. In some instances the same individual has his name spelt in two or three different ways.

With regard to the Rudston monolith, Mr. Clinch may be pleased to know that its depth underground has been demonstrated by excavation ; also, there were formerly four or more stones forming the Devil's Arrows near Borough-bridge (see 'Naturalist,' Feb., 1903, pp. 34-5). To his list of pre-historic antiquities in Yorkshire many additions might be made. This list is 'an attempt to put on record all the more important discoveries of pre-historic relics in Yorkshire.' Curiously enough, Mr. Clinch omits from this list some of the specimens he describes in the text. There are several localities in which important finds have been made not referred to in the list, and South Ferriby should not have been included, as the place is in Lincolnshire. The mark at Ferriby on the Yorkshire map containing Mr. Clinch's notes, therefore, does not apply, though

1907 June 1.

it might be allowed to stand to indicate the locality in which probably the finest bronze spear head in the north of England was found (which is in the possession of Canon Greenwell), but is not referred to in the list. The stone and bronze axes referred to under 'Hull' were not found at that place, but *near* Hull, as recorded in Sir John Evans' books, and the only pre-historic implement ever found at Hull, curiously enough, is not mentioned in the list, as it is in the Wilberforce historical museum, which presumably was not examined by Mr. Clinch when he visited Hull. We also notice that some forgeries have been included as though they were genuine. Perhaps the most unexpected omission from the list of place names is 'Ulrome,' where so many important discoveries at various periods have been made, some of which Mr. Clinch describes elsewhere.

These omissions, however, are of somewhat minor importance, and speaking generally, the account of Yorkshire's pre-historic remains as given by Mr. Clinch is very well done indeed.

Mr. A. F. Leach writes an account of the schools of the county, which is a very useful compilation, and appears to have been very conscientiously done. This contribution occupies 86 pages.

The concluding chapter (24 pages) deals with Forestry, and is written by the Rev. J. C. Cox, whose volume on Forests we recently had the pleasure of noticing in these pages ('Naturalist,' 1906, pp. 95-97). We then dealt at some length with Dr· Cox's researches amongst the old records relating to the Forests of Yorkshire; and in the present chapter these are somewhat extended. The name of Dr. Cox at the head of this important contribution is a sufficient guarantee of the thoroughness of the work.

———◆◆———

BIRDS.

Arrival of Migrants near York.—The following are the dates of the arrival of migrants near York. In most cases the dates are later than last year, the cold weather being no doubt responsible for the somewhat erratic arrival of the birds.

March 24.—Wheatear.	April 27.—Sandmartin.
April 19.—Willow Warbler.	May 1.—Cuckoo.
,, 19.—Sandpiper.	,, 8.—Whinchat.
,, 22.—House Martin.	,, 8.—Swift.
,, 22.—Swallow.	,, 10.—Landrail.
,, 23.—Whitethroat.	,, 13.—Nightjar.

SYDNEY H. SMITH, York.

HISTORY OF THE BRITISH MUSEUM (NATURAL HISTORY) COLLECTIONS.*

THE Trustees of the British Museum and the staff at Cromwell Road are to be congratulated upon the preparation of a valuable History of the Collections, two volumes of which have already appeared. The first deals with the libraries and the departments of Botany, Geology, and Minerals; and the second is devoted to the special accounts of the several collections of Zoology, written respective assistant keepers and assistants. The general by the History of the Zoological Department will be issued as a third volume shortly.

Too much stress cannot be laid upon the importance of placing upon record particulars of the origin and growth, not only of our national collections, but of all those in the museums throughout the country. As it is, a perusal of the volumes under notice proves that anything like a reliable record of even our national treasures is exceedingly difficult to obtain; in many instances the history of an important specimen is lost, and thus much of its value is also gone. Difficult as it undoubtedly has been to prepare the present history, there is no doubt that had it been longer delayed it would have been impossible to have got the information together.

The history of the British Museum dates from 1753, in which year an Act of Parliament was passed for the purpose of purchasing the well-known collection of Sir Hans Sloane. We observe no mention is made in this official guide of the subsequent Act which was passed authorising the raising of the funds required by means of a lottery! † As time went on the collections grew, and in 1860 it was decided that the Natural History Series be removed from the British Museum, and three years later the site of the International Exhibition at South Kensington was purchased for these collections. The volumes under notice chronicle the various changes that have taken place in the staff, arrangement, etc., of the Natural History Museum, as well as the yearly acquisitions. This last-named item is a very important one, and when it is remembered that in one department in one year no fewer than 116,000 additions were made, whilst most of the departments acquire thousands, or tens of thousands per annum, it will be seen that the question of

---

* Vol. I., 442 pp., 1904, sold at the Natural History Museum, Cromwell Road, 15/-. Vol. II., 782 pp., 1906, 30/-

† See 'Naturalist,' 1905, p. 120.

---

cataloguing the specimens is a very serious matter. In addition to this chronological record is an alphabetical list of the donors of the specimens to each of the departments, together with particulars of the objects obtained from them.

In the chapter devoted to a History of the Libraries is a 'List of Current Serial Publications Presented to the British Museum (Natural History),' which appears to be incomplete, as we know of one or two items regularly presented which do not appear in this list. It is also unsatisfactory to find so many of the sets of Transactions, etc., are incomplete. Surely the various societies referred to would complete their sets of Publications in the national museum were they approached—or, at any rate, if they could not do so, a small outlay on the part of the authorities would secure the desired publications and thus preserve them for all time. Strangely enough, we notice that the colonial and foreign publications presented are nearly all in sets commencing 'No. I.,' whereas from our own country complete sets are distinctly in the minority.

We have pleasure in drawing attention to these valuable publications—they should be in every reference library, and every naturalist will find them of distinct value. The authorities at the Natural History Museum are to be commended on the production of the volumes.

————◆◆————

The Council of the University of Leeds has decided to establish separate chairs of zoology and botany.

'This year rooks are said to be thicker than ever,' we read, 'and rook pie is therefore likely to be cheap.' A beautiful instance of 'caws' and effect.

On account of the illness of the Curator of the Haslemere Museum the publication of the *Museum Gazette* is suspended for the present. We trust that Mr. Swanton may soon be well again. It is hoped to re-issue the journal in January next.

As his presidential address to the Geologists' Association Mr. R. S. Herries took for his subject 'On the Constitution and Management of Scientific Societies.' It is printed in Vol. XX., pt. I., of the Society's Proceedings, just issued. This paper should be carefully read by all those concerned in the management of such societies.

In the new 'Lancashire Naturalist' we notice the Spring Vale Ramblers are to have two rambles 'from Clitheroe to Clitheroe,' which savours of a fishing trip. By the way, fishermen must be grateful to Mr. Alan R. Haig-Brown, who, in the same journal, makes the following contribution :—'Do fishermen juggle with the truth! I sometimes think they get the reputation thereof, because, being out often, and alone with Nature, they see sights which to the stay-at-home seem incredible.' The next time we hear our angling friends relating details of the incredibly big fish which they *nearly* caught, we must remember that, with the possible exception of a basket and flask, they were at the time 'alone with Nature.'

In Memoriam.

EDWARD HALLIDAY.

WE were very sorry to receive the announcement of the death of Mr. Edward Halliday, of Halifax, which event took place on April 5th last. He was in his eightieth year, and was almost the last of the old band of working men lepidopterists of a generation ago, being a contemporary of James Varley, William Talbot, George Liversedge, and many others. Halliday retained his interest in lepidoptera up to the last; and at the excursion and meeting of the Yorkshire Naturalists' Union at Hebden Bridge—one of his favourite localities—on June 11th, 1904, was quite enthusiastic over the species which occurred there.

G. T. P.

JOHN FRANCIS WALKER, M.A., F.G.S., F.Z.S.

WE regret to hear of the death, on May 24th, of John Francis Walker, M.A., F.G.S., F.Z.S., of Bootham, York, at the age of sixty-seven. He was a member of an old York family, and for many years took a keen interest in the York Philosophical Society, of which he was a vice-president. He was also a frequent donor to the York Museum. He was perhaps best known from the keen interest he took in geology—particularly in collecting. His series of fossil brachiopoda, over which he spent many years in getting together, is exceptionally complete, and has formed the subject of several papers in the Annual Reports of the York Philosophical Society, and elsewhere. At the recent York Meeting of the British Association he was elected Chairman of a Committee for the investigation of the Neocomian Beds at Knapton. Mr. Walker was a life-member of the Yorkshire Naturalists' Union.

T.S.

Birds I have known.—A. H. Beavan. T. Fisher Unwin, 256 pp. Second Edition, 2/- This appears to be a reprint, on thinner paper, of the work noticed in these columns for May, 1905 (pp. 158-9). The cover is certainly more attractive than that of the first edition, and at its present price it is a very cheap volume.

FIELD NOTES.

MAMMALS.

Note on Bats.—In the 'Naturalist' for January (p. 28) Mr. H. B. Booth has a note on 'the small quantity of air necessary to sustain life in a bat.' The following may be of interest to Mr. Booth and others who are studying these physiologically puzzling animals :—

W. Derham, in his 'Physico-Theology' (London, 1737), Book I., Chap. i, p. 28, gives the results of some experiments on many animals. 'Birds, dogs, cats, rats, mice, etc., when placed under the receiver of an air pump, died in less than half-a-minute, counting from the very first exsuction. A mole . . . . died in one minute . . . . A bat (although wounded) sustained the pump two minutes, and revived upon the re-admission of air. After that he remained four-minutes-and-a-half and revived. Lastly, after he had been five minutes, he continued gasping for a while, and after twenty minutes I readmitted air, but the bat never revived.' A frog lasted eleven hours, and certain invertebrates twenty-four hours. Besides suffering from lack of oxygen, these subjects would probably be damaged by exploded or distended vessels and tissues.

In a footnote on p. 71, in Vol. I. of the English translation of Cuvier's 'Animal Kingdom' (1834) there is an account of the observations of Dr. Marshall Hall on the lethargic sleep of several hibernating mammals which 'can bear with impunity the abstraction of atmospheric air.' This faculty 'enabled a bat to preserve life for eleven minutes when immersed in water.' —FRED STUBBS, Oldham, April 18th.

—: o :—

BIRDS.

Water-rail and Gannett at Mablethorpe.—About a week ago a water-rail entered, by the open window, the dining-room of Mr. J. H. Joyce, of Mablethorpe. On examination it was found to be almost foodless. About the same date a Solan Goose, or Gannet, was found dead from exhaustion about two miles from Mablethorpe.—J. CONWAY WALTER, Langton, Lincs., April 20th, 1907.

REVIEWS AND BOOK NOTICES.

How to Study Geology: a Book for Beginners, by Ernest Evans. Swan Sonnenschein, 1907, 272 pp. In this little book the Natural Science Master at the Burnley Municipal Technical School has presented, in a form very suitable for a beginner, a series of essays on geological phenomena, illustrated by a profusion of blocks—some old—mostly new. Mr. Evans has for some time been lecturer to the Co-operative Holiday Association, and presumably the present work is the result of his notes which have been put together in connection therewith. By a series of simple experiments, also, many geological phenomena are explained. We hardly expected to find 'Signs of the Ice Age in Great Britain' dismissed in less than half a page. Fig. 8, showing two sets of inverted strata on the sides of an intruded mass, is surely an impossible section; on Fig. 12 'left' should be 'right' and 'right' should be 'left'; we have never seen a brick wall built across a glacier, as apparently shown on Fig. 31; and on Fig. 50 we are informed that 'the amount of reduction of the figures is denoted by $\times \frac{1}{2}$, which means reduced one half,' whereas the amount of reduction, except in a very few cases, is *not* denoted; on page 53, line 8, 'size' should be 'side.' These, however, are minor matters, which can be put right in the next edition.

'In Starry Realms,' 372 pp.. 1907; **'In the High Heavens,'** 384 pp., 1907; and **'Great Astrononers,'** 372 pp., 1907.

Messrs. Sir Isaac Pitman & Sons, Ltd., are to be congratulated upon the way in which they are issuing some very useful and popular works on natural science at a reasonable rate. We have already referred to some of these publications in these columns recently. We have received the three well-known volumes named above, by Sir Robert Ball, F.R.S., cheap editions (3/6), which the firm have just placed on the market, and there is no doubt they will meet with a ready sale. In this way much is being done to popularise the natural sciences. In the first volume named Sir Robert has necessarily had to revise the chapters on 'Shooting Stars' and on 'Photographing the Stars.' He also anticipates that the recent discoveries in Radio-Activity 'will shortly cause some modification to be made in our present views as to the sustentation of the sun's heat.' In the second volume the discoveries recently made in reference to the satellites of Jupiter are referred to, and in this he also deals at some length with the nebular theory. Further consideration of this subject only makes Sir Robert feel more confident that the nebular theory does really express the law of nature. The additional matters in the third volume, as might be expected rom the nature of the volume, are not of much moment.

Thomas H. Huxley, by J. R. Ainsworth Davies, M.A. J. M. Dent & Co., 1907. 288 pp. Naturally in the excellent 'English Men of Science Series,' now being issued by Messrs. Dent, a volume devoted to Huxley must be amongst the first, and Mr. Davies has written a very interesting narrative of the life and work and hardships and successes of one whose name is a household word, though it must be admitted that to some he is perhaps better known as an 'agnostic' than as a man of science. But all who peruse the present small volume must marvel at the great work he did —too well known to all readers of this journal to require particularisation here. It is sufficient to state that it is a readable and concise account of the achievements of this great master, though there is much already familiar to those who have read the 'Life and Letters,' an admirable work which has admittedly been made good use of in the present volume. A useful feature are the appendices (*a*) Chief Biographical Sources (21 items): (*b*) List of Published Works (276 items); and (*c*) Classification of more Technical Works. There is a good index.

NORTHERN NEWS.

The April issue of 'Yorkshire Notes and Queries' contains the first instalment of a useful 'Bibliography of Yorkshire.'

At the request of the Executive Committee, Mr. H. Culpin, of 7, St. Mary's Road, Doncaster, has accepted the position of Treasurer to the Yorkshire Naturalists' Union.

The 'leading article' of a recent issue of 'The Country Side' is devoted to rabbit fleas, weasel fleas, dog fleas, cat fleas, fowl fleas (not lice), and other entertaining topics.

The seventy-third Annual Report of the Bootham School (York) Natural History, Literary, and Polytechnic Society is to hand, and contains evidence of a continued interest being taken in Natural History in the school.

Under the title 'Bibliotheca Pretiosa,' Messrs. Sotheran & Co. have issued a well illustrated catalogue of unusually choice books and manuscripts of literary and historical interest. The price is half-a-crown.

The opening address to the Geographical Section of the Birmingham Natural History and Philosophical Society, by Prof. W. W. Watts, F.R.S. ' has just been published by the society (55, Newhall Street, Birmingham) at one shilling.

In the May 'Entomologist's Monthly Magazine' Mr. E. A. Newbery describes *Enicmus fungicola*, Thoms., a species of Coleoptera new to Britain. This was obtained in some numbers in dry fungi on a tree at Edenhall, Cumberland.

Mr. Harper Gaythorpe favours us with a reprint of his paper on 'Pre-Historic Implements in Furness,' in which stone, flint, and bronze weapons are described. (Trans. Cumb. and West'd. Ant. and Arch. Soc. New Series, Vol. VI.).

At a recent meeting of the Bradford Natural History and Microscopical Society, Mr. J. W. Carter, F.E.S., the newly-elected president, delivered a presidential address, in which he reviewed the work of the Society, which was formed in 1875. Mr. Carter was one of its earliest members.

Our contributor, Mr. T. Petch, who is now the Government Mycologist of Ceylon, sends us his Report, as well as pamphlets dealing with ' Bud rot of the Cocoanut Palm,' 'Root disease of *Hevea brasiliensis*,' and 'Descriptions of New Ceylon Fungi.' We understand Mr. Petch is to visit England in October.

Drawn up by Prof. W. A. Herdman, F.R.S., is an admirably illustrated 'Guide to the Aquarium' of the Liverpool Marine Biology Committee at Port Erin Biological Station, Isle of Man. The illustrations have been principally prepared by Mr. H. C. Chadwick, the Curator of the Station, and the Guide is sold at the small price of threepence.

Newspaper Natural History. One of our dailies deplores the loss to shipping, due to the efforts of ' the worm, *Testudo* navalis'! A Welsh paper offers for sale a Shire Stallion, 'very muscular, good bone, *silky feathers.*' In drawing attention to the meteorological advantages of Scarborough, the *Scarborough Post* records that last year the total sunshine was 158,025 hours, an average of 430 hours of sunshine a day. No wonder Dr. H. R. Mill, at the British Association at York, gave us a warning as to the accuracy of meteorological records published at our holiday resorts! In Huddersfield a ' Chip Potato Plant' is offered for sale, and *Punch* suggests that in the same department may be found a Navy Cut Tobacco Plant, and a Stewed Celery Bed. In the *Huddersfield Chronicle* we notice the following displayed advertisement :—' Mr. ———, the old botanist from the ——— Society, will undertake to *name any local wild plant every Sunday evening* until further notice *John Bray*, proprietor, ——— Arms Hotel, ——— Street ——— Brewery Ales, Spirits, and Cigars of the choicest quality.' Huddersfield always has been a good botanical centre !

JULY 1907.

No. 606
(No. 384 of current series).

THE NATURALIST.

A MONTHLY ILLUSTRATED JOURNAL OF

NATURAL HISTORY FOR THE NORTH OF ENGLAND.

EDITED BY

T. SHEPPARD, F.G.S.,

THE MUSEUM, HULL;

AND

T. W. WOODHEAD, Ph.D., F.L.S.,

TECHNICAL COLLEGE, HUDDERSFIELD.

WITH THE ASSISTANCE AS REFEREES IN SPECIAL DEPARTMENTS OF

J. GILBERT BAKER, F.R.S. F.L.S.,
Prof. P. F. KENDALL, M.Sc., F.G.S.,
T. H. NELSON, M.B.O.U.,

GEO. T. PORRITT, F.L.S., F.E.S.,
JOHN W. TAYLOR,
WILLIAM WEST, F.L.S.

Contents :—

LONDON :

A. BROWN & SONS, LIMITED, 5, FARRINGDON AVENUE, E.C
And at HULL AND YORK.

Printers and Publishers to the Y.N.U.

PRICE 1/- NET. BY POST 1/1 NET.

THE LINNET: SPECIMEN PLATE FROM Mʳ CHARLES STONHAM'S
"THE BIRDS OF THE BRITISH ISLANDS" (Margins much reduced)

NOTES AND COMMENTS.

ANOTHER NEW MAGAZINE

DURING the past few years we have chronicled the appearance of many new magazines. Most have had a meteoric existence —came into the world with a great glare, as quickly 'fizzed out,' and left not a wrack behind. Others had a more lingering career ; but eventually sickened and died. One or two still linger, though apparently suffering from 'galloping consumption.' In most cases their fate was evident from the first. A few, a very few, seem still healthy. To these last had been added '.British Birds,' * which, judging from the first part just received, is most likely to be a success. The study of birds in recent years has been followed by an enormous number of serious students, as well as by a still greater number of drivelling dabblers, many of whom evidently consider that to buy (or borrow) a field-glass, to spend a week-end spying sparrows, and to have excess to a few monographs are the only qualifications for writing books and articles to periodicals. Page after page of the most blithering piffle are in this way printed. In 'British Birds,' however, there will be none of this. With such capable editors as Messrs. H. F. Witherby and W. P. Pycraft we can depend upon having nothing but the best, and 'Part I.' of the new publication certainly is evidence of this.

AND ITS CONTRIBUTORS.

The articles it contains are :—' Additions to the List of British Birds since 1899,' by Mr. Howard Saunders, 'A Study of the Home Life of the Osprey ' (with three excellent plates), by P. H. Bahr, 'Remarks on the supposed New British Tits of the genus *Parus*,' by P. L. Sclater, 'Nesting Habits observed abroad of some Rare British Birds,' by F. C. Selous, as well as notes and correspondence by W. Eagle Clarke, J. H. Gurney, J. L. Bonhote, and Charles Whymper.

It cannot be said that there is no room for such a periodical as 'British Birds.' It will unquestionably do good. And, with a possible exception, it will do no harm to its contemporaries.

BOOMING BEMPTON.

It is apparent that the attractions of the Cliffs of Bempton and Speeton are at last being recognised in the way they deserve.

---

* Vol. I., Pt. 1, June 1st.   32 pp., 1/- net. (monthly). Witherby & Co. London.

1907 July 1.

P

The North Eastern Railway Company has issued an admirable coloured poster of the Bempton Cliffs, after a painting by F. W. Booty. By the courtesy of the Company we are enabled to present our readers with a reproduction of the poster, in colours (Plate XXII.). From this it will be conceded that the North Eastern Company—which is so ready to grant every facility to the natural history and other societies using its lines, is also to the fore in issuing artistic posters. In addition, the Company has issued an attractive pamphlet, written by the Secretary of

Guillemot Eggs on Black Shelf, Bempton.

the Yorkshire Naturalists' Union, setting forth geological, ornithological, and other features of the area. This pamphlet, which can be obtained on application to the Chief Passenger Superintendant, N.E.R., York, is illustrated by reproductions of photographs by Mr. E. W. Wade and others.

WADE'S 'BIRDS OF BEMPTON CLIFFS.'

Messrs. A. Brown & Sons, Hull, the publishers of the 'Naturalist,' have issued a second edition of Mr. Wade's well-known pamphlet on the Birds of Bempton Cliffs. This contains

After painting by F. W. Booty.

The Bempton Cliffs

all that appeared in the first edition, with much additional matter, which extends the pamphlet to forty-one pages without the illustrations. The latter are now all printed on plate paper, which adds much to their value. Besides being issued in more handy form, the price of the present impression is one shilling, just half that of the first edition. It will doubtless have a ready sale.

Our attention has also been called to a handbill issued by one of the 'climbers' of Bempton, who modestly describes himself as 'known as the old Cliff Climber *throughout Great Britain!*' But of this the least said the better.

THE 'BIRDS OF YORKSHIRE.'

We are pleased to announce that by the time this is in our readers' hands the publication of the 'Birds of Yorkshire' will be an accomplished fact, and within a few days the subscribers will have received their copies. There has been delay, but when the work is seen in its two excellently printed and well illustrated volumes, the delay will be pardoned. In the issue of this work the Yorkshire Naturalists' Union is to be congratulated on the completion of still another of its monographs, the material for the present one having been accumulated during the past twenty years. Quite recently have been completed the Fungus Flora, the Algæ Flora, and the List of Lepidoptera of the County, as well as Baker's North Yorkshire. The present—the most formidable of all—adds to this excellent list, and the row of volumes is one of which any society might be justly proud. We hope to give an extended notice of the 'Birds of Yorkshire' in our next issue.

STONHAM'S 'BIRDS OF THE BRITISH ISLANDS.'

We have previously drawn attention to this sumptuous work in these columns. The illustrations unquestionably will attract the attention of those who are interested in birds. By the kindness of the publishers we are able to let readers of the 'Naturalist' see the nature of the plates by the reproduction of a drawing of one of the small birds (Plate XXIII.). This speaks for itself. The size of the plates, however, is about twice that of the specimen given. The publisher, Mr. E. Grant Richards, of 7, Carlton Street, Regent Street, S.W., will be glad to supply a detailed prospectus of the work on application.

DENTALIUM GIGANTEUM.

In our issue for Feb. (pp. 35-36), we gave a figure of some specimens of *Dentalium giganteum* on a piece of Lias. It was then stated that on a certain horizion in Yorkshire this species was extraordinary plentiful. This was demonstrated at the recent meeting of the Yorkshire Naturalists' Union, when huge slabs of rock, entirely covered with the shell, were seen strewn on the beach immediately north of Robin Hood's Bay. Fortunately a fair sample, which is figured herewith, was

Photo by *Godfrey Bingley.*

**Slab of Lias, with *Dentalium giganteum*.
Robin Hood's Bay.**

secured for a certain institution at Hull. It measures twelve by eight inches, and is about three inches thick. As will be seen from the photograph, the upper surface is thickly covered with the shell—which also occurs throughout the slab—several being visible on the underside.

SIR H. H. HOWORTH AND GLACIAL NIGHTMARES.

In the 'Geological Magazine' for June, Sir Henry H. Howorth is once again on the track of the glacialists. Of the general line of his argument we do not propose to deal, even were we capable of doing so. On one point, however, Sir Henry has apparently a convenient memory—or lack of it! It will be remembered that some time ago he put forward the theory that the Scandinavian erratics on our east coast were 'Viking anchors'! or ballast lost or thrown over from Scandinavian ships. A lengthy list of records of boulders of rhomb-porphyry, etc., at great depths below the surface, and at localities inland * caused Sir Henry to write to the 'Geological Magazine' (1897, pp. 154-5) as under :—'I am bound to say that . . . the reports of the British Association, and especially Mr. Sheppard's recent researches, make it *impossible for me to maintain any longer* the extreme position I took up in my controversy. I am convinced now that the rocks are certainly Norwegian, and that these Norwegian boulders have actually been found in undisturbed Boulder-clay and in its associated beds, and that, therefore, if they are to be explained, some other explanation than the one I gave must be forthcoming.' That was ten years ago. The same writer, who then found it impossible to maintain his theory any longer, now says :— 'I am more than ever convinced that a great proportion of the foreigners in the shingles, especially the *so-called* Scandinavian boulders, are derived from ballast, either from wrecks or discarded from ships, and are entirely misleading in their testimony.' Where are we now? Sir Henry has accomplished impossibilities!

MR. LAMPLUGH'S BRITISH ASSOCIATION ADDRESS.

Mr. Lamplugh's address to Section C of the British Association at York † is also referred to. 'That Mr. Lamplugh is right in his views about inter-glacial beds I have no doubt. . . . The conclusions which he now publishes, as if he was the first to generalise in their sense have been pressed for nearly thirty years in many papers and two big works by one Howorth. None of these publications are noticed in his address.' We have previously found Sir Henry complaining of his two big books being ignored, so that in this respect Mr. Lamplugh is not alone. Possibly there is a reason for it. As regards the Interglacial beds, we must congratulate Mr. Lamplugh on being in the same boat as Sir Henry Howorth. If they don't pull together they will at any rate 'row,'—towards Norway and the Maelstrom, surely! where 'the waters seamed

---

* 'Glacialists' Magazine,' 1895, pp. 129-131.
† See 'Naturalist,' 1906, pp. 304-317 and pp. 360-366.

and scarred into a thousand conflicting channels, burst suddenly into frenzied convulsion—heaving, boiling, hissing—gyrating in gigantic and innumerable vortices, and all whirling and plunging on to the eastward.' We can just imagine it.

BIRD MIGRATION.*

The nature of this valuable report, issued by the British Ornithologists' Club, was described in detail in our issue for May last year. The present report is on the same lines, and is similarly lavishly illustrated by maps showing the distribution of the various species, referred to. Messrs. C. B. Rickett and C. B. Ticehurst have been added to the Migration Committee. Mr. J. L. Bonhote is largely responsible for classifying the records, and Dr. N. F. Ticehurst has prepared the maps. These most useful reports of the ' B.O.C.' will unquestionably go a long way towards solving the complex problems of bird migration.

——◆◆——

A Text-book of Plant Diseases, by George Massee (Principal Assistant (Cryptogamic) Royal Herbarium, Kew). Duckworth & Co., 8vo. pp. xx., 472. Third edition, 6/- net.

This is the third edition of Mr. Massee's book, which speaks well for its popularity. We have had the book in constant use for several years, and proved its value to gardeners, nurserymen, and others. While the most important diseases are described and illustrated and the means of prevention and cure given, we find that the criticism of the average English gardener is that while several diseases not known in this country are described, a number of parasites frequent on plants in general cultivation, and some of these very destructive, are omitted. In a future edition it would be well to extend the types in this direction. This edition agrees closely with the second, except that eight pages have been inserted in front of the introduction dealing with the black scab of potatoes, American gooseberry mildew, and a cluster-cup disease of conifers ; and the price has been slightly increased.

The Gamekeeper's Manual, by Alexander Porter. David Douglas, Edinburgh. 140 pp., 3/- net. The fact that this is the third edition of this popular manual shows that the work of the Chief Constable of the counties of Roxburgh, Berwick, and Selkirk is appreciated, and serves a useful purpose. It deals with poaching, day trespass, preservation of game, poison, taking or destroying eggs, owners' and occupiers' rights, The Wild Birds Protection Acts, the powers and duties of a gamekeeper, etc., etc., and the author is naturally a well-qualified person to deal with the subject. It is interesting to note with regard to the Wild Birds Protection Act that 'there is no special power given to anyone to enforce the Act. *Any person*, on seeing an offence committed, may demand the offender's name and address, and if he refuses or gives a false name or address he renders himself liable for an additional penalty for so doing.' There is much in the book that might well be perused by others besides gamekeepers !

---

* *Bulletin of the British Ornithologists' Club.* Edited by W. R. Ogilvie-Grant. Volume XX. ' Report on the Immigration of Summer Residents in the Spring of 1906.' London, Witherby & Co., 1907, pp. 189. Price 6/-

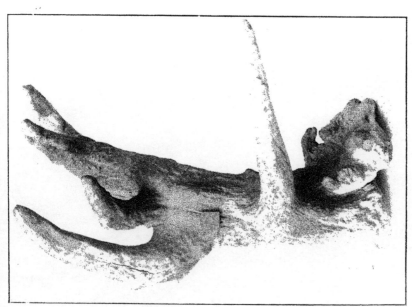

Two views of Malformed Antler of Red Deer from the Peat at Sutton-on-Sea, Lincs.

NOTE ON A MALFORMED ANTLER OF A RED DEER
(*Cervus elaphus*),
FROM THE PEAT AT SUTTON-ON-SEA, LINCS.

T. SHEPPARD. F.G.S.,
Curator of the Municipal Museums, Hull.

WE have recently received a curiously malformed antler of a red deer, which was found in 1906 by Mr. H. Bocock at Sutton-on-Sea, Lincolnshire. It occurred in the peat bed, generally known in these parts as the 'submerged forest,' immediately opposite Sutton. Mr. Bocock has previously found normal antlers of the red deer in the same deposit, but has never seen anything like the specimen now being described. I am indebted to a Lincolnshire friend for his interest on our behalf.

As will be seen from the photographs (Plate XXIV.), the antler is by no means normal; the tines branching out in a very irregular fashion. There are in all seven large tines remaining, and formerly there were nine, but two were cut away in order that the specimen might more readily fit against a passage wall to be used as hat pegs! Of the two tines cut away I am informed that one measured 7½ ins. and the other 4 ins. in length. Of those remaining, what is apparently the brow tine is 14 ins. in length, and 7½ ins. in circumference where it joins the main stem. The next tine to this in the ascending order is a little over a foot in length, and is 9 ins. in circumference at its base, and from this one of the branches referred to has been sawn.

In addition to the large tines, there are numerous small excrescences which are apparently partially developed tines. The 'horn' or shaft of the antler is, as will be seen from the photograph, unusually short and broad, and somewhat flattened. Its greatest measurement (round the corona) is 2 ft. 4 ins. Just above the base of the horn proper it is 12½ ins. in circumference. Immediately above the brow tine, at a distance of 4½ ins. above the last measurement, it is 1 ft. in circumference; 4 ins. higher still it is 1 ft. 2 ins.; a further 4 ins. and it is 12½ ins.; and immediately above the last tine it is 7 ins. in circumference, from which place for about 8 ins. the antler gradually tapers to a point.

The antler has evidently been cast, and appears to have two places of attachment to the skull; the principal one being somewhat the shape of a shoe-sole, $5\frac{1}{4}$ ins. long, $2\frac{3}{4}$ ins. in width at the broad end and, $1\frac{3}{4}$ ins. wide at the narrow end. Adjoining this, though apparently separated by a portion of the corona, is a smaller place of attachment, oval in shape, measuring $1\frac{1}{2}$ ins. by $1\frac{1}{4}$ ins.

The total weight of the specimen is 18 lbs. It is exceedingly hard, and dark coloured from its contact with the peat.

From the unusual thickness of the horn, from its irregular shape, the number of tines branching at all angles, and lastly from the nature of the places of attachment to the skull, it seems evident that it really represents a pair of antlers, which, owing to some accident or disease, have grown together as one.

Through the kind offices of Dr. A. Smith Woodward, F.R.S., of the British Museum (South Kensington), the specimen was recently exhibited at a meeting of the Zoological Society of London, and the Fellows did not remember having seen anything quite like it previously. Sir Edmund G. Loder, Bart., who was present, kindly informs me that he has known stags in Scotland which had only one antler each year. In each case the antler was exceptionally fine, in one instance having six points. In these instances there had never been an antler on the other side.

In the ' Proceedings of the Zoological Society ' * is printed a short note ' On some Abnormal Remains of the Red Deer (*Cervus elaphus*) from the Post-Pliocene Deposits of the South of England,' by Mr. Martin A. C. Hinton. In this the author places on record the discovery in various post-pliocene deposits in the south of England of certain remains of the red deer which present characters of abnormal nature. Three specimens are described : the first is a fragment of the frontal and antler of a very young individual, from the Pleistocene of Ilford ; the next is from a fissure deposit of the same age in the Isle of Portland ; and the third is from the Holocene alluvium of Moorfield, Slandon. The instances described show that whilst there is not the enlargement and thickness shown in the Sutton-on-Sea example, the antlers have unquestionably grown in an abnormal way. In the opinion of the author (p. 212) 'it is probable that these specimens belonged to individuals which had suffered injury to the testes at an early period of life, which resulted in making the retention of youthful characters possible for a longer period than is usually the case.' In the same journal, in 1905 (pp. 191-197), Mr. R. I. Pocock has a paper on ' The Effects of Castration on the

---

* 1905, Vol. I., pp. 210-212.

Horns of a Prongbuck.' In this, illustrations are given of deformed antlers in modern species due to the same cause, and from this it would appear that in all probability the extraordinary formation of the Sutton-on-Sea antler is due to some such injury to the animal at an early period of its life.

———◆◆———

Among the donations to a northern museum we notice 'a double shell-less hen's egg.'

Dr. F. A. Bather, of the Geological Department, South Kensington, represented the British Museum at the celebration recently held in Sweden in commemoration of the Bi-centenary of the birth of Linnæus.

At the Annual Meeting of the Leeds Geological Association, held a few days ago, the treasurer announced a balance in hand of over £6. The present membership of the Association is 87. Mr. F. W. Branson was elected President for the ensuing year.

We are pleased to acknowledge a further donation of two pounds from Mr. W. H. St. Quintin, J.P., a guinea from the Hon. Mrs. Carpenter, 5/- from Mr. W. J. Clarke, 5/- from the Leeds Co-operative Naturalists' Field Club, and half-a-guinea from Mr. D. Legard towards the Protection of Birds in Yorkshire.

The Liverpool Botanical Society is about to prepare a Flora of South Lancashire (*i.e.*, that portion of the county which lies south of the Ribble), and asks for the co-operation of botanists and others who are able to assist. Information is desired relating to any collections of South Lancashire plants, or old herbaria containing such, or private MS., notes or records dealing in any way with the botany of South Lancashire. Particulars of any plants will be of value, as the distribution of even the commonest species is imperfectly known in some parts of the area. Communications should be addressed to the Hon. Sec., South Lancashire Flora Committee, Mr. W. G. Travis, 107, Delamare Street, Liverpool.

In addition to the reports of the Society and its Museum for two years, the 'Transactions of the Natural History Society of Northumberland, Durham, and Newcastle-on-Tyne' (Vol. I., pt. 3) contains the following papers of interest ;—' On the Crustacean Fauna of a Salt-water Pond at Amble,' by Dr. G. S. Brady (with descriptions of new species); 'The Spiders of the Tyne Valley,' by Dr. A. R. Jackson (an excellent list); 'Notes on New and Rare Local Beetles,' by R. S. Bagnall and Prof. T. H. Beare (reprinted from another source) ; ' Derwenthaugh Land in Derwent Gut,' by Rev. A. Watts (with details of borings); 'The Landslip at Claxheugh, Co. Durham, September, 1905,' by Dr. D. Woolacott ; and 'Larval Trematodes of the Northumberland Coast,' by Miss M. A. Lebour. Most of the articles are illustrated by plates.

On the 22nd May, the Lincoln City and County Museum was informally opened to the public, and since then the attendance has been very fair. It will be remembered that the Lincolnshire Naturalists' Union has all along had for one of its aims the formation of a County Museum. This aim has now been achieved, and the Natural History Collections in the Museum are principally presents from the Union or its members. It is a great pity that the Lincoln Corporation lost the opportunity of making the existence of the county museum well known by having a public opening. There must be scores of valuable objects in the possession of private individuals who would be pleased to send them to a permanent home in the county, but as it is, few of these can possibly know of the existence of the Museum under Mr. A. Smith's charge. 'Tis sad to admit, but even Museums must advertise.

THE BIRDS OF THE FARNE ISLANDS.

R. FORTUNE, F.Z.S.
Harrogate.

THE Farne Islands are a group of rocky islets off the coast of Northumberland, having altogether an area of about eighty acres. Here sea fowl in countless numbers resort to breed.

I remember visiting these islands twenty years ago, when a sort of half protection was afforded the birds. This 'protection' really meant that those who were supposed to be looking after the birds did so chiefly for their own profit. The fishermen made ceaseless raids, taking large quantities of eggs, and collectors did likewise.

However, all this is now changed. the birds are looked after during the nesting season by a party of gentlemen banded together under the name of the Farne Islands Association. Four or five keepers are kept upon the Islands during the whole of the nesting season, and their sole work is to look after the birds. This is, of course, not done without considerable expense. There is no fixed subscription to the Association, each member gives what he likes or can afford. It is pleasing to find that there are several members of the Yorkshire Naturalists' Union in the Association.

To Mr. Paynter, of Alnwick, the Hon. Sec. of the Association, the thanks of all Naturalists are due for his untiring efforts on behalf of the birds.

The best way to approach the islands is from the little fishing village of Seahouses, where, at the Bamborough Castle Hotel, passes may be obtained. Every visitor must sign a form promising not to take eggs. Visitors are not allowed to stay very long on any of the islands, and especially is their time restricted where the Terns nest.

What with looking after the visitors and keeping their 'weather eyes' open for raids from the fishermen, the keepers in fine weather have their time fully occupied. In bad weather they have a fairly easy time, as landing upon the islands is practically impossible.

Altogether about fourteen species of birds find a sanctuary here. *The* bird of the islands is, without a doubt, the Lesser Black Back Gull. He is everywhere, and although undoubtedly a fine and handsome species, I cannot help thinking he is

R. Fortune.

Young Kittiwakes on Nests.

Kittiwakes on Nests.

R. Fortune.

Near view of the Pinnacles with their crowd of Guillemots.

far too numerous, as he is a sad rascal, robbing the other birds regularly both of their eggs and young. When the young Terns are hatching it is impossible for the keepers to drive the Gulls away until they have had their fill. Young of other species have to pay toll in a like manner. Although taking the eggs of other birds whenever an opportunity offers, they do not seem fo interfere with those of their own species.

Herring Gulls are present in small numbers. It is quite impossible to identify their nests and eggs without seeing the birds upon them, so closely do they resemble those of the Lesser Black Backs.

One of the sights is that of the renowned Pinnacle Rocks, a group of basaltic pillars rising to a considerable height. They have perfectly flat tops, and are tenanted with a huge crowd of Guillemots. It is a most interesting sight; from the main island the spectator can look right on to the heaving mass of birds, and one cannot help wondering whether by any chance a bird leaving its egg for a short visit to the sea could ever regain it again. Our own Yorkshire coast furnishes a most interesting spectacle of Guillemots in the nesting season, but nothing to approach the Pinnacle Rocks for large numbers in a small space. A few Guillemots nest on some of the ledges on the main island, and we found one or two laying their eggs amongst the Cormorants' nests on the Wamses.

One of the most pleasing results of protection is the great increase in the numbers of the charming Kittiwake, the most handsome and most gentle of all the Gull tribe. There can surely be no finer sight than a colony of Kittiwakes sitting upon their nests, their beautiful plumage affording a striking contrast to the dark rocks upon which their nests are placed, the smallest ledges being sufficient for a foundation for them. On the Farnes these birds are delightfully tame and confiding, and the visitor is thoroughly repaid by being allowed to enter into the mysteries of their home life. The nests are placed on the sides of the Pinnacles and on the cliffs of the Staple Island.

Razorbills are very few in number, only about half-a-dozen pairs nesting on the Farne Island. Very few birds nest on this island, so that it is not of great interest to the ornithologist. It has, however, an interest of its own.

Another gratifying feature is the great increase in the number of Eider Ducks. They are found practically everywhere nesting amongst the rocks, vegetation, or on the open beach. Some are wonderfully tame, even occasionally allowing

1907 July 1.

the visitor to stroke them upon the nest.   Should a duck leave
her nest in a hurry without covering the eggs with down, as
they sometimes do, it is a bad look out when she returns, as it
is a certainty that they will have been carried off by the Lesser
Black Back Gulls.   What astonishes me is that the Gulls have
never yet tumbled to the fact that beneath the heap of down
reposes a tempting array of dainties which they know well
how to appreciate.   The male Eider, in his striking plumage,
is a great contrast to the sober attire of the female ; he keeps
well out to sea, and rarely comes on land, but should the female
leave her nest for a little relaxation he is speedily in attendance.

Puffins are also very numerous on one or two of the islands.
Upon Staple Island the soil is riddled with their burrows, so
much so that it is almost impossible to walk over the surface
without ones foot sinking into a Puffin's burrow.   The Lesser
Black Back Gulls cannot get at the Puffin's eggs, as they are
deposited some distance in the burrow, but when the young
come forth they play havoc amongst them.   The Puffin is a
very quaint fellow, with a grotesque bill, the most gaudy
portion of which he sheds during the winter months.

Oystercatchers are fairly common, and may be seen on most
of the islands.   On the Staple Island there are the remains of
an old lighthouse, on the top of the old wall of which a pair
of Oystercatchers regularly place their nest.

Ringed Plovers nest on the shingly beeches of some of the
islands, but they are not abundant.

Cormorants were nesting in two colonies on the Wamses
and on the Harcar ; the Megstone, their former abode, being
deserted by them owing to their nests being continually washed
away by the sea.   I learn, however, that a few pairs returned
to the Megstone later in the season, and were able to rear
their broods.   Unfortunately, upon the Wamses and Harcar
are large numbers of Gulls.   When visitors approach the
islands the Cormorants take flight.   Immediately they go the
Gulls fly to their nests and clear out every egg in the nests in no
time.   At the first visit paid by Mr. Booth and myself, in June
1906, we thought it would be an impossibility for the Cormorants
to rear any young ; we were therefore greatly delighted upon
paying a second visit in early August to find young birds in
almost every nest.   A period of bad weather, during which
visitors could not land upon these islands, had allowed the birds
to stick close to their nests, thus protecting their eggs from the
ravages of the Gulls.

Eider Duck and Nest.

Eider Duck on Nest.

Lesser Black Back Gull.   *R. Fortune.*

**Kittiwakes and Guillemots
on the Staple Island.** ·

A Group of Puffins. *R. Fortune.*

It was upon the Wamses we had a good view of a so-called Iceland Gull, which, however, as we anticipated, turned out to be a variety of the Lesser Black Back.

A few Shags may be seen about, but they do not nest upon the islands.

Next to the Kittiwakes the Terns appear to me to have the most interest, and special attention is paid to their protection. Four species nest here, viz., the Common, Arctic, Roseate, and Sandwich Tern. By far the most abundant is the Arctic Tern, the Farnes being their southern breeding limit on the East Coast. Common Terns are not so numerous. It is practically impossible to distinguish them from their Arctic cousins when on the wing. An odd pair or two of Roseate Terns nest amongst the commoner species. They may be recognised when in the air by their cry, which differs from that of the other two. On the ground they are easy enough to detect, but unless the bird is seen upon the nest it is impossible to distinguish between the eggs of all three species. A few Arctic Terns nest on the Brownsman, but the headquarters of these birds are upon the Wide-opens and Knoxes. These two islands are quite different in character from the others. The Knoxes is a low island covered with sand and pebbles, with a somewhat higher rock formation at one end. This island is practically sacred to the Terns, which lay their eggs all over the place amongst the pebbles or on the sand in a most indiscriminate manner, usually without the slightest attempt at a nest. So close do they deposit their eggs to the edge of the water that quantities are destroyed every season when the wind brings the waves a little higher than usual. There is also a fine colony of Sandwich Terns, which seem to prefer to nest on a higher ground than the others, and form a very conspicuous object clustered together on the top of an elevated place.

Numbers of Terns nest on the Wide-opens also. When a visitor lands it is a perfect babel of harsh cries, for though these birds are dainty and handsome creatures, their cry is harsh to a degree. They have a curious habit of suddenly ceasing their outcry and taking a sweeping flight out to sea, returning to re-commence their clamour.

On the Wide-opens the Arctic Terns make a better attempt at nest building, but this may be on account of the materials being handier to obtain. Eider Ducks are very plentiful on the Wide-opens, nesting amongst some luxuriant vegetation ; on the Knoxes they nest on the bare sand.

·Of land birds the principal representative is the Rock Pipit, which nests commonly upon the suitable islands.

Space will only allow of a very short account of the birds upon these most interesting Islands. To any naturalist who contemplates paying them a visit I can promise a thorough treat. The immense numbers of the birds, and the great variety of species to be seen cannot be surpassed in any portion of our country.

In conclusion, one cannot help but be delighted with the efforts of the Farne Islands Association, who have provided a real sanctury for many interesting species. If the Lesser Black Back Gulls could be kept back somewhat, I am sure the rest of the birds would benefit considerably. One matter which no one can contemplate with any pleasure, and which tells considerably against our own county, is the fact that a lot of these dainty and charming Terns are, at a considerable expense, thoroughly protected throughout the nesting season, laying their eggs and bringing up their young in comparative security simply for the purpose of finding Sport (?) for a lot of, well, I can only call them unthinking, individuals upon our coast, who slaughter them unmercifully on their passage south.

DESCRIPTION OF PLATES.

Plate XXV.—Kittiwakes on Nests.
Young Kittiwakes on Nests.
Plate XXVI.—Near view of Pinnacles, with their crowd of Guillemots.
Plate XXVII.—Eider Duck and Nest.
Eider Duck on Nest.
Lesser Black Back Gull.
Plate XXVIII.—Kittiwakes and Guillemots on Staple Island.
A Group of Puffins.
Plate XXIX.—Puffins.
Cormorants on their Nests.
Sandwich Terns on their Nests.

———◆◆———

At the recent meeting of the Yorkshire Geological Society in the Isle of Man, Dr. Wheelton Hind read a paper on ' Dendroid Graptolites in the Carboniferous Rocks of Britain.' One of the specimens exhibited was found on Pendle Hill, Yorkshire, the other in the Isle of Man.

The police have at last got a conviction under the Wild Birds Protection Act! At Horncastle recently a boy caught a crow in a field. took it home, and wrung its neck. The bird had been tamed and taught to talk by a local doctor. The boy was ordered to pay the costs—half-a-crown. In Teesdale, on the other hand, Peregrines are allowed to be caught in Pole-traps without comment.

Puffins.

Cormorants on their Nests.

Sandwich Terns on their Nests.　　　　　*R. Fortune.*

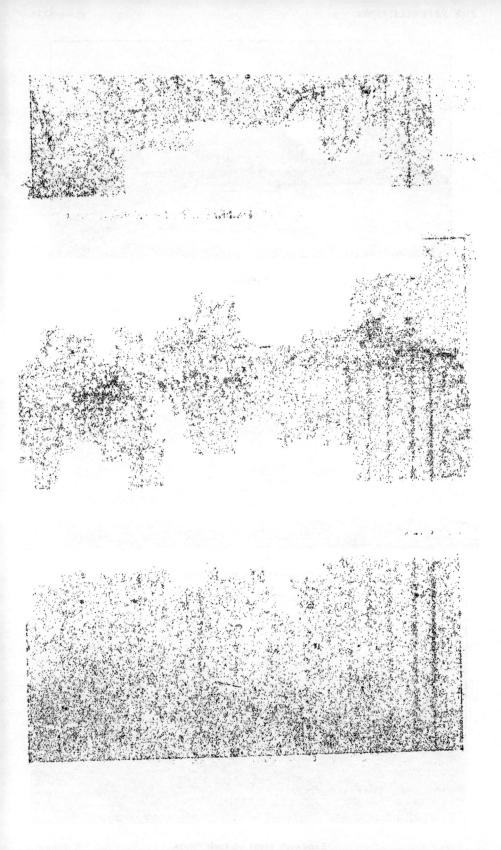

THE MARTEN IN LAKELAND.

EDWARD T. BALDWIN.

WHEN in Borrowdale again this spring I tried, somewhat unsuccessfully, to obtain records as to the 'Mart' (*Martes sylvestris*), subsequent in date to those sent this time last year (*Vide* 'Naturalist' for 1906, pp. 221-222).

However, I found that one (a very fine one, I believe) was got, in November or December, 1906, by a man called Gillbanks between the well-known Lodore Falls and the Bowder Stone. With this exception, I could not hear of any recent captures.

I had an interesting chat with Thomas Jackson, of Rosthwaite, who, in his time, was a noted 'Mart' hunter, but who now no longer 'follows it,' as the local saying is. It appears that the two young 'Marts' I mentioned as having been got last spring close to the Bowder stone were obtained by his nephew, and were not trapped (as I said), but worried by his terriers. Jackson says that the most he ever got in any one year was sixteen, but that he had twice got eight. He thinks that there is no doubt that they are plentiful still, and especially mentioned Eel Crags as a favourite locality. Eel Crags are a range of almost sheer precipice facing Dale Head (2473 feet) and Hindscarth (2285 feet). There, a good deal of foil is to be seen, and in winter the tracts are plainly visible in the snow. The track of the 'Mart,' by the way, is something like that of the hare. Another 'strong bield' is close to Eagle Crag, and here lately, amongst a quantity of mart foil, were found two small metal bands with numbers, such as are usually attached to the legs of carrier pigeons. Eagle Crag is the huge hump of rock which guards the entrance to the beautiful but wild and desolate Langstrath Valley, at the head of which towers the massive Bowfell (2960 feet). Jackson told me of a pair of peregrine falcons, which were successful this year in nesting and bringing off their brood. The locality I think it as well not to disclose. He considers the foumart, or polecat (*Mustela putoria*) virtually extinct in that district.

Wandering over these great grassy Cumberland fells, the absence of bird life strikes one very forcibly. Very little but a few meadow pipits and stone chats, with an occasional raven, carrion crow, or buzzard is, as a rule, to be seen. But at the time when the bracken clock hatches out (generally at this time of the year), numbers of the lesser black-headed gull (*Larus*

ridibundus) make their appearance ; this year, however, owing
to the cold and delayed spring, scarcely one was to be seen.   The
contrast between the teeming bird life in spring (*ex. gr.*, curlew,
peewit, golden plover) on many of the Scottish hills and the
meagre stock on these Cumberland fells is remarkable, and I am
not convinced that the absence of suitable food is the right
explanation.

Ornithological and Other Oddities, by Frank Finn, B.A., F.Z.S.
John Lane, 1907.   295 pp., plates, 10/- net.   The cry is still 'they
come !'. Another bird book—'Ornithological Oddities.'   In this volume the
late Deputy-Superintendant of the Indian Museum, Calcutta, gathers together his various contributions to numerous journals and dailies, and with
the aid of fifty-six illustrations produces a volume, bound in a light green
cover, with a bird upon the back.   As might be expected, the essays
principally deal with eastern species, though several of the chapters are of

Australian Barn Owl.

general interest, *e.g.*, 'The Study of Sexual Selection,' 'The Courting of
Birds,' 'Hybrid Birds,' 'Love among the Birds,' etc.   There is also a
chapter on 'Blushing Birds.'   Amongst the 'other' oddities are 'Monkeys I
have met.'   The book contains numerous ancedotes written in Mr. Finn's
familiar style.   The photographs (one of which we are permitted to
reproduce herewith) are mostly very good—that of 'Japanese monkeys;
father, mother, and child', (p. 246) is certainly funny—the parents are
busy with a usual occupation, whilst the 'child' sits by watching the
'hunt,' with an expression forcibly reminding one of that of little Oliver
when he wanted 'more.'

THEORIES OF EVOLUTION.

An Historical Outline.

AGNES ROBERTSON, D.Sc.

(*Continued from page 215.*)

IV.—EMBRYOLOGY.

Embryology is a source from which some help is derived in the study of evolution. Every animal and plant begins life as a unicellular organism, but whereas some continue unicellular to the end of their lives, others go through developmental processes of varying degrees of complication before they reach the mature form. Evolutionists believe that the whole organic world has been evolved from some very simple form of life, perhaps not differing greatly from certain unicellular organisms of the present day, so that in phylogeny (the ancestral history of the species), and ontogeny (the developmental history of the individual), the starting point and the goal are the same, which suggests the tempting view that the path between is the same also. An idea of this kind was put forward in 1810 by Oken, and later more definitely enunciated by Von Baer. The 'recapitulation theory,' as it is called, states that in the developmental stages passed through by an embryo we can recognise the different stages passed through in previous ages in the evolution of the species, or in the words of the epigrammatist, 'Every animal in the course of its development climbs up its own genealogical tree.' But this conclusion can only be accepted with great reservations, since at the best embryology only gives us a much abbreviated and altered picture of the phylogenetic history, for the embryo attains its end by many short cuts, and ancestral traits are often obscured by adaptations which fit the immature organism for the special conditions of its life.

V.—HEREDITY AND HYBRIDISATION.

It has always been a matter of common observation that children resemble their parents, but it is only of late years that scientific methods have been brought to bear upon the subject of heredity. In early days the mechanism by which characters

were transmitted was unknown because the nature of fertilisation
was completely misunderstood. For a long time after the discovery of the male fertilising element in the seventeenth
century it was believed that the spermatozoon itself developed
into the offspring, and that the only part played by the female
was in nursing and protecting it. We now know that the
essential thing in fertilisation is the union of two nuclei, the egg
nucleus of the female and the sperm nucleus of the male.

In investigating the laws of heredity experimentally it is
obvious that results will be more easily obtained if we choose
for the parents individuals which differ considerably ; that is to
say, we shall get more information by observing the manner in
which a character peculiar to one parent appears in the offspring
than by investigating a character common to both. For
instance, if we fertilise the ovules of a yellow-seeded pea with
pollen from a green-seeded pea, and find that the next generation bears only yellow seeds, we may infer that the seedlings
have inherited their seed-colour from the female parent, whereas
if we had used yellow-seeded peas for both parents we should
not know whether the yellowness in the next generation was
due to the influence of the male parent, or the female parent, or
both. The way in which the presence of the greatest number
of differentiating characters can be secured is by working with
hybrids, that is to say, crossing individuals belonging to
different races.

The most remarkable work that has ever been done on plant
hybrids is that of Mendel, and is of such extreme importance
that I must treat of his classical paper at some length.
Gregor Mendel (1822-1884) was the son of a peasant in Austrian
Silesia, and entered an Augustinian Monastery, of which he
eventually became Abbot. All his experiments were made in
the convent garden, and his great paper on pea hybrids was
published in an obscure local journal in 1865. It was quite
overlooked by biologists for more than thirty years, but since it
has been unearthed we might almost say that it has revolutionised
our ideas as to the nature of living things. Mendel began his
work by looking about for some species which should be as
convenient as possible to grow and to experiment with, and
the different races of which should be quite fertile when crossed
together, and should produce fertile hybrids. He finally hit
upon *Pisum sativum*, the ordinary eating pea. This species
has the great advantage of always fertilising itself when left
alone, so that there is no risk of accidental crossing. Twenty-

two different races of peas were chosen for the experiment. First of all the seeds of all these strains were grown separately for two years to make sure that they were genuine races and not mixtures. They were found to come quite true and constant from seed. In crossing his peas, Mendel concentrated his attention on certain definite differences between the races. He decided to investigate the inheritance of sharply defined pairs of characters by the possession of one or other of which the different races could be distinguished. These were :—

1. Difference in form of ripe seeds. Round or wrinkled.
2.   ,,   ,, colour of cotyledons. Yellow or green.
3.   .,   ,, colour of seed-coats. Coloured or white.
4.   ,,   ,, form of ripe pods. Inflated or constricted.
5.   ,,   ,, colour of unripe pods. Green or yellow.
6.   ,,   ,, position of flowers. Axial or terminal.
7.   ,,   ,, length of stems. Tall or dwarf.

A set of experiments was made with each of these pairs of differentiating characters. For example, in the first set of experiments plants bearing round seeds were crossed with plants bearing wrinkled seeds, and in the second set plants with yellow cotyledons were crossed with plants with green cotyledons, and so on. Seven sets of hybrid plants were thus produced. It might have been supposed that these hybrids would turn out to be intermediate in their characters between the parents (for instance, that the peas resulting from a cross between a round pea and a wrinkled pea would be intermediate in form between the two), but it was found that this was not the case. All the hybrid peas were round—no wrinkled ones were found at all. The same results were got with the other six pairs of differentiating characters—in each case one of the two characters had got the upper hand, and the other did not appear at all ; the offspring of the tall peas and the dwarf peas were all tall—the offspring of the peas with inflated pods and the peas with constricted pods were all inflated, and so on. It is convenient to distinguish the two opposing characters of each pair respectively as *dominant* (D) and *recessive* (R), the dominant character being that which appears in the hybrid to the exclusion of the recessive character which remains latent. On the hybrids being self-fertilised and their offspring again examined, a most curious result came to light. I will only describe the case of the round and wrinkled peas, as the other cases fall exactly into line with this. The first generation of hybrid peas were all round, but when these were sown and the flowers of the next generation self-fertilised, a mixture of smooth and wrinkled peas were obtained. There were no intermediates, and usually some of the peas in the same pod were round and some wrinkled. From 253 hybrid pea plants 7,324 seeds were obtained, of which 5,474 were round and 1,850 were wrinkled, *i.e.*, there were almost exactly three times as many round peas as wrinkled peas (2.96 : 1). This proportion was found to hold good for each ' of the seven pairs of characters worked with ; on an average the proportion in the seven trials was 2.98 : 1, which is practically 3 : 1.

To understand what these numbers mean we must follow the hybrids to further generations. Take first the one quarter which show the recessive character, *e.g.*, the wrinkled peas. These, when self-fertilised for a number of generations, continue to produce nothing but wrinkled peas, so that they may be regarded as a pure wrinkled race. But the remaining three-quarters which show the dominant character, roundness, when self-fertilised, do not all behave alike ; they break up into round and wrinkled peas in a constant numerical proportion.

Mendel required some theory to account for these remarkable numerical relations, and he put forward one which is extremely simple and complete. He supposes that *the germ cells are pure with respect to certain characters*, that is to say, in the case of . two opposing characters, such as roundness and wrinkledness of seeds, each egg and sperm can only transmit one or other attribute, and not both. He makes also the further postulate that on an average one half of the germ cells carry one character and one half the other character. The hybrid plants produced by crossing a round pea with a wrinkled pea would on this hypothesis bear eggs, half of which carry the character of roundness and half of wrinkledness, and in the same way sperm cells, half of which carry each character. When the hybrid is self-fertilised the eggs and sperms mate according to the laws of chance. Let us consider the case of four eggs and four sperms mating in this way :—

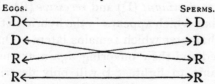

As the result of this mating we get

1DD, 2DR, 1RR,

but as the recessive character will be masked by the dominant one, the DR plants will, to all appearance, resemble DD, so that we shall get three apparent dominants to one recessive, which is the observed proportion. The fact that these dominants are not really all alike, but are made up of some pure dominants and some hybrids is shown in the next generation, in which, on Mendel's theory, the recessive (wrinkled) peas will produce only offspring constant to the recessive character, while the dominant (round) peas will give rise to a mixture of round and wrinkled. The whole thing may be represented diagrammatically :—

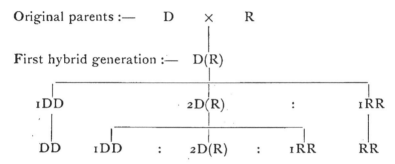

Original parents :— D × R

First hybrid generation :— D(R)

1DD   2D(R)   :   1RR

DD   1DD   :   2D(R)   :   1RR   RR

When the numerical proportions that would be expected on this scheme are worked out, it is found that they are closely in accordance with those obtained in Mendel's experimental work : from which we may infer that the theory of the purity of the germ cells has good claim to rank as a sound working hypothesis. ' In so far as Mendel's law applies, therefore, the conclusion is forced upon us that a living organism is a complex of characters, of which some, at least, are dissociable, and are capable of being replaced by others. We thus reach the conception of *unit-characters*, which may be re-arranged in the formation of the reproductive cells. It is hardly too much to say that the experiments which led to this advance in knowledge are worthy to rank with those which laid the foundation of Atomic laws of Chemistry.' *

Mendel's experiments have been confirmed and extended by later workers, and results such as he got have been found to hold true very widely both for plants and animals. In plants, such pairs of opposed characters as hairiness and glabrousness, sugary and starchy endosperm (maize), pinnate and palmate leaf-form (*Primula*), etc., have been found to fulfil Mendelian predictions, as also a great many characters in animals. There

---

* W. Bateson. Jr. Roy. Hort. Soc., 1901.

are some very curious pairs of qualities which work out according to Mendelian expectations.   For instance, when a race of wheat known to be highly susceptible to rust disease is crossed with one which is immune to the disease, in the first generation all the plants are susceptible, in other words, *susceptibility* is dominant.   In the next generation three-quarters are susceptible and one quarter immune !

The phenomenon of gametic purity is of far greater theoretical importance than that of dominance.   There are many cases in which the first generation from the cross, instead of showing the complete dominance of one or other character, shows something quite different, often an apparent reversion to a more primitive type.   For instance, the cross between an albino mouse and a black and white waltzing mouse has been found to resemble a wild mouse.   Among plants there is a white sweet-pea called ' Emily Henderson ' which has round pollen grains, while most sweet-peas have long pollen grains.   ' Emily Henderson ' is not, however, a pure race, since some of the plants, which in every external character exactly resemble their fellows, have *long* pollen grains.   If the long-pollened and short-pollened ' Emily Hendersons ' are crossed together the long pollen is dominant over the round, but, as regards flower-colour, instead of the whiteness which we should expect, we get a purple flower with a chocolate standard, which is supposed to be the ancestral form from which all sweet-peas have been derived !   It would seem that the present races of sweet-peas are a series of ' analytical dissociations ' of the characters of the original ancestor, and that when two of them are crossed we sometimes get the characters re-combining and giving an appearance recalling the ancestral form.

It is impossible in this extremely brief outline to do more than touch upon the fringe of Mendelian work.   I have left out of account everything but the simplest possible cases.   For further information I should like to refer the reader to a little book on ' Mendelism,' by R. C. Punnett (Macmillan), in which the present aspects of the subject are clearly discussed.

VI.—THE MUTATION THEORY.

According to the Darwinian theory the Origin of Species has been an extremely slow and gradual affair, depending chiefly upon the preservation and accumulation by natural selection of minute individual variations of a favourable kind, while similar minute variations of an unfavourable kind tended to the extinction of the creatures exhibiting them. The type of variability expressed in these minute individual differences is called 'normal' or 'fluctuating' variability, and is characterised by the fact that it conforms to the law of chance. As an instance, we may take some statistics given by Galton of the strength of pull of 519 men between 23 and 26 years of age.

Strength of pull.	Number of cases.
Under 50 lbs.	2 %
,, 60 ,,	8 %
.. 70 ,.	27 %
80 ,	33 %
,, 90 ,,	21 %
,, 100 ,,	4 %
Over 100 ,,	5 %

We notice from this table that the largest class have a pull of 70-80 lbs; that is to say the moderately strong individuals are more numerous than are the extremely strong or extremely weak, and we also notice that the numbers decrease gradually as we pass towards either extreme. In other words, 'normal' variability behaves just as we should expect if it conformed to the laws of probability.

In speaking of Darwin's work I have mentioned the difficulty of the 'uselessness of minimal variations;' an adaptation in its fully devoloped form may be very useful to the creature possessing it, and yet it is often difficult to see how its rudimentary beginning could be useful enough to be preserved if it were due to the gradual accumulation of the minute changes due to fluctuating variability. This difficulty, along with other considerations as to the nature and inheritance of normal variations which I cannot enter upon here, have led some biologists to the view that species have not arisen by fluctuating variations, but have progressed by jumps, separated by periods of stability; such jumps have been named 'mutations.' In discussing the possibility of mutations the analogy has been suggested of a polygon which is in a position of stable equilibrium when it is standing on one face. When it is tipped up it is not in a position of equilibrium at all, and either falls back on to the original face or forward on to the next face. In the second case it has, as it were, gone through a mutation and reached a second position of stability.

A number of cases have been reported in which a 'mutation' seems to have been completely stable. The Ancona sheep

appeared in Massachussetts in 1791 as a sudden variation; it
was kept and bred from because it had bent legs which prevented its jumping into neighbouring fields, and it has formed
a distinct and permanent race ever since.   In 1761 Duchesne
found a strawberry plant in his garden with simple instead of
ternate leaves; this bred true and is still in cultivation.   I
mention these cases because they are so very well marked, but
the advocates of the mutation theory do not demand such large
jumps as these for the evolution of species.   The question, 'What
is a species?' is one for which no biologist has yet found a
satisfactory answer, but still we have a working notion of what
we mean when we speak of a species.   We mostly use the term
in the sense of classifactory binomial unit of Linnæus [e.g.,
Veronica chamædrys=the Germander Speedwell].   But later
investigations have shown that many systematic species are not
really units at all, but comprise a number of distinct races,
which are found on cultivation to breed true indefinitely.   In
some extreme instances the number of races lumped together
under one specific name is very large.   *Draba verna*, the little
Whitlow-grass, has over 200 local races, which are uniform
and come true from seed!   These sub-divisions of a species
have received the name 'elementary species,' and the 'mutation'
of a species generally does not imply a greater jump than that
from one elementary species to another.

   The great worker on the subject of mutations has been Prof.
Hugo de Vries.   There is a kind of Evening Primrose, *Œnothera
Lamarckiana*, which is unknown in the wild condition.   Not
very long ago De Vries discovered it growing in thousands,
apparently as an escape, in a field near Amsterdam.   He
noticed that while the majority of the plants were of the true
Œ. Lamarckiana type, others differed widely from it in such
characters as height, leaf-form, leaf-surface, pigmentation, and
so on.   For the sake of more satisfactory observation he dug
up, in 1886, nine large rosettes of the type form, and transferred
them to his own experimental garden.   The seeds of these
produced 15,000 young plants, all but 10 of which were of the
true *Œ. Lamarckiana* type, while the divergent ten belonged
to two quite new types, to which De Vries gave new specific
names.   In the same way in later generations a great majority
of normal and a small minority of divergent types were produced.
Seven new forms were thus obtained, and were found to come
true from seed.   These new forms De Vries regards as elementary
species, and considers that in Lamarck's Evening Primrose we

have caught a species in the mutating period, in other words, in the act of giving birth to a number of new species. De Vries imagines that all species have periods of stability during which they only vary according to the laws of fluctuating variability, and then pass through a mutating period, to which a period of stability again succeeds.

De Vries supposes that the struggle for existence is not so much between individuals of a species as between different elementary species. When a group of elementary species arises by mutation, natural selection determines which of these elementary species shall survive.

On general grounds the chief point in favour of the mutation theory is that it does not require so long a period for the evolution of the organic world as that demanded by the Darwinian theory, and is thus more in harmony with the calculations of physicists as to the age of the earth.

De Vries' theory has been much criticised. A great deal depends on the previous history of *Œ. Lamarckiana*. De Vries assumes that it is an ordinary pure species, but it has been suggested that it is possibly a hybrid. The fact that it is unknown in the wild state, and the sterility of some of the pollen and ovules, lend support to this suggestion. The ' mutations' isolated by De Vries strongly recall the ' analytical variations' of Bateson, obtained by hybridising sweet-peas, etc. So it seems that the question of the real existence of mutations must remain an open one till further experimental evidence is brought forward.

VII.—CONCLUSION.

There are a number of important branches of evolutionary work which I have not even touched upon in this brief outline —amongst others the contributions of Weismann and the work of the modern biometrical school. I have confined myself to the briefest possible mention of those lines of research which seem to me most hopeful for the future. It is *experimental* work on heredity, such as that undertaken by those who have confirmed and extended the conclusions of Gregor Mendel, which gives the clearest promise of further light on these huge and complicated problems.*

---

* I should like to refer any reader who wishes to follow up the subject to 'Recent Progress in the study of Variation, Heredity, and Evolution,' R. H. Lock (Murray, 1906).

YORKSHIRE NATURALISTS AT ROBIN HOOD'S BAY.

MAY 18th to 20th, 1907.

(*Continued from page 202*).

GEOLOGICAL SECTION.—Mr. Cosmo Johns writes :—The policy. of the section in concentrating its .attention on some special geological feature in each district visited was abundantly justified during this particular occasion. The variation in thickness of the Dogger on both sides of the Peak fault—the significant fact that on one side it rested on the Alum Shales and was itself reduced to six feet, or occasionally even less in thickness, while just across the fault it was much thicker, and succeeded Liassic rocks in normal sequence, with the upper zone well represented—all lent weight to the view that we were here dealing with a most notable example of inter-formational or persistent faulting. It is rarely, however, that such movements can be so clearly demonstrated.

It is unfortunate that no fuller exposure of the Middle and Lower Lias is visible on the downthrow side of the fault, so it was not possible to determine whether the pre-Oolite movement had been preceeded by inter-Liassic dislocations. Quite as interesting as the normal movements were the evidences of the direction of movement having been reversed at some time, and the possible connection of this later adjustment with the bifurcating fault on the beach was discussed. What was clear, however, is that the earth movements of the Cleveland area present a series of problems as complex, and therefore as interesting, as any in Britain. That they are chiefly, if not entirely, due to gravitational stress makes it just possible that, when worked out, they might throw some light upon the structure of the Palæozoic rocks that lie buried at some unknown depth below.

It was this concentration .of attention on the earth movements that caused the Dogger to loom so largely during the meeting. A ferruginous sandstone—which had probably been a calcareous sandstone—with a few badly preserved derived fossils, which would otherwise have been hardly noticeable, became of great importance. Forming as it does the base of the Oolites, and carrying in itself the proofs of the pre-Oolitic movement, it was traced along the length of coast visited. It had been hoped that some definite evidence of the extent of the pre-Oolitic denudation would be obtained, but the worn condition of the fossils found made it clear that the time available during the excursion was insufficient for the purpose.

VERTEBRATE SECTION.—Messrs. H. B. Booth and R. Fortune write:—The members of the Vertebrate Section had a very enjoyable time.

Saturday and Sunday were devoted to working the coastline. Herring Gulls were found to be nesting on the whole range of cliffs between Robin Hood's Bay and Whitby, many of the nests being easily accessible from the top. Jackdaws and Starlings were also nesting freely on the cliffs. At Hawsker we found the Cormorants nesting, altogether about forty birds were seen, about half of which were intent on family matters ; apparently about one-third had not assumed adult breeding plumage. A small colony was also seen south of Robin Hood's Bay at the Peak. The return journey from Hawsker was devoted to working the various ravines, ideal places for warblers. The result was extremely disappointing, as they appeared to be almost devoid of bird life ; two nests of Carrion Crows were seen and one of a Magpie. The investigation of the large patches of gorse growing on the hill-sides was also disappointing. Linnets, which we expected to find in great abundance, were not at all plentiful. Although ideal places for the Stonechat, we did not see a single bird of this species, and the Winchat was comparatively scarce. The absence of the Corn Bunting and the Rock Pipit was also very noticeable. The Cuckoo was not seen until our return, when half-a-dozen were encountered together, apparently mating.

Monday was devoted to the investigation of the Fyling Woods, Ramsdale, and the moors above. Wood Warblers, Chiff Chaffs, and Garden Warblers were found to be common and in full song. Green Woodpeckers and Tawny Owls were comparatively abundant, and it was pleasing to hear the peculiar note of the Grasshopper Warbler. Both the Kestrel and Sparrow Hawk were seen, and Carrion Crows and Magpies appeared to have a firm footing in the district. The absence of Pheasants was remarkable, as was also the absence of the Spotted Flycatcher. The district is eminently suitable for both birds, and their apparent scarcity or non-existence was much commented upon. The Blackcap Warbler was both seen and heard, but, in common with most parts of Yorkshire, was not

anything like as abundant as the Garden Warbler. Three
species of Tits were seen, but this family was not as abundant
as might have been expected, and the apparent absence of the
Creeper was remarkable.

Dippers and Grey Wagtails were found frequenting the
stream, and a few Curlews and Red Grouse on the moors. It
was pleasing, also, to notice the Merlin, and disappointing not
to see the Ring Ouzel.

As we were in the vicinity, the opportunity was taken of
paying a hasty visit to Kettleness to see how the colony of
Herring Gulls was faring there. The Marquis of Normanby
has this year given his keepers instructions to prevent the raids
upon the eggs of these birds. This kindly forethought has,
however, not had the effect of putting a stop to the plundering
altogether, as there was abundant evidence that the eggs have
been taken in numbers. The Herring Gull nests in greater
numbers here than upon any other portion of the coast, and it
seems a pity that they should be so persecuted (see Plate XXX.).

Altogether, 61 species of birds were seen, many of them
nesting.

Only seven species of Mammalia were recorded. A Badger's
earth, with fresh footprints of the animals showing, was
inspected in Ramsdale. Apparently the species had formerly
been fairly abundant here. Circumstances caused their practical
extermination, and now they appear to be re-establishing themselves. From notes kindly supplied by Mr. Barry, and read
at the meeting, we learnt that Polecats are still found in the
neighbourhood, but we were not lucky enough to view one.
Moles, by their hills, were apparently much more numerous
on the moors above than on the lower ground. Long-tailed
Field Mice, Common Shrew, and Field Voles were noted.

The only specimens of Reptilea, etc., seen were Viper, and
Frog, and one Fish, the common trout, which did not appear
to be very plentiful.

MOSSES.—Mr. J. J. Marshall gives the following list :—

Orthotrichum pulchellum, on Elder.	*Gymnocybe palustris.*
Bryum capillare, Cloughton.	*Hypum palustre.*
Pogonatum aloides, side of Peat Bog.	*Rhacomitrium aciculare*,
Grimmia pulvinata.	*Hypnum exannulatum.*
Webera nutans.	

All are from Fylingdales woods or moor, except the one
from Cloughton.

R. Fortune.

Herring Gull and Nest at Hawsker.

FUNGI.—Mr. C. Crossland reports that the two Mycologists —C. H. Broadhead and himself—had a good time, which would have been still better could they have lingered longer in Ramsdale Wood, a typical place for their purpose, on the Monday. They were, however, repaid in other ways ; the route laid out by Mr. Barry, who personally guided the party on that day, proved to be full of interest from many points of view. Though we kept moving on from one point of interest to another, fungi were not altogether lost sight of. One desirable *Discomycete—Peziza sepiatra*—was found growing on some petrified moss we went to look at under a great overhanging, dripping rock near the waterfall ; and a spring *Entoloma—E. clypeatum*—on the ground in the wood corner. The very pretty, but not uncommon *Omphalia umbellifera* was met with on heathy ground in one of the glacial slacks visited.

On the two other days the search for fungi was conducted more leisurely. As predicted in the circular, many spring species, which do not appear at any other time of the year, were found. Mr. Fortune and Mr. Booth, while in quest of other game, came across a splendid group of St. George's mushrooms—*Tricholoma gambosum*—in a field near the head-quarters ; these were growing in half a ring four or five yards in diameter : the formation of a complete ring was prevented by the field wall. They were the largest specimens the writer has had the pleasure of seeing, many being over 5 in. across the pileus, with plenty of substance about them, as will be seen by the accompanying photo. Being one of the very best of edible species, they were not left to perish in the field, but gathered by two or three members who knew their real worth and parcel'd up for home consumption. Pastures are their home, but Mr. Fortune saw a few in one of the woods. The uncommon *Agaric —Bolbitius flavidus* (Bolton)—was found on cow dung in pastures on the cliffs in two places. There were about a score of other fungi belonging to various groups of the larger kinds, which are of fairly common occurrence.

Nearly two-thirds of the 74 species found belong to what are understood as micro species, although some which grow in colonies are prominent enough to the naked eye. Ten or eleven days were spent in working these out. There is one species new to Yorkshire—*Triposporium elegans,* a beautiful little brown mould with a *three-legged* arrangement of its conidia, hence its generic name. There are five first records for this (N.E.) division of the county—*B. flavidus,* mentioned above ;

Hypoxylon atropurpureum, on dead alder-branches; *Leptosphæria fuscella,* on dead furze; *Mollisia carduorum,* on dead thistle stems; and *Pseudopeziza rubi,* on dead bramble stems; all in or about Stoup Beck ravine. The last four, besides being new to N.E. Yorks., confirm hitherto solitary county records, and are perhaps as valuable as new ones; so far as records are considered to have any value. *Collybia tenacella,* on dead fir-needles in Fyling Hall shrubbery, and *Pseudopeziza sphæroides,*

Photo by   R. Fortune.

St. George's Mushroom.
Robin Hood's Bay.

on dead stems of rose campion, Stoup Beck, have also, up to the present, been met with but once before in Yorkshire.

The beautiful white *Dasyscypa virginea,* which grows on various decaying twigs and herbaceous stems, was definitely seen on decaying branches of furze, rose, bramble, woodsage, and honeysuckle within the space of a couple of yards, in a furze cover. The tiny *D. fugiens,* which grows on damp, rotting rushes, was seen in plenty ; unless great care be taken to keep the rush stems very moist, this delightful little Disco., $\frac{1}{300}$ to $\frac{1}{250}$ inch across its disc, disappears before it can be examined. Likely places proved most prolific in micro species when carefully investigated.

The rose rust *Phragmidium subcorticatum*—Æcidospore stage—was common on wild rose bushes.

(*To be continued*).

FIELD NOTES.

BIRDS.

Dunlin in Wharfedale.—Whilst watching several common Sandpipers feeding on the shore of a tarn on the moors above Burley, in Wharfedale, I had the pleasure of seeing an adult male Dunlin, in full summer plumage, feeding in the shallow water round the edge of the tarn. The bird at intervals uttered its characteristic love call, which was uttered by the bird whilst standing in the water. Mr. W. H. Hudson speaks of this call as being 'uttered in the air, or as the bird descends to earth with set, motionless wings and expanded tail.' On visiting the tarn two days later I found that it was still there, and I am inclined to think that there may have been a nest somewhere near ; but not being able to spare the time, I could not make a thorough search.—S. HOLE, Leeds, May 30th, 1907.

Quail in East Yorks.—A Quail was picked up on the 4th of June under the telegraph wires at Buckton, East Yorks.—E. W. WADE, Hull.

Note on the Cuckoo.—On May 24th, about 7 a.m., I saw a Cuckoo carrying an egg in her bill. I had her under observation for about an hour, during which time some Meadow Pipits were chasing her away from their nests. No doubt she was trying to deposit her egg in one of them, but the Pipits were successful in driving her away.—W. WILSON, Skipton-in-Craven.

—: o :—

LEPIDOPTERA.

Eupithecia coronata **in Yorkshire.**—On June 7th and 8th, in a large wood on the outskirts of Sheffield, I took three specimens of *E. coronata*. This species has hitherto only been recorded for Yorkshire as having occured at Scarbro' on the authority of Stainton's Manual. Its usual food plant,

Clematis vitalba, does not grow anywhere in the wood, which disposes of any probability of the moth having been introduced by artificial means.—L. S. BRADY, Sheffield, June 11th, 1907.

—: o :—

MOLLUSCA.

Clausilia bidentata m. dextrorsum in Lincs.—On April 18th Mr. Vernon Howard, M.A., visited Haugham Pasture, an old wood on the lower chalk near Louth, and there collected mollusca ; amongst those he kindly submitted to me was a good example of *Clausilia bidentata* m. *dextrorsum.*—C. S. CARTER, Louth.

—: o :—

FLOWERING PLANTS.

The Butterfly Orchis in East Yorkshire.—*Habenaria bifolia*, a plant rarely found in the East Riding, at least on or east of the Wolds, has been discoverd in a fine patch near Tibthorpe, Driffield (see remarks in 'Flora E. R. Yorks.')—(Miss) L. F. PIERCY, Tibthorpe, June 13th, 1907.

—: o :—

MOSSES.

Luminosity of Schistostega osmundacea.—A short time ago the luminosity of this plant was alluded to in the 'Naturalist.' As is well known, the spores of mosses produce a branched filamentous protonema on germination. The cells of this protonema (which contain chloroplasts) are of remarkable form, convex above and conical below. The light which enters these cells is first refracted and then reflected from the sides of the cone in such a way as to go across the conical parts of the cells, where it is again reflected from the sides, and on emerging into the air is again refracted. In this way some of the light which enters the cells leaves them again in the direction of the observer. As this moss always grows in cave-like places, it gets but a moderate amount of light for photosynthesis, and there is no doubt but that the peculiar morphological character of its protonemal cells is an ecological adaptation. I first saw this moss twenty-nine years ago in a cavernous place about three miles west of Buxton, and I shall never forget the strange sight.—WM. WEST, Bradford.

Inglebro' Mosses.—*Hypnum giganteum* Schp. This moss grows in considerable quantity in the rills joining the Fell Beck just above Gaping Gyll. This is a new locality for *H. giganteum*, which is rare in West Yorks.

Some other mosses not previously recorded for Inglebro' are :—

Oligotrichum incurvum Ldb.
Brachyodus trichodes Fürnr.
Blindia acuta B. & S.

Heterocladium heteropterum B. & S.
Hypnum ochraceum Turn.
Hypnum stramineum Dicks.

On the same excursion the Inglebro' records of the following montane mosses were confirmed :—

Dicranodontium longirostre B. & S.,
   var. *alpinum* Schp.
Gygodon gracilis Wils.
Tetraplodon minoides B. & S.
Amblyodon dealbatus P. B.

Bartramia ithyphylla Brid.
Plagiobryum zierii Ldb.
Bryum filiforme Dicks.
Pseudoleskea catenulata B. & S.

CHRIS. A. CHEETHAM, Armley.

—:o :—

FUNGI.

Volvaria parvula.—Mr. C. H. Broadhead, of Wooldale Nurseries, near Thongsbridge, has sent me several specimens of this most interesting little agaric. They were found growing on soil in one of his greenhouses this week.

C. CROSSLAND, Halifax, June 26th, 1907.

—: o :—

MISCELLANEOUS.

Yorkshire Earthquakes.—In an article under the title 'Yorkshire Earthquakes' in the 'Naturalist' for January, 1905, Dr. Charles Davidson intimates that any notices of the effects in Yorkshire of the great Lisbon earthquake of 1755, if any such were observed, would be of considerable interest to him. I have just noticed in a pamphlet entitled 'Notes on Old Peterborough,' by Andrew Percival (Published in 1905 by the Peterborough Archæological Society, p. 40), a quotation from 'an Ipswich paper' 'in 1755,' as follows :—'By a letter from Thirsk, in Yorkshire, we learn that very lately a terrible shock of earthquake was felt, inasmuch that several large rocks were removed to considerable distances ; several large grown elms were swallowed up by the earth so that no part of them remained to be seen but the uppermost branches. A man driving a cart near the place, the horses were so much frightened by the shock that they broke loose from the carriage and ran away. The horses seem to have behaved very sensibly.'—H. E. WROOT, Bradford.

The effects of the weather at Harrogate in 1907.—It is perhaps worth while to make a note of the disastrous results caused by the inclement and changeable weather this spring.

In this district, at the time of writing (the middle of June),
many ash trees are quite bare; Chesnuts have leaved and
flowered for a very short time. The leaves of many have, however, quite shrivelled up, as if by the action of frost. Rose trees
and many others are suffering in the same way. The chesnuts
are now putting forth a lot of new leaves, and look most
peculiar. The various flowering plants have had a very short
show, the blossoms being ruined almost as soon as they
opened.

The reverse is the case with fungi, which appear to be
unusually numerous. The St. George's mushroom is particularly plentiful.

Birds have suffered exceedingly, and one comes across
many deserted nests with eggs or dead young ones. Several
times I have seen a nest full of young birds, ground breeding,
washed out of the nests and drowned. Warbler's nests are
suffering very much, and the other day I even found a Tree
Creeper washed out.

There is a great scarcity of some species of migrants,
notably in Spotted Flycatchers, Whinchats, Grasshopper
Warblers, and Landrails, and also, I think, Cuckoos. The
arrival of migrants this year was most erratic.

Despite this depressing news, it is most gratifying to find
the House Martin nesting in some of the haunts which have
been without them for years.—R. FORTUNE.

'Collecting Lepidoptera in the Lake District in 1902, 1903, and in 1905,
1906' is the title of some notes by Mr. A. H. Foster in the June 'Entomologist.'

The York and District Field Naturalists' Society is issuing elaborate
excursion programmes, giving hints of species likely to be met with on the
rambles. But aren't *all* secretaries 'organising'?

An important paper on 'The Geology of the Zambezi Basin around the
Batoka Gorge (Rhodesia)' is contributed to the May 'Quarterly Journal of
the Geological Society' by Mr. G. W. Lamplugh, F.R.S.

We learn from the 'Quarry' that in recent Pompeiian explorations two
beautiful 'Blue John' vases were found—evidence that the famous 'Blue
John' mines of Castleton were worked some two thousand years ago.

We regret to hear of the death of Mr. C. Mossop, of Barrow-in-Furness,
late goods manager of the Furness Railway. Mr. Mossop took a keen
interest in natural history. He was President of the Barrow Naturalists'
Field Club in 1902, and his portrait appears in the Club's Report for that
year.

The Council of the British Association for the Advancement of Science
has nominated Mr. Francis Darwin, Foreign Secretary to the Royal Society,
and author of the 'Life and Letters of Charles Darwin,' to be President of
the meeting next year, when, for the fourth time in its history, the Association will assemble in Dublin.

LINCOLNSHIRE MITES.

RHYNCHOLOPHIDÆ.—(Continued).

C. F. GEORGE, M.R.C.S.
Kirton-in-Lindsey.

RHYNCHOLOPHUS. The members of this genus, though resembling *Erythræus*, will be seen to differ in many important particulars. They have longer legs, some of them very long ones, especially the hinder pair, and from this fact one of them has been called *longipes;* possibly it may be the one figured; of this, however, I am not sure. The internode above the tarsus of the hind leg is very long, much longer than any other. These mites are very active and good runners, and hence perhaps not easily caught. They are said to be common, but I have not found this to be the case. All the tarsal joints are flattened from side to side, those of the first pair are the largest (see figure 4), and are provided with a sort of hair pad on the under side, which gives them good foothold irrespective of the claws. I have one taken in this neighbourhood in 1877, which is mounted in Balsam for the microscope, and not very well displayed; I had, however, been furnished with a few specimens from Guernsey by Mr. Luff, which appear to be identical, and it is from these specimens, preserved and mounted by me, that Mr. Soar has made the drawings which illustrate this article. The mite is of a red colour, the body rather densely covered with short bristles, thickly pectinated (see figure 6). The palpi are shown much enlarged at figure 3. The fifth joint is pear-shaped and larger in proportion to the fourth than in *Erythræus;* the proboscis is also furnished at the tip with a very curious circlet of hairs, best seen in the living or unmounted mite. The eyes are very different from those of *Erythræus;* instead of a single ocellus on each side, there are two ocelli joined together on a kidney-shaped process (see figure 2), embedded in the skin. Another remarkable organ is the chitinous rod, which lies in the dorsal groove in the centre between the two eyes, and is furnished at the anterior end with a rather large and globular capitulum (figure 2), seen greatly enlarged at figure 5; it is furnished with several longish hairs, or spikes, which are pectinated, and also two stigmatic openings, each having a very fine and rather long tactile simple hair. The posterior end of the rod is also slightly enlarged, and likewise bears two stigmatic openings, with their accompanying tactile hairs—well shown in the figure. The mite figured is a female. The position of the vulva is evident, a point of some importance.

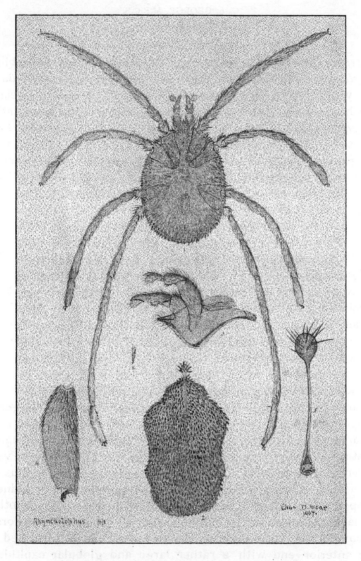

Fig. 1. Female *Rhyncholophus*, ventral surface.
Fig. 2.   ,,   ,,   dorsal surface.
Fig. 3. Palpi and rostrum, much magnified.
Fig. 4. Anterior tarsal joint.
Fig. 5. Chitinous rod with capitulum.
Fig. 6. Single pectinated hair from the body.

The Ægir, Gainsborough (Oct. 12th, 1904).
Showing the after-waves, locally called 'The Whelps.'

REVIEWS AND BOOK NOTICES.

·THE SHAPING OF LINDSEY BY THE TRENT.' *

In the 'Naturalist' for 1895, Mr. F. M. Burton contributed a paper on 'The Story of Lincoln Gap.' A year previously his presidential address to the Lincolnshire Naturalists' Union was entitled, 'How the land between Gainsborough and Lincoln was formed.' The fascinating line of research then formulated by Mr. Burton has since been followed up, and in his present attractive volume is given an exceedingly readable and clear account of the origin of the Lindsey landscape. Proper stress is laid upon the important part played by rivers in the formation of the present surface features of the area, and by the aid of diagrams and several excellent plates Mr. Burton indicates the old courses of some of the rivers, explains how and why these were changed, and deals in an interesting manner with the work being done by the present streams. Mr. Burton's book should do much to popularise the study of geology in north Lincolnshire, and this we know, from personal experience, is what the author wishes. Being produced at Browns' Savile Press, the 'get up' of the volume is all that can be desired. We are enabled to give our readers a specimen of the plates (see Plate XXXI.).

The Report and Proceedings of the Manchester Field Naturalists' and Archæologists' Society for the year 1906 is to hand. It contains a hundred pages, principally devoted to reports of excursions. The longest article is on the Society's excursion to the Winchester district. It is a pity the millinery, etc., advertisements are not kept on separate sheets of paper. They could then be torn out without the volume suffering.

We have received Part I. of **'The Book of the Open Air'** (1/- net, to be completed in 12 parts, Hodder and Stoughton), a book which promises well. It is edited by Mr. Edward Thomas, and is produced in a way which warrants every praise. The first instalment contains (1) An open-air diary for April, and 'Introduction,' by the Editor; 'In praise of rain,' by W. W. Fowler; 'The Otter's Holt,' by A. W. Rees; 'The Flowers of Early Spring,' by Rev. Canon Vaughan; 'Some English Butterflies,' by A. Collett; 'Birds as Architects,' by D'Esterre Bailey; 'The Venus Eve,' by G. A. B. Dewar; 'Ancient Ponds,' by W. Johnson; and 'The Making of Scenery,' by E. Clodd. These articles are mostly charmingly written—the last named being a perfect poem. Adding much to the attractions of the publication are some drawings and photographs, well reproduced by the three-colour process. These are mounted on thick tinted paper, and there will be fifty of them in the complete volume. If the remainder keep to the standard of those already issued, they will alone be worth the cost of the book. It would be ungrateful to find the smallest fault with this beautiful publication.

---

* By F. M. Burton, F.G.S., A. Brown & Sons, London and Hull. 60 pp. and plates. Price 2/6.

Mr. J. Donkin, F.R.I.B.A.. favours us with a copy of his **'Conservancy or Dry Sanitation** *versus* **Water Carriage'** (E. & F. Spon, London. Price 1/-). This is clearly written and illustrated by diagrams. It should be read carefully by all interested in sanitation.

Under the title of **'Douglas English Nature Books,'** Messrs. Bousfield & Co. have issued two admirable shilling books. The first contains 100 illustrations from photographs from life of the Shrew Mouse, the Dormouse, the House Mouse, the Field Mouse, the Meadow Mouse, and the Harvest Mouse. A fair sample of these excellent photographs is reproduced herewith by permission of the publishers. But of even more interest

The Dormouse.

and value are the 'Notes on some smaller British Mammals.' by Mr. English —which occupy 30 pages. No. 2 of the series contains a similar number of beautiful illustrations of bird life, and thirty pages of letterpress by Mr. R. B. Lodge. Both books should do much to popularise natural history, as well as serve as handy guides to amateur photographers.

Forty Years in a Moorland Parish. Reminiscences and Researches. in Danby in Cleveland, by Rev. J. C. Atkinson. Macmillan & Co., Ltd., 1907, pp. xlv. + 472, price 5/- net. We feel sure that it will be unnecessary to call the attention of readers of the 'Naturalist' to the nature of this. charming volume, one of the most popular of those written by the late Canon Atkinson, who has deservedly been styled the Yorkshire Gilbert White. We desire, however, to congratulate the publishers on the appearance for the first time of an excellent memoir of the author, written by Mr. G. A. Macmillan; together with two portraits, which are beautifully reproduced. To describe the contents of 'Forty Years in a Moorland Parish' would be almost an insult to the readers of this journal. Should there be any, however, who have not read the volume, let us recommend them to do so at once. And those who have already seen it will be glad to have the present edition from the additional matter and illustrations which it contains.

Birds' Eggs of the British Isles.—Messrs. Brumby & Clarke, the well-known colour printers, Hull, have drawn our attention to the volume they issued some little time ago under the above title. This contains the 24 beautiful plates of birds' eggs which appeared in the larger work, 'British Birds with their Nests and Eggs;' and in their present form will appeal to those who are more particularly interested in eggs than birds, or who do not care to pay the price of the larger work for the sake of the illustrations of the eggs. Mr. A. G. Butler has written a hundred quarto pages of descriptive letterpress, and the drawings are by Mr. F. W. Frohawk, whose work is so well known. The figures will particularly appeal to the collector from the wonderful varieties of eggs shown—some of these being most unusual; the eggs of the black-bird, for instance, being represented by eight extraordinary varieties. There are in all 472 illustrations, and the price of a guinea is not out of the way. In his introduction Mr. Butler gives some useful hints on the preparation of specimens for the cabinet.

Wild Life at Home: How to Study and Photograph it, by R. Kearton, F.Z.S. Cassell & Co., 204 pp. New and revised edition, 1907. Price 6/- A further edition of this well illustrated volume is evidence of its popularity.

Gulls Feeding.—A scramble for breakfast.
(From 'Wild Life at Home.')

The work of the brothers Kearton is well known to our readers. The present edition has been revised, and the illustrations, type, paper, and binding are uniformly excellent. We are permitted to reproduce one of the illustrations herewith.

Birds and their Nests and Eggs, by G H. Vos. London, G. Routledge & Sons, Ltd., 148 pp. and numerous illustrations. Price 1/.

This well printed and cloth bound volume is the cheapest of its kind that we have seen for some time, and will be very suitable for a prize book for young naturalists. The photographs are mostly good, though the method of sticking *badly stuffed* birds in their more or less natural surroundings and photographing them is not very desirable, especially as in so many cases any schoolboy can 'spot' the artificial twig upon which the bird is perched; on p. 41 the starling, which 'is a well got up and groomed gentleman at all times of the year,' certainly belies this description, though possibly the specimen figured has suffered through being carried in the photographer's pocket! The photographs of eggs at the end of the volume are of little value except for showing the relative sizes of the eggs.

NORTHERN NEWS.

Amongst the birthday honours we notice that Prof. E. Ray Lankester is created K.C.B.

Mr. E. E. Lowe, of the Plymouth Museum, has been appointed Curator of the Leicester Museum in place of Mr. Montague Browne, resigned.

No. 4 of the Bankfield Museum Notes (Halifax), deals with the 'Egyptian Tablets' in the collection, and has been written by Mr. Thomas Midgley, of the Bolton Museums. The pamphlet is sold for one penny.

An admirable case of representations of British Butterflies has been prepared by the 'Young Citizen,' of 12, Salisbury Square, London, E.C. This contains beautifully coloured copies of twenty-four of the most striking of our British Butterflies, which are cut out in cardboard and fastened in the case by pins, each species having a printed label. At the low price of half-a-crown, it is most suitable for schools, etc.

We much regret to record the death of Prof. A. Newton, F.R.S., whose 'Dictionary of Birds' is so well known. For half-a-century Prof. Newton has taken a leading position in ornithological science, and to him the present generation is largely indebted for his herculean efforts in the matter of the protection of birds, upon which subject he read a paper to the British Association so long ago as 1868. He was born in 1829.

Dr. C. F. George of Kirton Lindsey sends us a reprint of a useful paper, 'Hints on Collecting and Preserving Fresh-water Mites.' This is from the 'Hastings and East Sussex Naturalist.'

The following particulars of sales during 1906 of plumes, for the adornment of the (un)fair sex, seem incredible, but are guaranteed accurate :—

	Osprey Feathers. Packages.	Birds of Paradise.
February	327	8,508
April	260	7,188
June	289	11,841
August	242	3,948
October	485	5,700
December	265	3,600
	1868	40,785

[The printer refuses to set up our comments, which were, perhaps, rather too forcible, though expressing our feelings at the time—ED.].

Mr. E. K. Robinson, B.E.N.A., informs the readers of 'The Country Side' that he is 5 ft. 9 in. in height, weighs 13 stone, and is short and stout. He has not 'the same elegant sort of figure as the ex-Prime Minister, Mr. Balfour.' In view of this, possibly the readers of the 'Naturalist' may care to know that the beard of the writer of these notes is not at all like that of the King of the Belgians!

Speaking of Mr. E. K. Robinson reminds us that some little time ago he asked the opinion of the readers of 'The Country Side' on his 'proposed innovation' in substituting the word 'animals' for 'mammals.' We have watched the paper since, and apparently either Mr. Robinson's request has been ignored by his readers, or, what seems more likely, the opinions of his readers have been ignored by Mr. Robinson. We ventured to express our views on the matter, but presumably both the copies of this journal which were sent to him have miscarried. Or is it that in snatching this 'chestnut' from his well-stocked grate Mr. Robinson has burnt his fingers, and dropped it amongst the ashes, which must now be rather rapidly accumulating.

AUGUST 1907.

No. 607
(No. 385 of current series).

THE NATURALIST.

A MONTHLY ILLUSTRATED JOURNAL OF
NATURAL HISTORY FOR THE NORTH OF ENGLAND

EDITED BY

T. SHEPPARD, F.G.S.,
THE MUSEUM, HULL;
AND

T. W. WOODHEAD, Ph.D., F.L.S.,
TECHNICAL COLLEGE, HUDDERSFIELD.

WITH THE ASSISTANCE AS REFEREES IN SPECIAL DEPARTMENTS OF

J. GILBERT BAKER, F.R.S. F.L.S.,   GEO. T. PORRITT, F.L.S., F.E.S.,
Prof. P. F. KENDALL, M.Sc., F.G.S.,   JOHN W. TAYLOR,
T. H. NELSON, M.B.O.U.,   WILLIAM WEST, F.L.S.

Contents :—

LONDON:

A. BROWN & SONS, LIMITED, 5, FARRINGDON AVENUE, E.C.
And at HULL AND YORK.

Printers and Publishers to the Y.N.U.

PRICE 6d. NET. BY POST 7d. NET.

BOTANICAL TRANSACTIONS OF THE YORKSHIRE NATURALISTS' UNION, Volume I.

8vo, Cloth, 292 pp. (a few copies only left), price 5/- net.

ontains various reports, papers, and addresses on the Flowering Plants, Mosses, and Fungi of the county

Complete, 8vo, Cloth, with Coloured Map, published at **One Guinea.** *Only a few copies left,* **10/6** *net.*

THE FLORA OF WEST YORKSHIRE. By FREDERIC ARNOLD LEES, M.R.C.S., &c:

This, which forms the 2nd volume of the Botanical Series of the Transactions, is perhaps the most omplete work of the kind ever issued for any district, including detailed and full records of 1044 Phanero. ams and Vascular Cryptogams, 11 Characeæ, 348 Mosses, 108 Hepatics, 258 Lichens, 1009 Fungi, and 392 reshwater Algæ, making a total of 3160 species.

680 pp., Coloured Geological, Lithological, &c. Maps, suitably Bound in Cloth. Price **15/-** *net.*

NORTH YORKSHIRE: Studies of its Botany, Geology, Climate, and Physical Geography.

By JOHN GILBERT BAKER, F.R.S., F.L.S., M.R.I.A., V.M.H.

And a Chapter on the Mosses and Hepatics of the Riding, by MATTHEW B. SLATER, F.L.S. This Volume orms the 3rd of the Botanical Series.

396 pp., Complete, 8vo., Cloth. Price **10/6** *net.*

HE FUNGUS FLORA OF YORKSHIRE. By G. MASSEE, F.L.S., F.R.H.S., & C. CROSSLAND, F.L.S.

This is the 4th Volume of the Botanical Series of the Transactions, and contains a complete annotated list f all the known Fungi of the county, comprising 2626 species.

Complete, 8vo, Cloth. Price **6/-** *post free.*

HE ALGA-FLORA OF YORKSHIRE. By W. WEST, F.L.S., & GEO. S. WEST, B.A., A.R.C.S., F.L.S.

This work, which forms the 5th Volume of the Botanical Series of the Transactions, enumerates 1044 pecies, with full details of localities and numerous critical remarks on their affinities and distribution.

Complete, 8vo, Cloth.   Second Edition.   Price **6/6** *net.*

LIST OF YORKSHIRE LEPIDOPTERA. By G. T. PORRITT, F.L.S., F.E.S.

The First Edition of this work was published in 1883, and contained particulars of 1340 species of acro- and Micro-Lepidoptera known to inhabit the county of York. The Second Edition, with Supplement, ontains much new information which has been accumulated by the author, including over 50 additional pecies, together with copious notes on variation (particularly melanism), &c.

In progress, issued in Annual Parts, 8vo.

TRANSACTIONS OF THE YORKSHIRE NATURALISTS' UNION.

The Transactions include papers in all departments of the Yorkshire Fauna and Flora, and are issued in eparately-paged series, devoted each to a special subject. The Parts already published are sold to the public s follows (Members are entitled to 25 per cent. discount): Part 1 (1877), 2/3 ; 2 (1878), 1/9 ; 3 (1878), 1/6 ; 4 (1879), /- ; 5 (1880), 2/- ; 6 (1881), 2/- ; 7 (1882), 2/6 ; 8 (1883), 2/6 ; 9 (1884), 2/9 ; 10 (1885), 1/6 ; 11 (1885), 2/6 ; 12 (1886), 2/6 ; 3 (1887), 2/6 ; 14 (1888), 1/9 ; 15 (1889), 2/6 ; 16 (1890), 2/6 ; 17 (1891), 2/6 ; 18 (1892), 1/9 ; 19 (1893), 9d. ; 20 (1894), 5/- ; 1 (1895), 1/- ; 22 (1896), 1/3 ; 23 (1897), 1/3 ; 24 (1898), 1/- ; 25 (1899), 1/9 ; 26 (1900), 5/- ; 27 (1901), 2/- ; 28 (1902), 1/3 ; 9 (1902), 1/- ; 30 (1903), 2/6 ; 31 (1904), 1/- ; 32 (1905), 7/6 ; 33 (1906), 5/-.

HE BIRDS OF YORKSHIRE. By T. H. NELSON, M.B.O.U., WILLIAM EAGLE CLARKE, F.L.S.,

M.B.O.U., and F. BOYES. 2 Vols., Demy 8vo 25/- net. ; Demy 4to 42/- net.

nnotated List of the LAND and FRESHWATER MOLLUSCA KNOWN TO INHABIT YORK-

SHIRE.   By JOHN W. TAYLOR, F.L.S., and others.   Also in course of publication in the Transactions.

HE YORKSHIRE CARBONIFEROUS FLORA. By ROBERT KIDSTON, F.R.S.E., F.G.S. Parts 14,

18, 19, 21,.&c., of Transactions.

IST OF YORKSHIRE COLEOPTERA. By Rev. W. C. HEY, M.A.

HE NATURALIST. A Monthly Illustrated Journal of Natural History for the North of England. Edited

by T. SHEPPARD, F.G.S., Museum, Hull; and T. W. WOODHEAD, F.L.S., Technical College,
Huddersfield; with the assistance as referees in Special Departments of J. GILBERT BAKER, F.R.S.,
F.L.S., PROF. PERCY F. KENDALL, M.Sc., F.G.S., T. H. NELSON, M.B.O.U., GEO. T. PORRITT,
F.L.S., F.E.S., JOHN W. TAYLOR, and WILLIAM WEST, F.L.S.   (Annual Subscription, payable
in advance, **6/6** post free).

EMBERSHIP in the Yorkshire Naturalists' Union, **10/6** per annum, includes subscription to *The Naturalist*,
and entitles the member to receive the current Transactions, and all the other privileges of the Union.
A donation of **Seven Guineas** constitutes a life-membership, and entitles the member to a set of the
completed volumes issued by the Union.

Members are entitled to buy all back numbers and other publications of the Union at a **discount of 25
per cent.** off the prices quoted above.

All communications should be addressed to the Hon. Secretary,

T. SHEPPARD, F.G.S., The Museum, Hull.

ANTED: Rennie's Field Naturalist (Set). Proc. Geol. and Polyt. Soc., West Riding Yorks., Vol. I. British
Association Report, 1839-1840. Quarterly Journal Geol. Soc., Vols. I-XXI. Barnsley Nat. Socy's,
Quarterly Reports (Set). Good prices given.

Apply T. SHEPPARD, F.G.S.,

The Museum, Hull

NOTES AND COMMENTS.

THE YORK PHILOSOPHICAL SOCIETY.

THE Annual Report of the Yorkshire Philosophical Society for 1906 is of particular interest. From it we learn that an attempt has been made to grow a classified series of British plants in the Garden, and though the scheme was not very successful, it is hoped to carry it through in the future. Mr. Oxley Grabham has again given some valuable objects to the museum, including some 'unique Guillemots' eggs, respecting which we should have liked more details. The society is justly proud of the part it played in connection with the 75th meeting of the British Association, and prints 'probably the only permanant local record of this noteworthy event' in its report. Mr. H. J. Wilkinson gives a very useful Historical Account of the Society's Herbarium and the contributors thereto. This paper contains lengthy notices of James Dalton, Robert Teesdale, Christopher Machell, Samuel Goodenough, Wm. Jackson Hooker, Joseph Dalton Hooker, William Bingley, Samuel Hailstone, Richard Spruce, Henry Ibbotson, Henry Baines, and James Blackhouse—all honoured names in the botanical world. Mr. George Benson has an excellent illustrated paper on 'Some Relics of the Viking Period recently found in York,' Mr. J. E. Clarke deals with 'The Windrush at Biggin,' and there are some useful meteorological tables. The report certainly gives evidence of scientific activity at York.

A LAND-SHELL NEW TO THE FAUNA OF THE BRITISH ISLES.

MESSRS. John W. Taylor and W. Denison Roebuck, of Leeds, have been spending a week with the Irish Conchologists, on the occasion of the Cork Conference of the Irish Field Club Union. The molluscan work was very considerable, for Messrs. Robert Welch, A. W. Stelfox, and J. N. Milne, of Belfast, Mr. R. A. Phillips, of Cork, and Mr. Robert Standen, of the Manchester Museum, were all in the field—and the districts round Cork, as at Youghal, Macroom, Kinsale, Aghada, Blarney, etc., were very closely and carefully investigated. At the Conference Mr. Taylor read a carefully worked-out paper, in which he, for the first time, made public the addition of a fine well-marked and conspicuously distinct species of land mollusc to the Irish and British list. This was *Vitrina elongata*, Dp., found in fair numbers in Co. Louth by Mr. P. H. Grierson, and detected by Mr. Taylor on examining a large series of Irish shells sent for authentication by that gentleman. The species is of interest as occurring at nearly sea-level in Ireland, the nearest European localities being at elevations of 8000 feet in the Alpine and mountainous regions of the Continent. This occurrence exemplifies well the soundness of Mr. Taylor's views in regard to the position and discontinuous distribution of what may, for want of a better word, be described as 'weak'— *i.e.*, not dominant—forms of life. Mr. Taylor's paper is to appear in the next number of the 'Irish Naturalist,' with illustrations, and the problems involved are discussed fully.

The Microscope and how to use it. London, R. Sutton. 160 pp., 3/- net. This is the third edition of Mr. T. Charters White's well known handbook for beginners. It has been revised and enlarged, contains a chapter on 'Marine Aquarium,' and another, by Dr. M. Amsler, dealing with 'Staining Bacteria.' The book is thoroughly practical, and can be recommended. It is illustrated.

Beasts Shown to the Children, by P. J. Billinghurst, described by Lena Dalkeith, 103 pp. T. C. and E. C. Jack, Edinburgh, 2/6. This is one of the series already referred to in these columns, though the coloured plates in the present volume are not nearly so successful as those dealing with birds and plants. The drawing of the 'whale' in cotton-wool water might be a design for a frieze, and the sea gull's nest facing p. 84 is the nicest thing in sea-weed frills that we have seen for some time.

Introduction to Plant Ecology, by Prof. G. Henslow, M.A., F.L.S., &c., London. E. Stanford. pp. viii. and 130. 2/6.

Prof. Henslow opens this well printed little book by some rather unfortunate remarks. He points out that he wrote two books, one in 1888, and another in 1895, on Ecology without knowing it, 'for this term had not been invented and is still unknown to our teachers of botany in schools.' If we grant the first, it should be noted with reference to the second statement, that Haekel used the term in 1866; again the numerous books and papers since written, mostly by teachers of botany, go to disprove the third assertion. That it might, with advantage, be more widely known is admitted, and it is the object of this book to supply the deficiency. The first five chapters are devoted to a general consideration of ecological questions, in which Prof. Henslow criticises, unnecessarily we think, certain well-known books and teachers. These chapters also contain much that might well be left to the usual text books. If these features had been omitted, more space would have been available for fuller treatment of plant associations and their ecological characteristics, the chapters dealing with this part of the work being based on the well-known Scotch and Yorkshire papers. Too meagre are the accounts of a heather moor, water and marsh plants, the distinctive types of woodlands and their associations, the significance of zones of cultivation and the like. The chapter on plant surveying is too general, and reference should have been made to the more exact and detailed methods already published, and definite examples given. Too often exotic examples are quoted when equally good illustrations could be given from British species, while the chapter on floral ecology would have been improved by the inclusion of results obtained by Willis, Burkill, and others. The chapters on 'Natural Selection and Evolution' are characteristic of the author who shows himself an eager exponent of ecology and much of what he has to say is interesting.

PROMINENT YORKSHIRE WORKERS.

III.—REV. E. MAULE COLE, M.A., F.G.S.

(PLATE XXXII.)

FEW people have done more to popularise the study of the geology and antiquities of Yorkshire than has the subject of this sketch. It can truly be stated that the good he has done in this direction by far exceeds the usefulness of his numerous published works, valuable though the latter undoubtedly are. By writing to various journals, by lecturing in about every town in the county, but more particularly by conducting excursions over his beloved Yorkshire Wolds, Mr. Cole has undoubtedly done much more in furtherance of the study of archæology and natural history than can be estimated. One of the greatest treats that the writer knows is to be taken along the wolds and dales under the leadership of the Vicar of Wetwang, who, in spite of his more than three score years and ten, can still 'cover the ground' as well as most. Every chalk quarry, field, hill, and dale—nay, even every mound and earthwork has a history, and for each Mr. Cole has a fascinating story. On the Yorkshire coast, too (particularly on Flamborough Head), and along the Roman wall, he has conducted scores of parties, each member of which has benefitted by his store of knowledge and by his fund of ready wit. The Yorkshire Naturalists' Union has been much indebted to him in this way, and several of its members owe to him their first lesson in natural science.

The Rev. E. Maule Cole [*] was born at Dover in 1833, and was educated at Brighton, Tonbridge, and Rossall. At the last place he became captain of the school. In 1853 he went to Oxford, where he won the Goldsmith and Ludwell exhibitions, and took honours in classics. Whilst at Oxford he was in two college 'elevens' and 'eights.' In 1857 he was the first old boy to go back to Rossall as master.

When he came to Wetwang in 1865 he had an excellent opportunity of following up his interest in geology. Coming into contact with the brothers Mortimer, he accompanied them in their barrow-opening expeditions, and frequently described the pre-historic remains which were secured in his presence.

---

[*] For a portrait and account of his father, the Rev. Wm. Sibthorpe Cole, see Speight's 'Lower Wharfedale,' 1902, pp. 86-88.

Of the Yorkshire Naturalists' Union, the Yorkshire Geological Society, and the East Riding Antiquarian Society, Mr. Cole has been a member of many years' standing, and to their publications has contributed several papers. The first of sixteen papers printed by the Yorkshire Geological Society was on 'The Red Chalk of Yorkshire,' and was published so long ago as 1878. His first paper in the 'Naturalist' was on 'A peat deposit at Filey' (1891).

In more recent years Mr. Cole has taken a keen interest in the British earthworks and Roman roads of the eastern part of our county, and from his intimate knowledge of every part of the area, few are better able to speak on these subjects than he is.

That he may have health and strength to continue his work for many years to come is the sincere wish of every Yorkshire Naturalist.

LIST OF PAPERS, ETC., BY THE REV E. MAULE COLE.

Noah's Flood.   1865.   16 pp.
Scandinavian Place Names in East Riding of Yorkshire.   1879.
          35 pp.
Geological Rambles in Yorkshire.   1883.   112 pp.
Geology of the Hull and Barnsley Railway.   1886.   55 pp.
Modern Science and Revelation.   1889.   12 pp.
Papers in the Proceedings Yorkshire Geol. and Pol. Soc.—
      The Red Chalk.   1878.
      The Origin and Formation of the Wold Dales.   1879.
      The White Chalk of Yorkshire.   1882.
      Sections at Cave and Drewton.   1885.
      The Physical Geography and Geology of the East
            Riding.   1886.
      The Parallel Roads of Glen Gloy.   1886.
      Dry Valleys in the Chalk.   1887.
      Geology of Driffield and Market Weighton Railway.   1890.
      Ancient Entrenchments near Wetwang.   1891.
      A Lake Dwelling at Preston, Lancashire.   1891.
      Duggleby Howe.   1890.
      The Boulder Clay Cliffs of Filey.   1894.
      The Roman Wall and Vallum.   1898.
      The Danes Graves, Driffield.   1898.
      Distribution of Moors near York.   1899.
      The Site of the Battle of Brunanburh.   1899.

Papers in the ' Antiquary.'—
>   The Entrenchments on the Wolds. 1890.
>   Archæology in Provincial Museums : Driffield. 1891.
>   British and Roman Roads in East Riding. 1892.
>   Huggate Dikes. 1894.
>   A Pictish Burgh near Lerwick. 1895.
>   The opening of a Tumulus at Sledmere. 1897.
>   Norman Features in Wold Churches. 1901.
>   Tumuli on the Wolds. 1903.

Paper in ' Old Yorkshire.'—
>   Ancient Farm House at Fimber.

Paper in the Transactions Hull Sci. and Field Nat. Club.—
>   Waterspouts on the Wolds. 1901.

Papers in the Transactions East Riding Antiq. Soc.—
>   Danes Dike, Flamborough. 1893.
>   Huggate Dikes. 1894.
>   Notices of Wetwang. 1894.
>   Driffield Moot Hill. 1895.
>   Ancient Crosses on the Wolds. 1896.
>   Notes on Field Names. 1898.
>   Roman Roads in the East Riding. 1899.
>   Duggleby Howe. 1901.
>   Ancient Fonts on the Wolds. 1903.

Papers in the ' Naturalist.'—
>   A Peat Deposit at Filey. 1891.
>   Report on the Erosion of the Yorkshire Coast. 1892.
>   In Memoriam—Robert Mortimer. 1892.

Papers in ' Flamborough Village and Headland.' 1894—
>   Antiquities of Flamborough. 7 pp.
>   Geology of Flamborough. 6 pp.

Geology of Scarborough and East Coast of Yorkshire. 4 pp.
>   (in Marshall's Guide to Scarborough.)

Saga Book of the Viking Club. Vol. IV., pt. 1.
>   ' On the Place Name Wetwang.' 1905.

T. S.

————◆◆————

The Insect Hunter's Companion, by the Rev. J. Greene, being instructions for Collecting and Preserving Butterflies, Moths, Beetles, Bees, Flies, etc. West, Newman & Co. 120 pp. Fifth edition, 1907. Price 1/6. The fact that this little handbook has reached its fifth edition speaks for itself. Though out of date in many respects, it is a handy guide for the beginner, and also contains many hints of use to the experienced collector.

THE BIRDS, ETC., OF WALNEY ISLAND.

HARRY B. BOOTH, M.B.O.U.

THIS season the members of the Bradford Natural History and Microscopical Society arranged an Excursion further afield than usual, and by taking advantage of a day trip to Barrow-in-Furness, visited Walney Island on June 8th.

At Grange, we noticed the great increase of Shelducks, many of which could be seen on the shore and on the sea as we stood on the station platform. A pair of Blue Titmice was also seen industriously feeding a nestful of young ones down the centre of an iron gas-lamp in one of the streets of Grange.

Arriving at Barrow, a short journey on a light railway brought us to Piel, where we briefly inspected the well-equipped Marine Laboratory of the Lancashire and Cheshire Fisheries Board, under the superintendence of Mr. A. Scott. From here a short journey of two or three miles in a fishing boat enabled us to land on the south end of Walney, near to the Salt Works, where we were soon amongst the hordes of nesting birds. The watcher (a new hand who knew very little about the different species) had estimated the number of breeding birds to be between forty and fifty thousand.

The Black-headed Gull was found to be by far the most numerous species, comprising quite two-thirds of the total numbers present. Many acres of grassy sand dunes were almost covered by their nests. There was an extraordinary diversity in the ground-colour and markings of their eggs, which were noted in all stages of incubation, several just being chipped by the young chicks inside. Young birds just hatched and still damp were seen, and through every stage up to about a fortnight old, the larger ones running away in small flocks from the intruders. Many dead young birds were noticed, and no doubt the long-continued cold and wet weather had greatly contributed towards this excessive mortality. Numerous egg-shells were strewn about, chiefly the work of a small flock of Jackdaws and a few Lesser Black-backed Gulls; these discreetly sat upon the sand near to the sea during our visit, and no doubt were almost as much annoyed by our presence as were the parent gulls, who gave indications of their anger by constant harsh cries.

The Common Tern.—This appeared to be the most numerous of the Terns present, and were nesting in one large and several smaller colonies amidst the Marram Grass and within the gullery, but away from the shore and shingle. They had evidently only just commenced to lay, as most of the nests contained only one or two eggs—three being seen in only one nest.

The Artic Tern.—Not so numerous as the preceding species, and for the most part breeding separately from them. The largest colony was on the shingle, and one or two smaller ones were on the outskirts of the gullery. As with the Common Terns they had mostly only laid one or two eggs. Several of the birds of this species were bringing in small fish for their mates.

The Lesser Tern.—We were delighted to find so many pairs of this graceful little bird present. Quite twenty nests—or rather depressions in the shingle—were seen along a stretch of half a mile along the shore. Many of these contained three eggs, and some appeared to be incubated, so that this bird would seem to breed slightly earlier than the other terns—at least during this cold and wet season.

The Sandwich Tern.—We had always understood that there was a numerous colony of Sandwich Terns on Walney, but we were disappointed. We sought carefully for them, and it was not until our last quarter of an hour that we came across a single pair and discovered the nest (with a single beautiful egg) amongst the gulls, and not far from the shore. It is possible that others occurred in the vicinity, and had not commenced nesting, but we did not see them. It would be interesting to learn from some reader of the 'Naturalist' if the Sandwich Tern has really decreased to this alarming extent on Walney Island.

The Roseate Tern (?)—I noticed two birds which from the first I felt sure belonged to this species, both by their shape and flight, but they persistently kept at a much higher altitude than the other terns. Although I hid myself as much as possible under the wet and scanty cover, they disdained to descend like the other species. I fancied that they were not yet nesting, although they had evidently chosen a site from their continually flying around one spot—not far from the shingle. When I passed the same place again just before leaving, there was this characteristic pair of birds over the same spot, considerably above the other terns, and of course the chance of hearing their distinctive 'crake' was quite out of the question at such a distance, and amidst such a babel of noises. Possibly some future visitor will keep a look out for this species.

The Shelduck.—This conspicuous and distinctive species was very numerous, and no doubt many females were sitting in the rabbits' burrows in the sand dunes. Several stray birds were noted on our approach, which leisurely betook themselves to the edge of the water during our visit.

The Ringed Plover.—Several pairs frequented the shingle, chiefly on that portion occupied by the Lesser Terns. Their characteristic vagaries of nesting were noticed. Birds quite a week old, and also incompleted clutches of fresh eggs, were seen.

The Oystercatcher.—These birds were exceedingly numerous, more so than in any other locality I have visited during the breeding season. They had evidently only just commenced to nest, as most of their chosen nesting places only contained one or two eggs. At one place three lots of eggs were noted in the space of a few square yards.

The Stock-Dove.—There were several stray Stock-Doves, and in one case four birds were seen flying together. They evidently were nesting in the disused rabbit burrows.

During our short stay upon the island, our time was so much occupied with the maritime species that we had but little opportunity of noticing the passerine birds. The great number of skylarks present, however, was very noticeable, and once when the sun broke through the damp atmosphere, the air was full of their song. In spite of the exceptionally wet season, a nest containing three eggs was seen in a hole quite six inches below the surface. Many pairs of Wheatears and Meadow Pipits were seen, and also, at least, one pair of Rock Pipits.

Mr. J. Beanland writes :—"The visit to Walney being a purely ornithological one, no systematic effort was made to ascertain what plants were growing, and those noted were casually dotted down by the botanists as interesting from the point of contrast that one naturally makes from an inland district with the maritime flora. The following list was made as the plants were seen in walking to and from the gullery :—

Glaucium flavum.	*Draba verna.*
Plantago coronopus.	*Erodium cicutarium.*
Silene maritima.	*Crambe maritima.*
Cochlearia danica.	*Arenaria peploides.*
Viola sylvatica.	*Mertensia maritima.*

Eryngium maritimum.	*Scripus pauciflorus.*
Gentiana campestris.	*Ænanthe crocata ?*
Hyoscyamus niger.	*Trifolium dubium,* and everywhere
Botrychium lunaria.	the Marram Grass.

It is rather interesting to find that in the Herbarium of Mr. F. A. Lees at the Cartwright Hall, *Crambe maritima* and *Mertensia martima* are represented from the locality we found them still growing in, the former being gathered by Mr. Lees in August 1870, and the latter in August 1875."

Our special thanks are due to Messrs. W. L. Page, A. Hawridge, and W. Sargeant (of the Barrow Naturalists' Field Club) for the splendid arrangements which they had made for us; and for their guidance on the island. Our thanks are also due to Ed. Wadham, Esq., J.P., the agent of the Duke of Buccleugh, for kindly granting a special permit to visit the gullery. As the Duke of Buccleugh employs a watcher to protect the birds, it is useless trying to see them without having obtained permission first.

———◆◆———

FISHES.

Fox-Shark at Whitby.—On Friday last, the 12th inst., a fine specimen of the Fox-Shark, or Thrasher, *Alopecias vulpes*, was captured (in the salmon nets of John Hall, Fisherman), on the Skate Heads within the Whitby Rock-buoy. After a heavy struggle, and with the assistance of other cobles, it was with considerable difficulty eventually got into the coble, which the crew quickly vacated in consequence of the shark's severe struggles, snapping with its jaws, trying to bite the men and also to strike them with its long and powerful tail. It was at last killed and brought into Whitby and exhibited. When first brought to shore it measured in length from the snout to the end of its tail 15 feet; I measured it towards night, and found it to be but 14 feet 4 inches long, the pectoral fins measuring about 24 inches each in length, and on the morning of Saturday, the 13th, it only measured 14 feet, the shrinkage in length in 24 hours being exactly one foot.

In August, 1898, a shark of the same species was wounded by the ironwork of the wrecked steamer, "Glentilt," and captured on the rocks at Kettleness near Whitby. It measured 14 feet 6½ inches. — THOS. STEPHENSON, Whitby, July 15th, 1907.

GLACIAL SURVIVALS.*

FRANK ELGEE,
Middlesbrough.

THE question of Glacial Survivals is one of interest to geologists, zoologists, and botanists alike, because of its relation to the Ice Age which their investigations have established, and on account of the light it throws on the distribution of animals and plants.

It may at the outset, however, be advisable to define precisely what is meant by 'Glacial Survivals.' In pre-glacial times our country was inhabited by animals and plants similar to those of to-day, with the addition, however, of species now extinct or living in other lands. With the approach of the glaciers of the Ice Age these would either be driven out, exterminated, or contrive to exist on ice-free regions. In the following notes an attempt will be made to ascertain if survivals of the pre-glacial animals and plants may still be traced in our present fauna and flora.

In a former paper † the possibility of some species of insects and plants having survived the Ice Age on the unglaciated region of North East Yorkshire was suggested. In the present communication it is proposed to consider in more detail the conditions under which organisms may have survived the Ice Age within the British Islands, as well as to bring forward further facts and inferences referring to our own county. The following notes simply endeavour to point out *possibilities* with regard to Glacial Survivals which may render the study of certain areas much more fascinating and interesting.

Opinions respecting the fauna and flora of Britain during the Glacial Period must, of course, largely depend upon the views held regarding the temperature and condition of our islands during that geological episode. So far as the latter is concerned it seems necessary to accept the land ice theory. The evidence is all against the submergence hypothesis, and the great feature of Mr. Kendall's work in Cleveland lies in the fact that the land glaciation of Britain receives absolute proof in his wonderful extra-morainic lake phenomena.

Admitting, therefore, the former existence of enormous glaciers and ice sheets within the British Islands, it seems necessary to admit an arctic temperature. There is, however, some disagreement on this point, but I think that for the

---

* Read at the Robin Hood's Bay meeting of the Yorkshire Naturalists' Union.

† 'Naturalist,' April, 1907, pp. 137-143.

purposes of this argument we shall be on the safe side if we regard the temperature of Britian during the Ice Age as being a low one, and resembling that of the Arctic Regions at the present day.

The question of Glacial Survivals is not new. In this connection Dr. R. F. Scharff stands pre-eminent, and in his various works has laid much stress on this survival as explaining many curious facts of distribution. But it does not appear to have been very precisely specified *where* or *how* organisms battled through the Ice Age; in many instances only sheltered nooks and corners are hinted at in a vague way.

In the first place it is incontestable that terrestrial species could *not* survive the Ice Age on land covered either by ice hundreds or thousands of feet thick, or by the waters of extra-morainic lakes. The only possible places where they could survive were the regions which we know from geological evidence to have been unglaciated.

At first these might not be thought very extensive, but when it is borne in mind, as Mr. Kendall reminds me, that ' the whole country south of the Thames, much of the Pennine Range south of the Aire Valley, the large area of North East Yorkshire, and several patches elsewhere' were unglaciated, it will be seen that the regions on which animals and plants may have existed are neither small nor unimportant.

Only on these areas could any species survive the Glacial Period. Those that did manage to exist thereon must surely have been principally Arctic forms, and the possibility of temperate and southern species managing to live on them seems small. It must not be forgotten, however, that the approach of the Ice Age would be gradual, and that many temperate and even southern forms might, in the course of generations, become adapted to the slowly increasing rigours of an Arctic climate, at the same time competing against the on-coming northern fauna and flora. This consideration goes some way towards explaining the strange mixture of animals of southern and northern habitats so frequently observed in British Pleistocene deposits.

However this may be, glacial survival can only be postulated where there were unglaciated areas on which species could exist.

The exact location of animals and plants in relation to glacial deposits is therefore of vital importance as differences in the fauna and flora of two adjacent areas, one covered with drift and the other driftless, apart from differences of soil, etc., may be perceptible. There are possibly indications of such differences in North East Yorkshire. The driftless region is essentially Arctic in the character of its fauna and flora, whereas the drift covered lands chiefly support Germanic forms, though both have invaded the territory occupied by the other.

The isolated occurrence of species on or near unglaciated areas which were surrounded by ice sheets is strongly suggestive of survival, be they Arctic, temperate, or southern forms, and in my former paper I quoted many examples. To these may be added the single Yorkshire record of the beetle *Pterostichus lepidus*, a ground species, obtained by the Rev. W. C. Hey on Sawdon and Ebberston Moors on the driftless area. In addition all plants ranging into the Arctic Circle and Greenland which live on the unglaciated moors may be included. It might be objected that these species are the relics of an Arctic fauna and flora which was once general in Britain, but has since largely disappeared owing to the amelioration of the climate and the advance of the Germanic section of our fauna and flora. But it seems probable that as the Arctic types came from the north at the commencement of the Ice Age, most of them lived on the ice free country throughout its duration, and have since stayed there, and not been driven to it by the Germanic section. This latter has doubtless, however, taken place in other districts.

To the insects already mentioned must be added the Dung Beetles (*Aphodius*), which live in the excreta of cattle, horses, and sheep, and are very numerous everywhere. One of them, *Aphodius rufipes*, lives in Siberia, the Caucasus, the High Alps, and Arctic Europe. It is of special interest as it also lives in Tropical Africa, a fact tending to show that the same species can become adapted to extreme climates ; from which it may be inferred that many southern and temperate forms became adapted to and survived the climate of the Ice Age on ice-free oases. Dr. Scharff, who gives the above instance, also states that * 'no less than six other species of *Aphodius* frequent alpine heights above 7000 feet, and a few ascend the region of permanent snow.' This genus may therefore be added to our list of glacial survivals, as suitable habitats would occur in a land where the Musk Ox, Reindeer, and other Arctic animals thrived, the remains of which have been found in the district.

(To be continued).

---

* ' European Animals,' p. 137.

RECENT GEOLOGICAL DISCOVERIES AT SPEETON.

T. SHEPPARD, F.G.S.

Perisphinctes lacertosus.—The specimen herewith figured is a large and unusually perfect example of *Perisphinctes lacertosus* Dum et Font, one of the characteristic Ammonites of the lower beds of the Upper Kimeridge in Filey Bay. These beds have very rarely been seen, but recent exposures of them above Reighton and Hunmanby Gaps showed that they contain

Perisphinctes lacertosus from the Kimeridge Clay, Speeton.

myriads of ammonites, so crushed, however, that only in the nodules can the collector hope to get even a fragment well preserved. The present specimen sent by Mr. C. G. Danford to the Hull Museum may therefore be regarded as an exceptionally fortunate find. It measures $12\frac{1}{2}$ in. across, is in some places slightly crushed, and the last two whorls have been partially squeezed asunder; otherwise the form is well preserved, with the sculpture remarkably sharp, and the trumpet-shaped mouth indicative of full growth.

The nodules of this part of the Kimeridge, though much smaller, are far more pyritous than those of its higher horizons, and the difficulty of extracting a fossil from such a matrix must be experienced to be appreciated. In the working out, various small ammonites were found, evidently belonging to the genus *Hoplites*, and probably to the species *pseudomutabilis* Loriol.

In 'Argiles de Speeton et Leurs Équivalentes,' A. Pavlow et G. W. Lamplugh, Moscow, 1892, p. 110, this species is referred to : (Translated) 'The specimen here figured belongs to the Geneva Museum, and is preserved in Pictet's Collection under the name *Ammonites biplex* loc. Speeton. Its characters perfectly correspond to the description of *Ammonites lacertosus* Fontannes, which description we do not reproduce. The horizon of the specimen in the Speeton Section is not indicated, but Mr. Lamplugh possesses a less well-preserved specimen of the same species, which he found near the outcrop of the Upper 'F.' Shales. In the shales themselves one often finds crushed ammonites suggestive of this species, but difficult to determine with precision.'

In the above the work the synonymy of the species is given as *Perisphinctes lacertosus* Dum et Font, 1876. *Ammonites (Perisphinctes) lacertosus* Dumortier et Fontannes, Crussol, p. 100 : and 1877 ; *Ammonites (Perisphinctes) lacertosus* Loriol, Pl. XV., fig. 1, Baden, Pl. VI., fig. 1, p. 50.

The species now under consideration is apparently that known in this country under the name of *Ammonites biplex*, and as the specimen now obtained is probably the best that has come from the Filey Bay deposits, it is thought advisable to place its present location on record.

Crocodilian remains.—Another interesting find was made by Mr. Danford in an exposure of the lower part of the Upper Kimeridge Clay, about 400 yards north of Reighton Gap. In this case several vertebræ, a femur, ribs, scutes, etc., were obtained, and on these being submitted to Dr. A. Smith Woodward he identified the remains as of *Steneosaurus*. On the completion of the examination of other similar remains in the British Museum, it will be possible to give the specific name. A typical dorsal vertebra from Speeton measures $2\frac{1}{4}$ inches in length, and $1\frac{1}{2}$ inches across, laterally. The bony scutes are particularly interesting, as nothing of the kind appears to have been obtained here previously. They are somewhat irregular in shape, about

2 inches in length, ½ inch in thickness, and the upper surface is covered by characteristic pittings varying from an eighth to a quarter of an inch across.

Remains of *Steneosaurus* do not appear to have been recorded previously from Filey Bay, and judging from the 'Catalogue of British Fossil Vertebrata,' by Smith Woodward and C. D. Sherborn, it would seem that the genus principally occurs in the Lias and in the Great Oolite. One example is recorded from the Kimeridge Clay of Kimeridge Bay. This is a snout, now in the British Museum, and is identified as *Steneosaurus* (*Teleosaurus*) *megarhinus*, Hulke. The specimen in question was described J. W. Hulke in the 'Quarterly Journal of the Geological Society' in 1871, and by Dr. Smith Woodward in 'The Geological Magazine,' in 1885. Since the 'Catalogue of the Fossil Vertebrata' appeared, numerous remains of *Steneosaurus* have been obtained from the Oxford Clay at Peterborough. One of the scutes from the Oxford Clay has been kindly placed in the Hull Museum by Mr. H. C. Drake for the purpose of comparison with the Speeton specimens.

Fish remains.—Another discovery of some importance, also made by Mr. Danford, is in a large nodule from the Kimeridge Clay. On the outside of this traces of bone were noticed, and after much labour these were carefully chiselled out, and revealed the bones of the head, a fin, a tooth, and other remains of a large fish, probably allied to *Caturus*, a genus well represented in the Liassic rocks further north at Whitby. On submitting this to Dr. Smith Woodward, he states that it resembles another species, undescribed, which is in the British Museum. That specimen, according to Mr. Lamplugh, is probably from the Neocomian zone *Belemnites jaculum*. If this is so, it would appear that the species is represented on the east coast in the Lias, Kimeridge, and Speeton Clays.

———◆◆———

There is an interesting paper 'On the Existence of the Alpine Vole (*Microtus nivalis*) in Britain during Pleistocene times,' by Mr. M. A. C. Hinton, in the Proceedings of the Geologists' Association (Vol. XX., Pt. 2).

In the same publication Mr. E. T. Newton has a 'Note on specimens of "Rhaxella Chert" or "Arngrove Stone" from Dartford Heath.' After very careful examination he concludes that specimens of Rhaxella Chert found in gravels at Cromer were derived from the Scarborough district.

In Cornwall recently a large stone of great archæological interest, the Giant's Quoit, which figures prominently in the legendary and historical records of the county, has been blasted and used for road metal! Were the perpetrators of the crime of suitable material, we could wish them a similiar fate. But they would be unsuitable—even for mending roads.

VITREA CELLARIA IN SHELL-MARL, NEAR HALE, WESTMORLAND.

J. WILFRID JACKSON,
Manchester Museum.

DURING a recent visit to the neighbourhood of Hale and Burton-in-Kendal, Westmorland, I had the good fortune to discover *Vitrea cellaria* (Müll.) in the lacustrine deposits so extensively developed there.

Its occurrence there is very interesting, as it adds another locality to the few already mentioned by Mr. J. W. Taylor in his Monograph (part 14, p. 36), some of which, for the sake of comparison, I give below :—

"Isle of Wight.—Prof. Forbes : Lacustrine beds at Totland's Bay, near Yarmouth.   Essex.—Mr. French : Alluvial

Section in Shell Marl near Hale.

shell-marl at Felstead ; Miller Christy : Rarely in shell-marl at Chignal, St. James. Yorks.—Mr. H. H. Corbett : In old lake deposit at Askern, near Doncaster. Ireland.—Marl deposits at Marlfield, near Clonmel, South Tipperary; at Drumcliff Crannoge, Co. Clare."

Along with *V. cellaria* and other more common species, I found a number of slender elongate *Limnæa truncatula*, which I submitted to Mr. Taylor, who refers them to the var. *lanceata*, saying he has had similar specimens from the black earth deposits in Nottinghamshire.

The accompanying photo shows a good section of the marl deposit near Hale.

NOTES ON THE CARBONIFEROUS ROCKS OF THE KETTLEWELL DISTRICT.

COSMO JOHNS, M.I.Mech.E., F.G.S.

In these notes it is proposed to give a brief sketch of the geology of the area to be visited by the Yorkshire Naturalists' Union this month in order to indicate the general features of the district, and to suggest, for the consideration of geologists attending the meeting, particular points towards which their attention might be directed. The difficulty is not so much to find interesting features as to select from the many, just those to which the most profitable attention might be given. It is perhaps fortunate that, owing to the recent work that has been done in determining the faunal sequence of the Lower Carboniferous Rocks of Britain and in establishing life zones in them, it will be possible to suggest a line of work not only interesting, but also serving to make somewhat clearer the relation of the various sections visited to those in other parts of the Pennines. The stratigraphy of the area is clear, and though the Great Scar Limestone and Yoredale Series do not stand out with the diagrammatic vividness of Ingleborough as seen from Chapel-le-Dale, yet their relation can be made out without much difficulty.

It is unfortunate that nowhere in the area to be visited is the base of the Carboniferous rocks to be seen, and the mantle of drift which rests, for some considerable height, on the sides of both Littondale and Wharfedale, increases the difficulty of working out the sequence. The occurrence of the Silurian Grit boulders in the drift below Kilnsey has suggested that pre-Carboniferous rocks were exposed in Upper Wharfedale in pre-glacial times. This is very probable, and would be in harmony with such paleontological evidence as is available. Despite the absence of a visible base and the occurrence of drift, it is possible to make out fairly clearly the faunal sequence of the Great Scar Limestone as far as it is exposed. This would seem to be the most promising work that could be undertaken during the few days of the meeting, and very possibly would also be the most interesting.

Before mentioning the particular zones which can b> made out in one area, it is desirable to call attention to some of the stratigraphical features. The Great Scar Limestone, which is so well exposed in Kingsdale, Ingletondale, Ribblesdale, Littondale, and Upper Wharfedale, is clearly a huge limestone

1907 August 1.

plateau some hundreds of feet thick, out of which the dales mentioned have been carved, exposing, in the case of the three first mentioned, the pre-Carboniferous rocks on which the Carboniferous series rests. There are no stratigraphical difficulties here. There is a sameness, too, about the Yoredale rocks that rest on this great limestone plateau, and at first sight this series of limestones, sandstones, and shales appear very simple. But the simplicity is only apparent, and when a large area is investigated it becomes very evident that the Yoredales are a very changeable series, and that these changes possess great significance. John Phillips, whose classic description of the Mountain Limestone district, in his 'Geology of Yorkshire,' is still the best account of our area, was fully aware of this, and devoted much attention to it. His most important conclusion was that the typical Yoredale series practically disappear on the western face of Great Whernside, and passes into the Shale Series which lies between the massive limestone and millstone grit of the region to the south east. The significance of this important change in the lithology of the Yoredale series cannot now be discussed, but interest lies in the fact that it takes place in the neighbourhood of Kettlewell.

To return to the Great Scar limestone and its faunal sequence: so far as it can be made out in the area under discussion, two zones can be determined. Neither are complete, but are still clearly distinguishable. One is characterised by the presence of *Productus giganteus* and the other by its absence. The first is the *Dibunophyllum* zone—the highest in the Lower Carboniferous rocks. The second one is the *Seminula* zone, the base of which is represented by the Carboniferous basement conglomerate of Ingletondale. Taking the *Seminula* zone first it will be found that only the upper sub-zone, or S_2 is well exposed, but it will be possible to obtain the characteristic fossils, though the beds of this age are not very fossiliferous compared with the rich collecting grounds in the D_2 beds of Wensleydale and other districts. In a small quarry on the left-hand side of the road from Kilnsey to Kettlewell, and near the last named place is a very interesting exposure of these S_2 beds, and *Seminula ficoides, Chonetes papilionacea, Productus corrugato-hemisphericus,* and *Syringopora* sp. can be obtained. To those interested in the zonal classification the most interesting piece of work will be to investigate in detail, starting from a known faunal horizon, a section affording a series of exposures to the top of the Great Scar limestone. They will thus be able to notice the change of fauna as the *Dibunophyllum* zone is entered, and to estimate the value of field evidence on which the classification is based.

This change of fauna occurs in a typical section at the first maximum of *Productus giganteus*, and the following fossils can be expected.

Dibunophyllum, sp.	*Productus giganteus.*
Carcinophyllum.	*P. hemisphericus.*
Syringopora cf *ramulosa.*	*P. corrugato-hemisphericus* (Mut D).
Cyathophyllum murchisoni.	*Chonetes* aff *comoides.*
Campophyllum aff *murchisoni.*	*Cyrtina septosa.*

This is by no means an exhaustive list, but it includes forms which have been obtained at this horizon in the Yoredale Province, and it is interesting to note that they can be compared with specimens from the same horizon in the Bristol and South Wales areas, similiar both in kind and stage of advance. The correspondence of the faunal sequence in the Carboniferous rocks of our area, so far as they are exposed, with that of similiar horizons in the Bristol area is very marked.

We have received parts I. and II. of '**The Hastings and East Sussex Naturalist,**' the journal of the Hastings and St. Leonard's Natural History Society. It is edited by Mr. W. Ruskin Butterfield.

The Annual Report and Transactions of the North Staffordshire Field Club for 1906-7 contains a portrait and memoir of the late J. Ward, F.G.S. In the same publication Dr. Wheelton Hind has some interesting 'Speculations on the Evolution of the River Trent.'

Mendelism. R. C. Punnett. Second edition. Cambridge, Macmillan & Bowes, 1907, 85 pp., price 2/- net. In view of the excellent summary of the theory of Gregor Mendel, which Dr. Agnes Robertson recently gave in these columns (July, pp. 242-246), our readers will be glad to know that in the present well printed volume is a very lucid and carefully written account of 'Mendelism.' At the low figure of two shillings the book deserves a large sale.

Peat : Its use and manufacture, by P. R. Björling and F. T. Gissing. London, J. Griffin & Co., 1907. 173 pp. 6/- net.

The volume under notice is the outcome of a suggestion of the late Sir Clement Le Neve Foster, who placed a number of valuable notes at the authors' disposal. Its main object is to indicate the most economical methods of digging and preparing peat for fuel. By the aid of sixty illustrations the various processes of its manfacture are clearly described— most of these, as might be expected, being from Swedish sources. The book deals with the formation, growth, and distribution of peat ; specific gravity and analyses; methods of digging, cutting, and dredging: drying ; peat fuel manufacture ; nature and uses of peat as a fuel ; and uses of peat otherwise than as fuel. There is a very useful bibliography. In the chapter on the distribution of peat, we learn that ' *At Holderness, near Hull,* there is peat, with trees, two feet deep, and at Hornsea, near Hull, beds are seen at low tide which contain peat and black root beds six feet deep.' But, strangely enough, no mention whatever is made of the extensive deposits at Goole and Thorne, which are worked so largely !

YORKSHIRE NATURALISTS AT ROBIN HOOD'S BAY.

(Continued from page 255).

In the following list of fungi the initials R. H. B. = Robin Hood's Bay; M. B. = Mill Beck ravine; S. B. = Stoup Beck ravine; and R. = Ramsdale.

AGARICACEÆ.

Armillaria mellea. Last year's mycelium on dead alder stump, M. B.

Tricholoma terreum. On the ground in a wood, R.

Tricholoma gambosum. Abundant in pastures; also in one of the woods, R. H. B.; R.

Collybia velutipes. On decaying stumps, S. B.

Collybia tenacella. Among decaying fir needles, R. H. B.

Omphalia umbellifera. On peaty ground on the moors.

Omphalia fibula. Among moss, R.

Entoloma clypeatum. On the ground in a wood, R.

Pholiota præcox. Among grass, roadside, R. H. B.

Galera tenera. In pastures on the cliffs, etc., R. H. B.

Galera hypnorum. Among moss in a wood, R.

Bolbitius flavidus. On cow dung in two places in pastures on the cliffs, R. H. B.

Agaricus campestris. In meadows, R. H. B.

Stropharia semiglobata. On dung in in the fields.

Psilocybe sarcocephala. On the ground, margin of wood, R.

Psathyrella gracilis. Among grass in pastures, R. H. B.

Psathyrella disseminata. On rotting stump among moss, R.

Marasmius oreades. In rings in pastures, R. H. B., etc.

POLYPORACEÆ.

Polystictus versicolor. On dead wood.

Poria vaporaria. On rotting branches, R.

HYDNACEÆ.

Hydnum niveum. On rotting branch of Ulex, S. B.

THELEPHORACEÆ.

Stereum hirsutum. On piled logs, Fyling Hall grounds.

Corticium calceum. On decaying branches, R.

Hymenochæte corrugata. On dead alder branches, S. B.

TREMELLACEÆ.

Dacryomyces stillatus. On old palings, R. H. B.

UREDINACEÆ.

Coleosporium sonchi. Uredospores on Coltsfoot, R.

Uromyces poæ. Æcidiospores on *Ranunculus repens* and *R. ficaria*, S. B.

Uromyces ficariæ. On *Ran. ficaria*, garden border, Thorp, R. H. B.

Puccinia poarum. Æcidiospores on Coltsfoot, R.

Puccinia suaveolens. On *Carduus arvensis*, on the cliffs, R. H. B.

Phragmidium subcorticatum. Æcidiospores. Common on wild rose stems.

Triphragmium ulmariæ. Primary uredospores on *Spirea ulmaria*, M. B.

USTILAGINACEÆ.

Ustilago violacea. In the anthers of *Lychnis diurna*, S. B.

PYRENOMYCETES.

Hypocrea rufa. On rotting wood near S. B. farm.

Hypoxylon atropurpureum. On decaying alder branches, S. B.

Rhytisma acerinum. On last year's sycamore leaves, Fyling Hall.

Diatrypella verrucæformis. Common on dead alder branches, S. B.

Diatrype stigma. On dead branches, S. B.

Eutypa lata. On dead branches, S. B.

Leptosphæria fuscella. On dead furze branches, S. B.

Heptameria doliolum. On dead herbaceous stems, S. B.

Pleospora herbarum. On dead herbaceous stems, S. B.

Sphærella rumicis. On living leaves of *Rumex obtusifolius.*

PERISPORIACEÆ.

Podosphæria oxyacanthæ. The *Oidium* stage on living hawthorn leaves, S. B.

HYSTERIACEÆ.

Hypoderma virgultorum. On dead bramble, S. B.

Gloniopsis curvata. On dead stems of wild rose, S. B.

DISCOMYCETES.

Peziza reticulata. On the ground in a moist copse, S. B.

Peziza sepiatra. On pieces of petrified moss, R.

Dasyscypha virginea. On decaying furze, rose, bramble, woodsage, and honeysuckle, S. B.

D. hyalina. On dead wood, S. B.

D. fugiens. On dead rushes, S. B.

Helotium cyathoideum. On dead umbellifer stems, S. B.

Mollisia cinerea. On dead wood in several places.

M. fusca. On decaying furze, S. B.

M. lividofusca. On decaying furze, S. B.

M. atrocinerea. On dead herbaceous stems, M. B.; S. B.

M. atrata. On dead stems of *Epilobium hirsutum*, S. B.

M. carduorum. On dead thistle, M. B.

M. dilutella. On dead stems of *Epilobium hirsutum*, S. B.

Pseudopeziza sphæroides. On dead stems of *Lychnis diurna*, S. B.

P. rubi. On dead bramble, S. B.

P. benesuedo. On dead alder twigs, S. B.

PHYCOMYCETES.

Mucor mucedo. On sheep dung on the moors.

DEUTEROMYCETES.

Phoma longissima. On dead stems of some Umbelliferous plant, S. B.

HYPHOMYCETES.

Penicillium glaucum. On rotting leaves, S. B.

Botrytis cinerea. On decaying herbaceous stems, S. B.

Periconia podospora. On dead furze, S. B.

Stachylidium cyclosporum. On dead herbaceous stems, S, B.

Cladosporium herbarum. On dead herbaceous stems, M. B.; S. B.

Triposporium elegans. On rotting wood, S. B.

MYXOMYCETES.

Stemonitis Friesiana, S. B.

Reticularia lycoperdon, M. B.

Trichia fragilis, S. B.

Tilmadoche nutans, R.

The last four all on dead wood.

T. S.

YORKSHIRE NATURALISTS AT SOUTH CAVE.

IT cannot be said that the visitors to South Cave on June 22nd were in any way inconvenienced by the lengths of the routes chosen by the local leaders. Within the more immediate vicinity of the railway station and village are charming tracts of country, which were open to the investigation of the members.

The well-wooded wold dales were equally agreeable to botanist and zoologist, whilst the geologists spent most of their time on the railway sections near the station, which, by the aid of sledge-hammers, proved even more productive of good things than usual. Millepore Limestone, the Kellaways Rock, Kimeridge Clay, Red Chalk, and White Chalk were all examined, though most attention was devoted to the Kellaways. From this several species of Ammonites were obtained, as well as some very fine gasteropods and lamellibranchs. The phragmocone of an ususually large belemnite, and some spines of an echinoderm were amongst the principal 'finds.' Messrs. J. W. Stather, Mr. G. W. B. Macturk, and the Hon. Secretary took charge of the formidable geological section, Councillor F. F. Walton and Mr. W. H. Crofts piloting the afternoon party ; and the botanists had the advantage of the President (Mr. C. Crossland), Dr. W. G. Smith, Mr. J. F. Robinson, and Dr. J. W. Wilson. Mr. W. Denison Roebuck represented the conchologists, and Messrs. H. Ostheide, J. W. Boult, and T. Stainforth were with the entomologists. Mr. E. W. Wade represented the vertebrate zoologists.

After tea at South Cave a hurried meeting was held in the Guild Hall, at which representatives from fourteen societies were present. In the absence of the president, who had to leave early, Mr. Cosmo Johns occupied the chair. Votes of thanks were passed to Mrs. Barnard, Col. Harrison Broadley, M.P., and the Hull and Barnsley Railway Company for the facilities given, and two new members were elected. Reports on the work accomplished were also made by the officers present, particulars of which follow.

In view of the February-cum-April weather which we had been experiencing for some time prior to the excursion, it was satisfactory to find that June 22nd proved to be one of the old-fashioned typical June days.

VERTEBRATE ZOOLOGY.—Mr. E. W. Wade, writes :—The South Cave district, comprising the Wold Valleys adjacent, many of them well wooded, the Houghton Woods, and the old Cliff

Warren, is far too large for a visit of three hours to produce any results worthy of record. Being the only representative of the ornithological section, my observations were confined to the part least known to me, viz., Cave Castle grounds. Twenty-seven species only were observed, of which the Great Tit and Spotted Flycatcher were feeding young in the nest, whilst the Blackcap, Willow Wren, Coal Tit, Blue Tit, and Chaffinch had young on the wing. Of the rest, perhaps the Chiffchaff and Tree Creeper were the most interesting. When I mention that this district contains, among the Mammalia, the Badger, and among the birds, Redstart, Nightingale, Lesser Whitethroat, Gold-crested Wren, Wood Wren, Grasshopper Warbler, Goldfinch, Hawfinch, Lesser Redpoll, Jay, Common Nightjar, Great Spotted Green Wood- peckers, Barn, Long-eared, and Tawny Owls, and Turtle Dove as breeding species, all but the Nightingale and Grasshopper Warbler being of regular occurrence, it will be seen that for this part of Yorkshire it is one of exceptional interest. No doubt the sanctuary accorded to the birds at Houghton, where all but Carrion Crows are strictly protected, has a good deal to do with the number of rarer birds to be found there. A dead example of the Red Field Vole (*Arvicola glareolus*) was picked by the conchologists near the station, and identified by Mr. Roebuck.

For the CONCHOLOGICAL SECTION, its Secretary, Mr W. Denison Roebuck, F.L.S., reported that he had devoted his whole afternoon to investigating the beech woods of Cave Castle, thus meeting with a number of beech-loving shells, but he had not the opportunity of working the rest of the area. The beech leaves swarmed with *Pupa cylindracea*, and *Buliminus obscurus* was also very common. *Helix rotundata, H. caperata, Hyalinia fulva, Arion minimus, Clausilia laminata*, and *Agriolimax agrestis* also occurred. *Clausilia rugosa* was found at Weedley Springs by Mr. Porter, and *Helix hortensis* by Mr. Wakefield, and *Arion ater* was brought in by another member; the total amounted to three slugs and eight land shells—eleven altogether. Ineffectual search was made for water shells in the lake at Cave Castle, in which water-lilies were abundant.

BOTANICAL SECTION.—Mr. J. F. Robinson writes:—The route taken by the phanerogamists and mycologists was via Drewton-, Weedley-, and East-dales, over South Cave Wold and down by ' Dicky Strakers' lane into the ancient little town of South Cave. Near the Railway Station, on the sandy outcrop of the middle oolite, there were noticed some plants characteristic of the district. Very fine was *Cerastium arvense*, a pronounced xerophile, commoner in the East Riding of Yorkshire than we have seen anywhere else. A few yards down Drewton lane a sedge that we do not often gather hereabout—*Carex muricata*—made its appearance. Five or six old walnut trees still flourish and fruit near the Manor House. The marshy springs in Weedley dale yielded several more sedges, including *C. paludosa* and *C. rostrata*, together with the allied cotton grass *Eriophorum latifolium*. On the chalk banks near the springs some good finds were made, notably *Bryonia dioica*, *Galium mollugo*, *Atropa Belladonna* (in flower), *Verbascum Thapsus*, and *Hyoscyamus niger* (Henbane, also in flower).

Ascending the Wold and going southward, a few of the first flowering plants of *Campanula glomerata* were noted. *Epipactis latifolia* grew in the beech wood traversed; whilst 'Dicky Strakers' Lane was festooned with the first flowering sprays of *Rosa canina*. *Rosa arvensis* was there in even greater quantity than the former species of wild rose, and was abundantly budded, although not a single flower was seen open. Amongst the briars and hawthorn of the hedges the shining green leaves and flowers of the black bryony (*Tamus communis*) were conspicuous.

Amongst cultivated forms it was a treat to visit the well-ordered grounds and rose-embowered gardens at Cave Castle, the guidance of the genial head-gardener, Mr. Curtis, being much appreciated. In the pond, the yellow and white water-lilies were flowering profusely, but not yet the small water lily-like plant *Limnanthenum peltatum*.

Fungi.—Mr. C. Crossland writes :—The route laid out by Mr. J. F. Robinson and Dr. Wilson proved a most excellent one. Evidently they had calculated the limited time at the disposal of the members. There was no need for rush; it suited the mycologists admirably; in fact, it might have been a mycological excursion, for each one took an interest in this branch, and picked up something or other. In the bottom, at the Springs near the railway, there was a typical collecting ground. A few very interesting species were met with, including *Clasterosporium fungorum*, a little black mould which lives on species of Corticium; this had only one previous record for Yorkshire. There are six records new to the S.E. division marked * in the list below. Most of the species found are of fairly common occurrence; still, it is well to know they were seen here also. In all, forty-seven species were noted, There is no need to specialise localities, South Cave will answer for all.

GASTROMYCETES.

Sphæroblus stellatus. On rotting stems of *Epilobium hirsutum.*

HYMENOMYCETES.
Agaricaceæ.

Mycena acicula. Among decaying leaves, twigs, etc.

* *M. hiemalis.* On decaying bark in moist shaded place.

M. tenerrima. On rotting chips.

Hypholoma lacrymabundum. On the ground among grass.

Psathyrella gracilis. Road - side among grass.

Coprinus comatus. On waste ground.

C. ephemerus. Among dead vegetation in hedge bottom.

Hygrophorus miniatus.

H. conicus.

H. nitratus.
All in pasture land.

Marasmius oreades. In 'fairy' rings.

POLYPORACEÆ.

Polystictus versicolor. On dead wood.

P. abietinus. On dead stump.

Fomes annosus. On stump.

Poria vaporaria. On rotting sticks.

HYDNACEÆ.

* *Grandinia granulosa.* On rotting sticks.

THELEPHORACEÆ.

Stereum hirsutum. On dead log.

Corticium calceum. On dead stump.

Hymenochæte rubiginosa. On decaying log.

TREMELLACEÆ.

Hirneola auricula - judæ. Last year's growth, on elder.

Dacryomyces stillatus. On decaying wood.

UREDINACEÆ.

Puccinia poarum. Æcidiospores on coltsfoot.

P. caricis. Æcidiospores on nettle.

P. suaveolens. On *Carduus arvensis.*

P. malvacearum. On *Malva rotundifolia.*

PYRENOMYCETES.

Xylaria hypoxylon. On dead stump.

* *Diatrype disciforme.* On dead beech branches.

Eutypa lata. On dead branches.

Lasiosphæria ovina. On rotting wood.

Lophiostoma caulium. On dead stems of *Epilobium hirsutum.*

Heptameria doliolum. On dead stems of *Epilobium hirsutum.*

DISCOMYCETES.

Peziza vesiculosa. On soil in stack garth.

Humaria granulata. On cow dung.

Dasyscypha virginea. On dead stems of *Epilobium hirsutum,* and on decaying wood.

D. hyalina. On decaying wood.

D. calycina. On living larch.

* *Helotium citrinum,* var. *pallescens.* on dead wood.

H. cyathoideum. On dead nettle stems.

Mollisia cinerea. On dead wood.

M. atrata. On dead stems of *Epilobium hirsutum.*

Ascobolus furfuracens, On cow dung.

PHYCOMYCETES.

Cystopus candidus. On living Shepherd's Purse.

HYPHOMYCETES.

Cladosporium herbarum. On dead herbaceous stems.

* *Clasterosporium fungorum.* On living *Corticium calceum.*

MYXOGASTRES.

* *Didymium farinaceum.* On dead wood.

Badhamia panicea. On decaying elm bark.

Tilmadoche nutans. On dead wood

To this list may be added *Mavasmius graminium,* on dead grass : this was accidently omitted, August 1894.—T. S.

In Memoriam.

JOHN HARRISON, F.E.S.

BY the death of John Harrison, a native of Barnsley, which occurred July 11th, Yorkshire has lost one of its principal lepidopterists. At an early period he took up the study of Natural History, and along with four other of his fellow townsmen, three of whom still survive, founded the Barnsley Naturalists' Society, Feb. 4th, 1867. As Secretary of the new society he worked hard to establish it on a sound footing, and lived to see it with a membership roll such as few had dared to hope for. Although he had ceased to be a member several years before his death, he had given evidences of a continued kindly interest in its prosperity. His inclinations led him to take up the study of Lepidoptera during the later years of his life, more particularly to the Micros. A reference to Mr. Porritt's 'List of Yorkshire Lepidoptera' will show how valuable were his contributions to that work. Always of a modest and self-depreciating cast of mind, it might almost be said that he rarely sought the society of his fellows; at any rate he never pressed himself upon them. At the same time he was ever ready to assist his brother entomologists, not forgetting those who were studying other orders than Lepidoptera. Thus on one occasion he brought the writer several specimens of an insect which his experienced eye recognised as an uncommon species. It proved to be the rare Longicorn Beetle *Gracilia minuta*. He was elected a F.E.S., 3rd April, 1889, and became a frequent exhibitor at the meetings of the society. His illness was of short duration, less than a week in fact, and was due to pleurisy and pneumonia. He died in his 74th year, and was interred in the Barnsley Cemetery, July 15th, 1907.

E. G. B.

Note on a blackbird's nest.—On June 13th my gardener shot a hen blackbird as she flew from her nest, and on going to the place shortly afterwards found the nest tenanted by a male thrush, who sat out the blackbird's eggs until they were hatched, about two days later. Unfortunately the young birds died on the cold wet night of Sunday 23rd. The lining of the nest was partly that of the blackbird and part the mud lining of the thrush.—T. R. CLAPHAM, Austwick Hall, Lancaster.

FIELD NOTES.

BIRDS.

Bramble Finch, &c., near Halifax.—The bramble-finch has been in Luddenden Dean near Halifax for some time. It was first observed about June 10th, and remained in the same neighbourhood in full song until July 1st, after which date it was not seen or heard again. As some shooting was heard in the wood, it is feared that someone shot it. The female, according to one observer, was also seen.

On June 8th I found a water-hen's nest, containing eggs, in a tree fully six feet above the level of the Aire near Bingley.— H. Waterworth, Halifax.

Red Breasted Flycatcher in East Yorks.—On June 4th, 1907, I heard the note of a bird quite unknown to me. It was singing a low warbling note of very little power. I saw the bird close to me, very low down in a thick hawthorn hedge skirting a beech plantation. Presently it flew up higher into the branches of the hedge, and was joined by another, evidently of the same species, for when the second bird joined the first the singing ceased and the male bird commenced flirting with the hen. The movements of the male were those of the Robin as often seen at pairing time : he raised and lowered his head and tail as if making most elaborate bows, and I feel certain the birds were nesting somewhere near.

They were about the size of a Willow Warbler, and when seen were in a bad light, being in the hedge under the shade of the branches of large beech trees, but the colour of both was a uniform light brown, the striking marking of the male being a bold red patch under the throat, extending partly down the breast. The birds were under observation for six or seven minutes, and my sisiter, who was with me, noticed the red throat referred to above, and also the peculiar warbling note.

The birds were, in my opinion, Red Breasted Flycatchers, and were seen at Thearne Hall, near Beverley. Mr. Haworth Booth informs me that he saw a male this spring at Hull Bank House about a mile-and-a-half, as the crow flies, from where I saw the pair. He saw the bird on May 20th, and reported it in the *Field* on the 25th.—Harold R. Jackson, Grosvenor House, Hornsea.

THE BIRDS OF YORKSHIRE.*

AT last! For a quarter-of-a-century has the writer patiently awaited the appearance of a monograph dealing with the birds of our greatest county. For the last five years he has been receiving prospectuses and reading notices and advertisements of the *forthcoming* 'Birds of Yorkshire,' and for the last five days he has been revelling in an advance copy of the two substantial volumes in which all that is known of Yorkshire Birds is printed.

The Yorkshire Naturalists' Union, of which Mr. T. H. Nelson and his colleagues are worthy members, has done more than any other private society towards recording particulars of the fauna and flora of its area. In the present work we have perhaps the most substantial account of the avifauna of any county extant. Not only have the authors, each of whom is well known in the ornithological world, given us of their best, but in addition to their own notes and knowledge, they have had unlimited access to the records of a whole corps of ornithologists and nature lovers, as well as the most useful records of Thomas Allis, John Cordeaux, and numerous others. The 'Naturalist,' 'Zoologist,' 'Field,' and similar publications have also been conscientiously searched, the particulars they contain have been most religiously examined and sifted; the grain has been properly placed in this store-house, and the chaff (and there was much of it) has gone to the winds. In addition, full use has been made of Howard Saunders 'Manual,' Sir Ralph Payne Gallwey's work on Decoys, etc.

The method adopted with regard to each species is most commendable, and in future the comparative rarity of any species can be learned at a glance, and it is to be hoped that we have now heard the last of the 'only record for Yorkshire,' which we have seen so frequently in the public press in recent years. After the common and scientific names of a bird, its status in the county is given. Then follows the first record, and the record given by Thomas Allis in his report presented to the York Meeting of the British Association in 1844. Allis's records, now printed for the first time, are an exceedingly valuable feature of the book. Details of migration, nidification,

---

* 'The Birds of Yorkshire,' by T. H. Nelson, M.B.O.U., W. Eagle Clarke, F.R.S.E., F.L.S., and F. Boyes. 2 vols. 899 pp., 164 plates. 8vo., price 25/- net; 4to., price 42/- net.

R. Fortune.

Young Grey Wagtails in Nest.

"The Birds of Yorkshire."

varieties, unusual habits, and distribution are given, as well as the 'local' names, folklore, etc. In this direction, however, the records are not always consistent.

That the 'Birds of Yorkshire' will at once take its place in the front rank of British works on ornithology there is no doubt, and for many years to come it will be the constant companion of every naturalist having an interest in Yorkshire or in British birds. The language used is such that its meaning can readily be grasped by the beginner, and certainly no one will complain of a super-abundance of 'fine writing.'

To the numerous illustrations we cannot give too much praise. They alone are worth a good proportion of the price of the volumes. There are some coloured—there are representations of early Yorkshire records, of unusually rare species, of famous nesting sites, of peculiaraties in the birds, and of nests and eggs and young of scores of interesting members of the avifauna of the county. The choice of the two hundred or more illustrations has been most happy, and no one can say that a single illustration is poor or that it is out of place. They form the most graphic account of the birds of any district that we have seen for some time. Very largely they are the work of that expert of bird photography, Mr. Riley Fortune, F.Z.S., who has in this way contributed greatly to the success of the work. Among other well-known naturalist photographers who have helped with the illustrations, we observe the names of Mr. T. A. Metcalfe, Mr. H. Lazenby, Mr. T. H. Nelson, and James Backhouse.

As appendices are given the Birds Protection Orders for the three Ridings, lists of 'Ancient Records,' etc., and the Indices (occupying 65 pages) are unusually complete and valuable.

From the Editorial we gather that the Hon. Secretary of the Yorkshire Naturalists' Union (Mr. T. Sheppard, F.G.S.) has had his finger in this (as in almost every other recent Yorkshire) pie, and doubtless this has resulted in the present work being unusually free from misprints or other errors.

In conclusion, we know not which to congratulate most—Mr. Nelson, the Yorkshire Naturalists' Union, the publishers, or the ornithological world generally on the appearance of these two volumes. Messrs. A. Brown & Sons have done their work thoroughly, and their excellent taste and good workmanship is exhibited on every page. They have certainly not spoilt the work by hurrying it through the press!

M.B.O.U.

1907 August 1.

REVIEWS AND BOOK NOTICES.

Lotus Land : Being an account of the country and people of Southern Siam, by P. A. Thompson. London, T. Werner Laurie. 312 pp.

In this well written and excellently illustrated volume Mr. Thompson gives the impressions he obtained during his three years' residence in Siam, principally amongst the peasantry. By the aid of his pen, pencil, and camera he has given a graphic and valuable record of Siamese life and lore. The introductory chapter gives a historical sketch of the country, and others deal with the Buddhist Religion, Temples, Images and Symbols, the Pees and Charms, Siamese Art, Camp Life, etc. Ethnology receives special attention—several of the photographs being of particular value in this direction.

'**A History of the Parish of Penistone** | embodying not only interesting particulars relating to its fine old Church and the Parish generally from the earliest times as contained in Hunter's History and Topography of the Deanery of Doncaster and other records, but also separate histories | of | the ancient grammar school of Penistone founded A.D. 1392; | of | the old markets of Penisale and Penistone established respectively A.D. 1290 and 1699, and the old Agricultural Society for a wide district around Penistone established A.D. 1804; | and of | the oldest pack of hounds in the world, viz., the Penistone Harriers or 'Olde Englyshe' Northern Hounds, probably in existence before the Conquest to prevent the ravages of wolves and other wild animals from the great forests of Hordern, Wharncliffe, and Sherwood, and the vast moors, wilds, and fastnesses of the district. | With illustrations and most interesting [*sic.*] local and general information, including educational, agricultural, and sporting gleanings, scraps, notes, etc., etc., made and collected from many sources during the last thirty years as well as my own recollections of Penistone and the districts around during the past fifty years | By | John N. Dransfield | Penistone | James H. Wood, the Don Press.'

The preceding is a copy of the most extraordinary title page we have seen for some time. Mr. Dransfield evidently believes in showing all his goods in his shop window! Immediately following, in the preface, he truthfully remarks, 'The title page of this book without anything further I think fully explains its subject matters.' He then equally truthfully admits 'the book is somewhat of a medley or hotchpotch.' This is our opinion, and as the author shares it we have no hesitation in expressing it. In its 569 closely printed large pages the author appears to have recorded and reprinted every possible scrap of information and tradition that he has seen or heard. If it were in MS it would be just what we should have expected the writer of a local history would have compiled as the basis of his book. Very little news of any sort appears to have been omitted. But we would suggest that were the whole contents of this heavy volume to be carefully digested, summarised, and re-written, *de novo*, with something like method, a really readable and useful work would result. From a prospectus we learn the volume 'will be taken up again and again and brighten *many* a spare or weary hour as a portion of its contents from their nature and from the unconventional way they are recorded, *are not intended to be, and cannot well be* grasped all at once.' The sentence is a bit awkward and difficult to understand, but we quite agree with the parts we have italicised! There are also included 'addresses and statements by Prof. Huxley, Ernest Renan, J. A. Froude, Sir Frederick Treves, Lord Napier of Magdala, George Muller, D. L. Moody, *and Buffalo Bill!*' The volume is illustrated, and in view of the quantity of printed matter alone is very cheap at half-a-guinea net. But, as a frontispiece, it contains a photograph and autograph of the author, and whenever we see a book embellished in this way, it 'kind of goes against the grain.' Mr. Dransfield may be a very kind and affable gentleman, but in his portrait he wears a somewhat stern and savage expression ; in fact, he looks as though he had just read this review !

Glimpses of Ancient Leicester, in six periods, by Mrs. T. Fielding Johnson. Second edition. Clarke and Satchell, Leicester, 440 pp.

In this well-written and scholarly volume a member of an old Leicester family gives a concise account of the main events in the history of the meeting place of the British Association for the present year, an event which makes the appearance of this edition peculiarly appropriate. The authoress deals with (1) Roman Leicester, (2) Leicester under the Anglo-Saxons and Danes, (3) Leicester under its Norman and Plantagenet Earls, (4) Leicester in the Sixteenth Century, (5) the Siege of Leicester, and (6) Leicester at the end of the Eighteenth Century. Perhaps the first chapter is of more general interest, containing as it does an account of the Roman occupation, several important evidences of which are extant—notably the mosaic pavements, the Jewry Wall, and the Roman Milestone, said to be the oldest stone inscription in Britain. The inscription upon this reminds one of the well-known advertisement boards on the line-side towards London, where, on large hoardings setting forth particulars of soap or pills, one sees in small characters, 'London . . . miles.' The following is a free translation of the Roman milestone :—' To the Emperor and Cæsar the august Trajan Hadrian, son of the divine Trajan, surnamed Particus, grandson of the divine Nerva Pontifex Maximus, four times invested with Tribunitial power, thrice Council. From Ratæ, Two Miles.'

The Days of a Year, by M. D. Ashley Dodd. London : Elkin Matthews, 1907. 173 pp. Price 2/6 net. This well-printed little volume contains a beautiful thought of a poet-naturalist for each day of the year. Each has clearly been penned on the date it bears, in the surroundings it describes ; and it is evident that with the author ' To see things in their beauty is to see them in their truth.' To peruse a few of the pages is quite refreshing. The reviewer has just returned from a ramble on the Yorkshire Wolds at South Cave, and turns to the date, June 22nd :—' Cool woods on the shady hillside, still and calm in the mid-day heat that quivers all around. The delicious earth-scent from dewy mosses and growing green things rises, penetrating and sweet, and in the thick trees an unseen wood-pigeon flutters, and flaps slowly away. The call of a late cuckoo comes faintly across from the outside far-off sunshine ; then again there is silence, till the woodland dreariness sinks deep into the soul, till the present slips away, and nothing is real but the memory of a few past happy hours.' In some such words might our ramble have been described. But it is not every one that can so well express one's thoughts !

European Animals : their Geological History and Geographical Distribution, by R. F. Scharff, Ph.D. Constable & Co. 258 pp. 7/6 net.

This work is based upon the Swiney Lectures on Geology which Dr. Scharff delivered at the Victoria and Albert Museum last year. In its pages the author has gathered together an enormous number of interesting facts bearing upon the past and present distribution of numerous animals and plants. An examination of the more recent geological deposits has enabled him to trace the probable direction of distribution of many interesting species. The present relative abundance of species of various genera is also shown to be indicative of the original source of these genera. By the aid of maps the distribution of many species can be seen at a glance. These are rendered of further value by the insertion of photographs or drawings of the species referred to. There are also ma ps showing the probable land connection at different periods between England, Ireland, and the Continent. Naturally, the many problems of distribution presented by the fauna and flora of Ireland receives attention. As indicated on the maps, the distribution of several species is indeed remarkable. Dr. Scharff's volume concludes with a very good bibliograpy and an excellent index. It is also an admirable illustration of the value of ' mere lists ' of species occurring in given areas. Were it not for such material it would have been impossible for this book to have been written. After perusing ' European Animals ' no one can reasonably say that lists have no value.

NORTHERN NEWS.

The Yorkshire Wild Birds Protection Committee begs to acknowledge the receipt of £2 2s. from the Royal Society for the Protection of Birds.

The July 'British Birds' contains an excellent portrait of the late Alfred Newton, F.R.S., together with an appreciative memoir by Dr. Bowdler Sharpe.

A schoolboy, writing an essay on wild beasts, wound up by saying that fortunately none were now at large in Britain, but they could be seen in safety in the Theological Gardens!

We notice that Mr. E. Kay Robinson vainly appeals for hints on 'How to make Garden Parties Entertaining for Elderly Guests.' Why not get someone to read extracts from 'The Country Side'?

In addition to reliable natural history, 'Country Side' occasionally gives its readers similarly reliable information of general interest. It recently referred to a famous motor run of 510 miles in 470 SECONDS. We hope a 'B.E.N.A.' badge was awarded to the chauffer.

A recent writer in *The Yorkshire Post* points out that 'the dangers of cliff-climbing' as carried on by the agricultural labourers under Bempton conditions are greatly exaggerated, and that young women have done what the 'Old Cliff Climber' and 'his gallant men' dare do.

After the Annual Dinner of the Museums Association held at Dundee recently, Dr. W. E. Hoyle, of the Manchester Museum, and President of the Zoological Section of the British Association, gave some useful hints on the methods of capturing rare animals for museums. Dr. Hoyle explained that these ideas were not original, but were obtained at a Congress of Zoologists at Breslau. :—

(1) To capture a Raccoon. The Raccoon, as is well known, is a very cleanly animal, and this characteristic is taken advantage of by the crafty huntsman. Near the haunts of the animal a bad sixpence is thrown on the ground. This the Raccoon sees and takes to a shop to buy soap. The shopkeeper detects the base sixpence, and the Raccoon is arrested for trying to pass counterfeit coin.

(2) To capture an Ape. The Ape is a great imitator. This characteristic is taken advantage of by the crafty huntsman, who takes a printing press into the forest, which he sets up and begins to print. He then leaves the press, but no sooner is his back turned than the Ape begins to set type and print. Proceedings are then taken against it for infringement of copyright.

(3) To capture a Lion in the forest. The Lion is the king of beasts. This fact is taken advantage of by the crafty hunstman, who takes a large cage with him into the forest. Various letters of the alphabet, mounted on cardboard, are then strewn about the ground, until the Lion has learned to read. The words 'no admittance' are then put over the cage door. The Lion, on seeing this, says 'No admittance! am I not the king of the forest!' and straightway walks into the cage. He is then caught.

(4) To capture a Lion in the desert. The desert, as is well known, consists principally of sand. This fact is taken advantage of by the crafty huntsman, who arms himself with a sieve and carefully passes the sand through it. What is left is the Lion!

(5) To capture a Camel. In order to capture a Camel it is necessary to be provided with an American millionaire, a balloon, and a needle. The balloon is filled, the millionaire placed in the car, which is released, and the needle is stuck into the sand. The Camel, seeing this, says to himself, before the millionaire gets to heaven I can walk through the eye of this needle, and proceeds to walk through. When half way through, the crafty hunstman ties a knot on his tail, and there he has him.

Other information of a similar kind was given, but the above will demonstrate the practical advantages of these Conferences! -

SEPTEMBER 1907.

No. 608
(No. 386 of current series)

THE NATURALIST.

A MONTHLY ILLUSTRATED JOURNAL OF
NATURAL HISTORY FOR THE NORTH OF ENGLAND.

EDITED BY

T. SHEPPARD, F.G.S.,
THE MUSEUM, HULL;

AND

T. W. WOODHEAD, Ph.D., F.L.S.,
TECHNICAL COLLEGE, HUDDERSFIELD.

WITH THE ASSISTANCE AS REFEREES IN SPECIAL DEPARTMENTS OF

J. GILBERT BAKER, F.R.S. F.L.S.,   GEO. T. PORRITT, F.L.S., F.E.S.,
Prof. P. F. KENDALL, M.Sc., F.G.S.,   JOHN W. TAYLOR,
T. H. NELSON, M.B.O.U.,   WILLIAM WEST, F.L.S.

Contents :—

LONDON :
A. BROWN & SONS, LIMITED, 5, FARRINGDON AVENUE, E.C.
And at HULL AND YORK.

Printers and Publishers to the Y.N.U.

PRICE 6d. NET. BY POST 7d. NET.

POST FREE.

(Of a few of these there are several copies.)

1. **Inaugural Address,** Delivered by the President, Rev. W. FOWLER, M.A. in 1877. **6d.**
2. **On the Present State of our knowledge of the Geography of Britis** Plants (Presidential Address). J. GILBERT BAKER, F.R.S. **6d.**
3. **The Fathers of Yorkshire Botany** (Presidential Address). J. GILBER BAKER, F.R.S. **9d.**
4. **Botany of the Cumberland Borderland Marshes.** J. G. BAKER, F.R.S. **6d**
5. **The Study of Mosses** (Presidential Address). Dr. R. BRATHWAITE F.L.S. **6d.**
6. **Mosses of the Mersey Province.** J. A. WHELDON. **6d.**
7. **Strasburger's Investigation on the Process of Fertilisation in Phænerc** gams. THOMAS HICK, B.A., B.Sc. **6d.**
8. **Additions to the Algæ of West Yorkshire.** W. WEST, F.L.S. **6d.**
9. **Fossil Climates.** A. C. SEWARD, M.A., F.R.S. **6d.**
10. **Henry Thomas Soppitt** (Obituary Notice). C. CROSSLAND, F.L.S. **6d**
11. **The Late Lord Bishop of Wakefield** (Obituary Notice). WILLIA WHITWELL, F.L.S. **6d.**
12. **The Flora of Wensleydale.** JOHN PERCIVAL, B.A. **6d.**
13. **Report on Yorkshire Botany for 1880.** F. ARNOLD LEES. **6d.**
14. **Vertebrates of the Wertern Ainst.** (Yorkshire). EDGAR R. WAITE F.L.S. **9d.**
15. **Lincolnshire.** JOHN CORDEAUX, M.B.O.U. **6d.**
16. **Heligoland.** JOHN CORDEAUX, M.B.O.U. **6d.**
17. **Bird-Notⁿ from Heligoland for the Year 1886.** HEINRICH GÄTKE C.M., **1s.**
18. **Coleopte: of the Liverpool District.** Part IV., Brachelytra. JOH W. El L.R.C.P. **6d.**
19. **Coleoptera f the Liverpool District.** Parts V. and VI., Clavicorni and Lam licornia. JOHN W. ELLIS, L.R.C.P. **6d.**
20. **The Hydradephaga of Lancashire and Cheshire.** W. E. SHARP. **6d.**
21. **The Lepidopterous Fauna of Lancashire and Cheshire.** Part I. Rhophalocera. JOHN W. ELLIS, L.R.C.P. **6d.**
22. **The Lepidopterous Fauna of Lancashire and Cheshire.** Part II. Sphinges and Bombyces. JOHN W. ELLIS, L.R.C.P. **6d.**
23. **Variation in European Lepidoptera.** W. F. DE VISMES KANE, M.A. M.R.I.A. **6d.**
24. **Yorkshire Lepidoptera in 1891.** A. E. HALL, F.E.S. **6d.**
25. **Yorkshire Hymenoptera** (Third List of Species). S. D. BAIRSTOW F.L.S., W. DENISON ROEBUCK, and THOMAS WILSON. **6d.**
26. **List of Land and Freshwater Mollusca of Lancashire.** ROBER STANDEN. **9d.**
27. **Yorkshire Naturalists at Gormire Lake and Thirkleby Park. 6d.**

WANTED.

Barnsley Naturalists' Society Quarterley Reports, Set.
Trans. South Eastern Union of Scientific Societies, Vol. I.
Yorkshire Lias, Tate and Blake.
Phillip's ' Life of William Smith.'
British Association Reports, 1839 and 1840.
Yorkshire Geol. and Polytechnic Society, Vol I.
Quarterly Journal, Geological Society, Vols. 1-21.
Reports Yorkshire Phil. Soc. for 1823, 1843, 1856, 1871, and 1882.
Reports Scarborough Phil. Soc. for 1830-3, 1837, 1839, 1845, 1854-5, 1861, 186
    1868-9, 1881-2.

Apply—Hon. Sec., Y.N.U., Museum, Hull.

NOTES AND COMMENTS.

THE BRITISH ASSOCIATION AT LEICESTER.

ANOTHER ' Parliament of Science ' has assembled, and dispersed. Leicester, which had never previously entertained the British Association, certainly did its best to make up for the past. On every side was the comfort of the members considered. An amount of over three thousand pounds was subscribed locally, and we hear that an even greater sum was personally spent by the Mayor, Sir Edward Wood, to whom, in no small way, the success of the gathering was due. The visitors were agreeably surprised with Leicester, its cleanliness, its fine streets, open spaces, and absence of smoke. The beautiful Abbey Park, and the Museum and Grounds, were transformed into something like what 'once upon a time' we thought Fairyland to be. The weather was most favourable. Whilst no extraordinarily startling scientific discovery was announced, each section studiously adhered to its work. There seemed to be fewer of the picnicking fraternity.

THE ATTENDANCE.

In connection with the York meeting a year ago, we pointed out that the disappointingly low figure on that occasion was in all probability due to the unusually early date at which the meeting had been fixed. At Leicester it was yet earlier, the proceedings starting on July 31st. As might well have been prophesied, the record at Leicester was even lower than at York. The attendance of course affects the amount of money available for scientific research, and few were the Committees that were not disappointed at the way in which their applications for financial aid were dealt with. But, perhaps, of more import than the question of grants, at any rate to several of the members, was the absence of many familiar faces, an absence entirely accounted for by the unfortunate date selected for the meeting. It may be that the British Association meets for the purpose of furthering scientific research, and it may be that most of the members assemble year after year with that object, but we cannot get away from the fact that the great charm and value of the gathering lies in the opportunities afforded of meeting and conversing with workers in the same field, from all parts of the world. If there is a continued choosing of early dates for the meetings, the result will unquestionably be detrimental to the interests of the Association. We are glad to learn that next year, at Dublin, the meeting will be held in September.

1907 September 1.

HANDBOOKS.

In the matter of handbooks the Members were well treated. Each was presented with a copy of Mrs. T. F. Johnson's 'Glimpses of Ancient Leicester' (noticed in our August issue), and a useful 'Handbook to Leicester and Neighbourhood,' prepared under the editorship of Mr. G. Clarke Nuttall. This, besides containing a description of the Topography and Antiquities of Leicester and its environs, has a paper on the 'Geology of the District,' by Mr. C. Fox Strangways; the 'Pre-Cambrian Rocks,' by Prof. W. W. Watts; 'Botany,' by Mr. W. Bell; 'Zoology,' and 'Cryptogamic Flora,' by Mr. A. R. Horwood; 'Entomology,' by Mr. F. Bouskell; 'The Charnwood Forest,' by Mr. J. B. Everard; and 'The Stone Roads, Canals, and Railways of Leicestershire,' by Mr. C. E. Stretton. A valuable Bibliography (geology, natural history, archæology,. etc.) by Mr. A. R. Horwood is included. But, as at York, we missed the useful sets of excursion handbooks, which have, in recent years, been given to the members. By the aid of these it was an easy matter for a stranger to chose the excursion most likely to interest him.

THE PRESIDENT'S ADDRESS.

The Address of the President, Sir David Gill, K.C.B., was one of the most brilliant Presidential Addresses that we have listened to for some time. Though dealing with 'The Science of Measurement,' 'The Solar Parallax, 'The Stellar Universe,' etc., etc.—technical astronomical subjects, he used simple and forcible language in such away that surely everybody was thoroughly interested throughout the Address. And this could not be said for every Presidential Address to the British Association !

THE VALUE OF A STANDARD.

An interesting illustration of Sir David's method occurred in his reference to one of Clerk Maxwell's Lectures in the Natural Philosophy Class at Aberdeen, which the President attended in 1859. Clerk Maxwell stated that 'A standard, as it is at present understood in England, is not a real standard at all ; it is a rod of metal with lines ruled upon it to mark the yard, and it is kept somewhere in the House of Commons. If the House of Commons catches fire there may be an end of your standard. A copy of a standard can never be a real standard, because all the work of human hands is liable to error. Besides, will your so-called standard remain of a constant length ? It certainly will change by temperature, it probably will change by age (that is, by the rearrangement or settling down of its component molecules), and I am not sure if it does not change according to the azimuth in which it is used. At all events, you must see that it is a very impractical standard—impractical because, if, for example, any one of you went to Mars or Jupiter, and the people there asked you what was your standard of measure, you could not tell them, you could not reproduce it, and you would feel very foolish. Whereas, if you told any capable physicist in Mars or Jupiter that you used some natural invariable standard, such as the wave-length of the D line of sodium vapour, he would be able to reproduce your yard or your inch, provided that you could tell him how many of such wave-lengths there were in your yard or your inch, and your standard would be availabe anywhere in the universe where sodium is found.' In this way Clerk Maxwell impressed great principles upon his students. Sir David adds, ' We all laughed before we understood; then some of us understood and remembered.' The vote of thanks to the President was proposed by Lord Kelvin, whom to see and hear was alone worth the journey to the Midlands.

THE GROWTH OF BOTANICAL SCIENCE.

In the course of his Presidential Address to the Botanical Section, Professor J. B. Farmer mentioned that the problems that confronted them as botanists were far more numerous and far more complex than formerly. They were attached to a science that was rapidly growing, and this advance was carrying with it a process of corresponding differentiation. In rating highly the value of maintaining a physiological attitude of mind towards the phenomena presented by the vegetable kingdom, one was mainly influenced by the logical necessity which such a position carried with it of constantly attempting to analyse their problems, as far as might be possible, into their chemical and physical components. He believed it was only by the help of the elder branches of science—chemistry and physics—that the accurate formulation, to say nothing of the final solution, of the problems in which they were engaged would be analysed.

THE HISTORY OF THE EARTH.

To the Geological Section, Prof. J. W. Gregory delivered a scholarly Address in which many interesting topics were touched upon. After referring to the geology of the inner earth, he said that the modern view of the structure of the earth added greatly to the interest of its study, for it recognised the world as an individual entity of which both the geological structure and the history had to be considered as a whole. Once the earth was regarded as a mere lifeless, inert mass, which had been spun by the force of gravity, that hurled it on its course into the shape of a simple oblate spheroid. Corresponding with this astronomical teaching as to the shape of the world was the geological doctrine that all its topography was the work of local geographical agents, whose control over the surface of the earth was as absolute as that of the sculptor's chisel over a block of marble.

THE SHAPE OF THE EARTH.

In his Presidential Address to the Mathematical Section, Prof. A. E. H. Love asked, if the ocean could be dried up, what would be the shape of the earth? By means of interesting diagrams he demonstrated his theory of ' gravitational instability,' accounting for the existence of the oceans, and the suggestion that without the oceans the sphere would be deformed into a sort of irregular pear-shaped surface, with the stalk of the pear in the southern part of Australia, and containing Australasia and the Antarctic continent. In attempting to estimate the bearing of his theory on geological history he was guided by the consideration that the earth is not now gravitationally unstable. From observations of the propagation of earthquake shocks to great distances, they could determine the average resistance to compression, and they found that this resistance was now sufficiently great to keep in check any tendency to gravitational unstability.

CEPHALOPODS.

Dr. W. E. Hoyle, of the Manchester Museum, in his Presidential Address to the Zoological Section, dealt with ' The small and economically unimportant group of the Cephalopoda,' a subject of which he has made a special study, and in giving the results of his own researches Dr. Hoyle unquestionably acted wisely. The classification of the Cephalopoda was the President's theme, and whilst it is admittedly not a subject which can be made interesting to all, his contribution can be looked upon as a clear statement of the present position of a far too neglected branch of Zoology. Dr. Hoyle spoke as a specialist, as a systematist, ' one whose main work has been the discrimination and definition of genera and species.'

NATURAL MONUMENTS.

Before a joint meeting of the Geological, Geographical, and Botanical Sections, Prof. Conwentz, the Prussian State Commissioner for the Preservation of Natural Monuments, delivered an interesting Address on the subject which he has made his own. He explained that the phrase 'natural monuments' was new in Germany as well as in England, but we should recognise that there could be monuments of nature as well as of art. The constant inroads of cultivation and industrial undertakings upon primitive nature have led, and are leading, to the disappearance of scientifically interesting and even unique natural objects and types of scenery. A widespread feeling has arisen that as much as possible should be done to prevent such destruction, and this has recently led not only to much local effort directed to this end, but in Prussia to the institution of a special State department, under the Minister of Education, for the purpose of directing and co-ordinating such efforts. In the opinion of Professor Conwentz, however, procedure by Government department is not the right method in this country ; we should rather depend upon voluntary effort. We would point out, however, that many of the valuable suggestions made by Professor Conwentz have been anticipated by Professor G. Baldwin Brown in his book on 'The Care of Ancient Monuments,' which was reviewed in these columns for November, 1906 (p. 387).

OTHER ADDRESSES.

The Presidential Address to the Chemical Section was by Prof. A. Smithells, who took 'Flame' for his subject. Mr. G. G. Chisholm addressed the Geographical Section on 'Geography and Commerce.' Mr. W. J. Ashley discoursed on the past history and present position of political economy to the Economic Science and Statistics Section; Dr. Silvanus P. Thompson, as President of the Engineering Section, referred to the Development of Engineering and its Foundations on Science ; Mr. D. G. Hogarth dealt with 'Religious Survivals' in his Address to the Anthropological Section ; Dr. Augustus D. Waller addressed the Physiologists 'On the Action of Anæsthetics ;' and Sir Philip Magnus addressed the Educational Science Section on 'The Application of Scientific Method to Educational Problems.'

PLANKTON INVESTIGATIONS.

· Professor W. A. Herdman, one of the Secretaries of the Association, gave a report of the Plankton Fishing Investigation carried out under his direction in the Irish Sea off the Isle of Man during April, 1907, with the object of testing different kinds of open and closing tow-nets and of gathering information as to the detailed distribution of the organisms according to the length, depth, and date. Examples were given of very different results both quantitative and qualitative to those from quite similar casts taken not far apart either in space or time. Sudden variations in horizontal distribution of the Plankton were discussed, and seasonal changes were also considered. The necessity of numerous gatherings in well-chosen restricted areas was emphasised. His conclusions were (1) That they must investigate their methods before they attempted to investigate on a large scale. (2) That they must find out much about their gathering of organisms before they could consider them as adequate samples ; and (3) that they must make an intensive study of small areas before they drew conclusions in regard to relatively large regions such as the North Sea or the Atlantic Ocean.

MIMICRY IN INSECTS.

At one of the evening meetings at the British Association, Dr. F. A. Dixey gave a discourse on " Recent Developments in the Theory of Mimicry." Dr. Dixey, at the outset, observed that the remarkable resemblances that existed between certain insects belonging to widely different orders, as, for instance, the likeness borne by some of the "clear-wing moths" to wasps and hornets, had long been known to naturalists. They were interpreted by the older observers as cases of "repetition," and "analogy" in nature. Kirby and Spence were the first to attempt a rational explanation. These authors got so far as to suggest that one species might gain an advantage by resembling another ; but the first really scientific account of the matter was given by Bates, who pointed out that certain kinds of butterflies in South America escaped attacks from birds by mimicking the appearance of other conspicuous species which were immune from persecution on account of the possession of distasteful qualities. This resemblance to a distasteful model he considered had been gained by a gradual increase of forms tending in the necessary direction. Bates' theory of mimicry, which was at once accepted by Darwin, and met with general approval, marked an important step in advance, but left certain facts unexplained. The lecturer then discussed the further contributions of Fritz Muller, Meldola, Poulton, and others, and illustrated his remarks by a beautiful set of lantern slides. In proposing a vote of thanks to the lecturer, the President, Sir David Gill, pointed out that Bates was a native of Leicester.

VIKING RELICS AT YORK.

On this subject, Dr. G. A. Auden read an interesting paper to the Anthropological Section. During the autumn of 1906 excavations for building purposes in the city of York, a few yards from the left bank of the Ouse, had revealed a number of objects which may with certainty be referred to the Viking period. The area in question is situate at the junction of Nessgate and Coppergate, and contiguous to the site in which a large number of objects, dating from the Scandinavian occupation, were found during excavations for the Public Library and Friends' Meeting House in 1884. Several objects are enumerated which have not been previously reported in England, and amongst these the chief interest centres in a bronze chape of a sword scabbard, exhibiting an open zoömorphic interlacing design terminating in a conventionalised animal head which attached the chape to the material of the scabbard. The zoömorphic *motif* is further illustrated by several portions of contemporaneous stonework which have been found from time to time in York, slides of which were shown. A consensus of opinion upon the objects described attributes them to the first half of the tenth century—a period which saw the Scandinavian power in York rise to its zenith.

HOLDERNESS GRAVELS.

To the Geological Section, Messrs. T. Sheppard and J. W. Stather gave some Notes on a New Section in the Glacial Gravels of Holderness. In these they pointed out that the North-Eastern Railway Company had recently been making some extensive excavations in a hill situated between the well-known Kelsey Hill and Burstwick Gravel Pits, in central Holderness. At the present time the section exposed is probably the finest of its kind in the country. The cutting is made through the heart of the hill, and is 1300 ft. long, and 45 ft. high in the centre, from which the section gradually slopes. The sides of the hill are flanked by boulder clay, and irregular masses also occur at intervals in the gravel. There are two types of boulder clay visible, the upper or Hessle clay, containing a preponderance of Cheviot rocks, and the purple or middle boulder clay with its Carboniferous Limestones and basalts. The gravels are somewhat similar to those described by Mr. Clement Reid at Kelsey Hill as interglacial, but the present authors consider them to be merely part of the terminial moraine of the North Sea ice-sheet. In addition to the far-travelled boulders, a lengthy list of marine shells, mostly of an Arctic type, has been compiled, and the species *Cyrena* (*Corbicula*) *fluminalis* (a freshwater form), also abounds. An interesting collection of mammalian remains has been secured, and includes bones of *Elephas primigenius*, Rhinoceros, Walrus, Red Deer, *Bison priscus*, Horse, and *Bos*. Some of these bearevidence of having been gnawed by the Hyæna. It is thought that the shells and mammalian remains have been caught up by the moving ice mass, and in this way incorporated in the moraine.

IRON ORE SUPPLIES.

Mr. Bennett H. Brough pointed out to Section C. that in Great Britain the principal iron-ore producing districts are Cleveland, in North Yorkshire, which in 1905 yielded 41.0 per cent. of the total output of the kingdom ; Lincolnshire (14.8 per cent.), Northamptonshire (13.9 per cent.), and Leicestershire (4.7 per cent.), together yielding 33.4 per cent. of the total output ; Cumberland (8.6 per cent.) and North Lancashire (2.7 per cent.), Staffordshire (6.1 per cent.) and Scotland (5.7 per cent.). The Cleveland iron ore occurs in a 10-foot bed in the Middle Lias, and contains about 30 per cent. of iron. It is worked by underground mining. In Lincolnshire, Northamptonshire, and Leicestershire the brown iron-ore beds form part of the Inferior Oolite, and contain about 33 per cent. of iron, the workings being mostly opencast. In Cumberland and North Lancashire the red hæmatite occurs in irregular masses in Carboniferous limestone. It contains more than 50 per cent. of iron, and is worked by underground mining. The ironstone in Staffordshire and in Scotland is mostly obtained from mines that also produce coal. Such, in brief, are the home deposits from which the British supply of 14,590,703 tons of iron ore, valued at £3,382,184, was obtained in 1905. Even that enormous output did not meet the consumption, and 7,344,786 tons were imported.

MARINE PEAT.

Mr. J. Lomas reported that during excavations in the Union Dock on the Mersey Docks and Harbour Board Estate, in the South End of Liverpool, a very remarkable peat band was discovered.   Reckoning downwards from a datum line three feet above Old Dock Sill, a section showed :—

Sand with black carbonaceous bands . .	4 feet.
Peat	6 inches.
Blue clay with rootlets	4 feet.
Sand with thin bands of peat . . .	2 feet 10 inches.
Boulder clay	3 feet 2 inches.
Bunter pebble beds	8 feet +.

The upper peat was entirely composed of marine plants, laminaria predominating.   On the fronds were numerous encrusting organisms, such as polyzoa, hydrozoa, the fry of young molluscs, etc.   The lower peat, while consisting mainly of marine plants, contained a few drifted pieces of oak and other land plants.   The sands accompanying the peat resemble those of the Mersey Bar, and besides the quartz which makes up the bulk of the deposit, contain zircon, garnet, tourmaline, dolomite, kyanite, rutile, staurolite, orthoclase felspar, biotite and muscovite, shell fragments, foraminifera, sponge spicules and polyzoa.   The deposit was probably accumulated in a sheltered bay in the old estuary of the Mersey.   The chief interest lies in the fact that peat may be formed from marine as well as from land plants.

YORKSHIRE FOSSIL PLANTS.

In the report on the Life-zones in the British Carboniferous Rocks Committee, Dr. Wheelton Hind stated that he was fortunate enough to secure a fine collection of plants obtained in an abortive attempt to find coal at Threshfield, near Grassington, in the Valley of the Wharfe.   The shales are stated to be those which occur below a bed of Millstone grit.   Mr. Kidston has examined the specimens, and the following list is the result :—

Sphenopteris elegans Bgt.	*Sphenophyllum tenerrimum* Ett. sp.
Calymmatotheca stangeri Stur.	*Lepidodendron* sp.
Rhodia moravica Ett. sp.	*Lepidostrobus* sp.
Sphenopteris sp.	Small Lycopodiaceous bract.
Calamites ostraviensis Stur.	*Rhabdocarpus ?* sp.
Calamites sp.	

Mr. Kidston states, with regard to the horizon : 'I have not the slightest doubt that the bed these specimens come from is on the horizon of the Upper Limestone group of the Carboniferous Limestone series of Scotland.   At any rate we know the Lower Limestone group of Scotland has a fauna which indicates the Upper *Dibunophyllum* zone.

FOSSIL FISH FROM THE CHALK OF NORTH LINCOLNSHIRE.

Dr. A. SMITH WOODWARD, F.R.S.
(British Museum, Natural History).

THE fossil fish* obtained by Mr. H. C. Drake from the *Rhynchonella cuvieri* zone ·in the chalk pit at South Ferriby is in a fragmentary condition, and lacks the skull ; but the parts preserved agree so precisely with the corresponding parts of *Elopopsis crassus*, that the specimen may almost certainly be referred to this Elopine species. It agrees with the known examples of *Elopopsis crassus* even in size. The bones of the opercular apparatus are remarkably thin, large and smooth, showing only a slight waviness parallel with the margins. The preoperculum is much expanded at the angle and in the lower limit, and bears marks of slime canals which radiate backwards from the main slime canal of its anterior border. The vertebral centra of the caudal region are about as long as deep, and they are strengthened by a few longitudinal ridges which extend between the stout anterior and posterior rims. The neural and hæmal arches are much flattened from side to side, and sharply inclined backwards. The scales are very thin, large, and deeply overlapping. They exhibit numerous small rounded pittings on their exposed portion, but are otherwise smooth, and only display their fine concentric lines of growth when abraded. A few of the stout pectoral fin-rays are smooth and undivided for a long distance at their proximal end.

The type skull of *Elopopsis crassus* was found in a Turonian zone at Southeram, near Lewes, Sussex. More satisfactory specimens of the fish have been obtained by Mr. G. E. Dibley, F.G.S., from the zone of Rhynchonella cuvieri in Peter's Pit, Wouldham, Kent, and will shortly be described in the Palæontographical Society's Monograph of the Fossil Fishes of the English Chalk.

REFERENCES.—*Osmeroides crassus*, F. Dixon, Geol. Sussex (1850), p. 376. *Elopopsis crassus*, A. S. Woodward, Proc. Zool. Soc., 1894 (1895), p. 759, pl. 43, fig. 1.

◆◆

'**Bulbs,**' by S. Arnott, and '**Weather,**' by the Hon. H. A. Stanhope, forms Nos. 11 and 12 of the 'One and All,' garden books, issued at a penny each. They are exceedingly useful publications.

---

* Hull Museum Specimen, No. 24.05.

NOTES OF CHANGES AMONG ANIMALS IN FYLING-DALES, NORTH-EAST YORKS., WITHIN THE LAST FIFTY YEARS.

J. W. BARRY.
Fylingdales.

MAMMALIA.

The Badger has much increased in numbers and also in range of late years. This is doubtless from the protection it has received on this estate. It has bred in the woods about a quarter-of-a-mile from the house ever since I remember, though the fact was not generally known until recently. In 1903, the Badgers there had become so numerous and, in one part, had so mined the side of the ravine with their burrows that I feared landslips. At the same time they had spread widely in the neighbourhood and began to annoy the farmers, one of my cottagers losing the whole of the cabbages in his garden through them. Under these circumstances it was necessary to make a thinning. Five were trapped, and not long afterwards the landslip which I chiefly feared took place, apparently exterminating all that were left in that spot. The rest seem to have scattered; but this year a pair or two have re-appeared in the same woods.

About twenty-three years ago, the somewhat unusual circumstance occurred of finding and catching a Badger asleep in broad daylight. This was by my then steward in an open part the woods. The Badger was a young one, of course. My gardener, who has seen a good deal of them (as for many years he did my gamekeeping), has only seen them two or three times when the sun was well up. He says, however, that, in a great number of instances, the Badgers lie not at the bottom of their holes but near the entrance, as he has heard them bolting on his approach. Elsewhere they have generally discovered him, he has found by scent rather than by hearing.

Sir C. W. Strickland's keeper has found that Badgers make their raids on wasp's nests by alternate sallies and retreats. When in another situation he once heard a strange cry from some unknown animal as if in distress, and saw a Badger rolling itself on the ground and trying to free itself from wasps. He then saw it go back, evidently to the comb, for it returned after a while with a fresh supply of wasps, from which it freed itself in the same manner as before. This process was constantly repeated.

The Polecat used to appear on the gamekeeper's lists of this estate ; but I never succeeded in seeing one until 1903, when (at the end of January or the beginning of February) my gardener caught a fine adult male in a trap set for rabbits, inside a burrow in the oak woods about three furlongs from the house. Its leg was broken, but it was kept for about three months in the ferret hutch of the garden. During this period it became slightly tamer, allowing itself to be touched when at rest, but it was uncertain. At last, owing to its unbearable stench and the large amount of raw meat which it required, it was destroyed. After its capture, the tracks of another Polecat were seen round the gardener's cottage, so that it would seem that they had been hunting in couples.

A few years ago, someone wrote to ' The Field ' to say that, when traversing Fylingdales Moor by the Whitby and Scarborough Road, he had seen five Polecats together somewhere in the neighbourhood of Jugger Howes. There is no more likely place for their headquarters than the Jugger Howes gill. I think that the captured Polecat already mentioned must have come down from the moor under stress of weather, tracks in the snow having been seen some time before in High Moor plantation.

The Squirrel was quite unknown on this estate and in this dale until about twelve years ago, when it suddenly appeared and multiplied so rapidly that, on seeing the damage which it was beginning to do to the young larch and spruce, and knowing what it had done elsewhere, as in Scotland, I reluctantly declared war upon it.

I do not doubt that it came from the Sneaton plantations across the moors. That is the nearest point at which it was found before, and then, as in the Newton House Woods below them, it had been abundant for a long time. One of the first that I saw was in the high part of the Ramsdale Woods, and the distance between that point and the nearest point of the Sneaton plantations is only about one mile and a quarter. With so short an interspace it may seem strange that the Squirrels should never have come before, but then it must be remembered that all of this is open moor, and that, no doubt, was the natural barrier which proved so efficient for such a number of years. It may also be asked why the Squirrel should have come when they did. I can only conjecture that the heavy falls of timber which have been going on in the Newton House Woods for the last thirty years, and the destruction in the early eighties (I think) of some hundreds of acres of Scots pine in the Sneaton plantations, in consequence of a sort of tornado, reduced the food supply and induced migration.

On the credit side of the Squirrel account I ought to mention that since their advent seedling hazels have spread in the woods in a way that they never did before.

The Black Rat was common in this house and the drains when I was a boy. I have seen three or four trapped together, and my earliest associations are as much with black rats as with brown.

Since I have had possession of the property, what appear to be hybrids between the black and brown rat have from time to time been killed. The last were caught three or four years ago; but my servants say that they are still in the house, and that those about the kitchen offices are exclusively of this sort. They are glossy black on back and hindquarters, dark grey on the sides and light brown underneath.

This kind of rat has long been confined to the house itself and the outbuilding next adjoining. The stables, the best feeding ground, have, for a great number of years, been monopolised by the Norwegian invader.

The man who has trapped them says (rather oddly), that he has found these rats much fiercer when caught than the brown rats.

The Wolf must have survived in this neighbourhood down to the fifteenth century. There is an entry of 1395 in the Abbey Roll of Disbursements for 'tewing' fourteen wolf-skins.

(To be continued).

To the list of papers by the Rev. E. M. Cole, given in our August issue, should be added: 'Roman Remains at Filey' ('Proc. Yorks. Geol. Soc.,' 1906), and 'Presidential Address to the Malton Field Naturalists' and Scientific Society,' printed in the Society's Report for 1905.

In the 'Memoirs and Proceedings of the Manchester Literary and Philosophical Society' there is a note 'On a Confusion of Two Species (*Lepidodendron Harcourtii*, Witham, and *L. Hickii*, sp. nov.) under *Lepidodendron Harcourtii*, Witham, in Williamson's XIX. Memoir; with a description of *L. Hickii*, sp. nov.,' by D. M. S. Watson. In this paper the author shows that under the one name Williamson really described two separate forms, and to the second the name *Lepidodendron Hickii* is now given.

We are glad to receive the Annual Report of the Louth Antiquarian and Naturalists' Society, and it augurs well for the Society's work to find that its rooms are inadequate, and that better accommodation is required. There have been several interesting exhibits at the indoor meetings, and field work has not been neglected. It also speaks well for the study of nature and archæology to find that the President of the Society, Alderman S. Cresswell, has recently attained the age of 90 years, and was presented with an illuminated address.

NOTES ON THE LAPWING.

F. STUBBS,
Oldham.

IT is well known that for the first few days of its existence the young lapwing closely resembles a fragment of earth or a lump of horse dung. On a bare fallow the bird is quite in harmony with its environment, and is well protected. The colours do not agree, however, with those of green pasture ; but, as it cannot be inconspicuous, the young bird finds safety in mimicking an unpalatable and common substance. I do not think that this instance of interchangeable protective colouration has been up to now appreciated by field naturalists, although very few observations will show the beautiful arrangement of colours necessary for this adaptability to different surroundings. As the bird grows bigger, the black and green feathers of the adult appear, and destroy the resemblance to horse dung. Then the young lapwing enters what may with propriety be called the "cow dung stage" of its existence, for everyone must have noticed the resemblance of the half-grown (or even full-grown, but flightless) young to a portion of a patch of vaccine droppings. On the fallows these colours still remain in harmony with the bare soil. Perhaps a brood of lapwings may be divided between two fields, the one a bare fallow, and the other a pasture. In one situation they provide an example of pure protective colouration, and in the other an excellent instance of mimicry. Early in life they mimic one substance, but when they grow out of this livery, they don a coat that enables them to mimic another substance dissimilar in appearance, but equally unattractive to birds or animals of prey.

We have here a possible explanation of the growing abundance of the lapwing in all parts of the country. The mere word "adaptability" does not quite explain the matter. The bird is well protected on the pasture lands, and multiplies, so that the number of pairs may exceed the number of suitable nesting sites. The surplus attempt to breed somewhere, although the effort to rear their progeny may be a vain one; but on the fallows the birds find that the colours of the eggs and young are well fitted to the new and quite dissimilar surroundings.

The increase of agriculture, the bane of so many wading birds, has provided the lapwing with countless nurseries. Game-preservation, also, has left this curious bird with little more than a single enemy, and even he is impotent during the most important part of the year. On the grass-lands the species has multiplied with our cattle, while elsewhere the plough has ousted all rivals, only to provide a safe shelter for the lapwing's eggs and young.

From these two great mutually-supporting reservoirs the birds overflow in all directions, and we find them nesting in the most incongruous situations. For extreme cases we need only think of salt marshes, rabbit warrens, chalk downs, peat bogs, maritime sandhills, and meadows. The lapwing's first requirement at the nesting season seems to be open ground, preferably free from tall herbage. If the land is under the plough there will always be fields lying fallow, and providing excellent hiding places ; if not, the presence in the locality of a few horses or cattle provides the bird with all the shelter it requires. Away from the influence of man it is just a rather aberrant wader struggling for existence with a host of other birds. It is not the fashion to seek for instances of Commensalism or Mutualism in the higher ranks of life, or one of these terms might be applied to this association of *Homo sapiens* and *Vanellus vulgaris*.

Although the great majority of breeding female lapwings have white chins, some authors state that the sexes are alike. Others say that old hens develop black chins, but never have long crests. A specimen in the Oldham Museum has a black chin and a crest as long as any male's, although the wing formula (1st primary as long as 4th) proves the bird to be a female. Before the youngsters are able to fly the sexes may be readily separated by the shape of their wings.

———◆◆———

In a paper recently read to the Geological Society on the Flora of the Inferior Oolite of Brora (Sutherland), Miss M. C. Stopes pointed out that most of the species found are identical with those of the Inferior Oolite of Yorkshire.

With the July number the 'Bradford Scientific Journal' commences its second volume. Mr. A. Whitaker gives some interesting notes on the 'Scenes and Habits of Bats,' Mr. T. Sheppard writes on 'Museums and their Functions,' Mr. W. H. Whitaker gives an account of the 'Fishes of Upper Airedale,' and Mr. W. P. Winter refers to the 'Mediterranean Plants in the Bradford Botanical Gardens.' There are some useful extracts from the Note-book of the late W. Cudworth. These are chiefly geological. A new feature is 'Science and Nature Notes.' These add to the interest of the journal and are certainly preferable to the queries and answers which formerly appeared. These Science Notes, however, are rather scrappy and would be more useful if more local in character. We were surprised to find Mr. A. Whitaker referring to 'the thirty-five wild animals indigenous to this county,' and throughout the article he appears to substitute 'animals' for 'mammals.'

NOTE ON THE DISTRIBUTION OF *DIATOMA HIEMALE* IN EAST YORKSHIRE, ETC.

R. H. PHILIP,
Hull

IN a sequestered dell, hidden away among the swelling Wolds, the springs of Weedley come to daylight at the base of the chalk, picturesquely overshadowed by hawthorn bushes at the source, then winding serpentine fashion through dense masses of willow herb and lush grasses, till passing under the Hull and Barnsley Railway the little stream forms the valley of Drewton dale. The Wold Springs, of which there are so many in this district, are some of the most interesting habitats of our fresh water diatoms, and in them are found many of our finest and rarest species. But we were not destined to find anything either fine or rare on the occasion of the visit of the York-

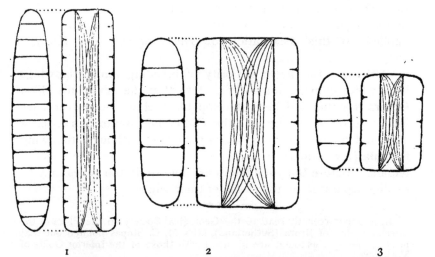

1   2   3

Fig. 1. *Diatoma hiemale* Sertigthal, Switzerland.
Fig. 2.   ,,   ,,   Weedley Springs, near Hull.
Fig. 3.   ,,   ,,   var. *mesodon* Chiavenna, Italy.

shire Naturalists' Union on June 22nd. The margin of the brook was indeed fringed for all its length with flocculent brown streamers, indicating the copious presence of diatoms, but subsequent examination proved that this consisted almost exclusively of a single species—*Diatoma hiemale* (Lyng.), Heib. And yet there are some interesting facts connected with this form which it may be worth while to put on record. Fifty years ago, when the late George Norman was working this district, this species was not known to exist here. It is not included in his list, and having had the advantage of examining many hundreds of his slides, I am able to say that I have not found it in any of his East Riding gatherings. Notwithstanding, it is a common and widely distributed species. I have found it in Switzerland, Italy, Scotland, North of Ireland, Wales, and the English Lake District, as well as in many places in the north and west Ridings of Yorkshire (Y.N.U. excursions to Sedburgh, Bowes, Hebden Bridge, Buckden, Ilkley, Farnley Tyas, etc.). A gathering I took from Weedley Springs, in the spring of 1897, shows no trace of it, and its first appearance in the East Riding to my knowledge was in Sept., 1899, when I found it plentifully in Newbald Springs. Since then I have found it in Weedley Springs, in Stream Dyke, Hornsea Mere, and in flooded fields after heavy rains off Beverley Road and Sculcoates Lane. In the last year or two it seems to have increased enormously. At the Pocklington excursion it was plentiful in the Great Givendale Springs, and also last year at Weedley. This year it appears to have ousted almost everything else from Weedley Springs. It will be interesting to see if anything of the kind has occurred in other Wold Springs, and also to frame a theory to account for the rapid and successful invasion. The species is somewhat variable, and in addition to the drawing of the form found at Weedley, I have drawn two others, the longest, with 10 to 14 costæ, being found in the Sertigthal, an Alpine Valley near Davos, and the shortest, with only 3 or 4 costæ (variety *mesodon*), coming from a mountain torrent at Chiavenna, North Italy. Our English forms are chiefly intermediate, but nearer to the latter.

———◆◆———

From July 8th to July 12th the Museums Association visited Dundee. The following Northern museums were represented :—Bootle, Sheffield, Bolton, Stoke-upon-Trent, Burnley, Newcastle, Stockport, Carlisle, Manchester, St. Helens, Warrington, Bolton, Huddersfield, York, Hull. Amongst the many papers read and discussed, the following were of particular interest and value :—' Methods of Collecting and Exhibiting English Pottery and Porcelain ;' ' Methods of Utilising Wall Spaces ;' ' Notes on the Attitude of Birds ;' ' Centralising Museum Work ;' ' Museums of Industrial Art ;' ' Museums in Higher Grade and Secondary Schools ;' ' Circulating School Museums ;' ' Museums Illustrating Town History ;' ' The Sunday Opening of Museums ;' ' Civic Museums ;' and ' An improved Method of Exhibiting Coins.' An exceedingly valuable feature of this Conference was the opportunity afforded of examining famous collections of furniture, china, &c., which exist in the district. Amongst the places visited in this way were ;— Rossie Priory, the seat of Lord Kinnaird ; the University College, Dundee ; the Perth Museum ; Glamis Castle, the seat of the Earl of Strathmore ; and The University Museum, Castle, and Marine Laboratory at St. Andrews.

1907 September 1.

GLACIAL SURVIVALS.

FRANK ELGEE,
Middlesbrough.

(*Continued from page 276*).

As the unglaciated area of North East Yorkshire comprises the deep, large sheltered valleys lying to the south of the anticline (Rosedale, Farndale, etc.,) their natural history must be particularly interesting. Unfortunately, however, very few records therefrom are known. The evidence, however, that many kinds of widely distributed Land and Fresh-water Molluscs lived in them throughout the Ice Age is decidedly strong. There is direct evidence that many of the most abundant freshwater shells can withstand a very severe climate, and would therefore be capable of surviving the Ice Age.

One of these is *Physa hypnorum*, the shell of which is so fragile as to need most careful handling. It has been noticed on the peninsula of Taimyr in North Siberia in 73° 30' N. lat., where the mean annual temperature is below 10° F. with a range of from 40° F. in July to − 30° F. in January.* If a species so delicate can exist under a frost of 62° it ought to have been able to survive on our driftless area.

Again, those common Pond Snails *Limnæa stagnalis*, *L. peregra*, and *L. truncatula*, as well as *Planorbis albus*, live at the present day in Greenland, whilst *Limnæa auricularia* has been known to survive after having been subjected to a frost of 34°. *Paludina vivipara* and *Anodonta anatina* have resisted a temperature of 23° F., and the former produced young shortly after being thawed out of ice. † In fact, most of the British land mollusca, except the operculate species, are extremely hardy, and may therefore have existed in the large sheltered dales throughout the Ice Age.

In confirmation of this glacial survival of Mollusca, the Arctic Freshwater Bed of the Norfolk coast contains, associated with typical Arctic plants such as the Dwarf Birch and Willow, *Succinea putris*, *S. oblonga*, and *Pupa muscorum*, and the wing cases of beetles. ‡ Altogether at the present time 40 species of terrestrial and freshwater molluscs inhabit circumpolar regions;

---

* Cam. Nat. Hist., Vol. 3, p. 24.
† *Op. cit.*, p. 14.
‡ Geikie, ' Text Book of Geology,' Vol. II.

of these several live in Britain, and appear to have survived the Ice Age.

If this existence of Mollusca during the Ice Age on the driftless region be accepted, we have a probable explanation of the distribution of two of our most characteristic northern shells, viz., *Acanthinula lamellata* and *Margaritana margaritifer*. Existing in Britain in pre-glacial times these two species might have lingered on within the driftless area during the Glacial Period. After its close they would spread out to the surrounding districts. In the case of *A. lamellata*, this extended across the North Sea (then probably a land area) to Sweden, Denmark, and North Germany ; in the case of the Pearl Mussel, to Scandinavia and Northern Germany.

During the Glacial Period, therefore, the Cleveland driftless region probably supported a fauna and flora, chiefly of an Arctic character, though several temperate forms may have also existed. After the ice sheets melted it formed a centre of distribution from which animals and plants radiated to the surrounding bare lands.

What has been here stated concerning North East Yorkshire is applicable, with modifications according to conditions, to other driftless areas. In this respect the peculiar flora of Teesdale is worthy of attention in relation to the glaciation of that valley. Dr. A. R. Dwerryhouse, in his valuable essay on the Ice Age in Teesdale,* shows considerable non-glaciated regions. May not an Arctic fauna or flora have existed here during the Ice Age, including the rare northern plants for which the district is famous ? A careful study of the flora and fauna (especially *Coleoptera*) in relation to these areas might reveal some interesting features.

It seems probable that the biological study of unglaciated areas will decide the problem of glacial survivals, temperate or Arctic. That the whole of our pre-glacial fauna and flora was destroyed during the Ice Age seems improbable, and in considering that of any ice-free country, discrimination must be made between the purely pre-glacial species (e.g., *Mutilla europœa*, the Heaths, *A. lamellata*, *M. margaritifer*) ; the Arctic species (e.g., *Miscodera artica*, Alpine Beetles, the Crowberry, Dwarf Cornel, etc.) ; and the species which have invaded the district since the Ice Age. At best, however, only probabilities and possibilities can be asserted, unless well-preserved fossils contemporaneous with glacial beds be found in any particular district.

* 'Quart. Journ. Geol. Soc.,' 1902.

NATURAL HISTORY OF THORNE WASTE.

THE 204th meeting of the Yorkshire Naturalists' Union was held at Thorne on Thursday, July 11th, 1907, for the investigation of Thorne Waste and the adjacent country.

Mr. H. H. Corbett writes that the very cold and wet season had militated against the prospects of some of the sections of the Union, and no doubt had also an adverse effect upon the numerical strength of the meeting. Notwithstanding this, between fifty and sixty members and associates risked a wetting and the after effects, and were rewarded by a really charming summer day. The whole of the district visited is low lying country, consisting either of re-claimed marsh or uncultivated peat bog, with small areas of slightly raised gravel and sand. The fields and lanes are for the most part bordered by broad ditches, and these, together with the canal, furnished both conchologists and botanists with plenty of material for study. Entomologists, who were strongly represented, found the results of their labours but poor, having regard to the very good repute of the district and the lengthy list of 'good things' tabulated on the official programme.

The outlook for geologists did not appear to be very exciting, but nevertheless they found good work to do, especially in the matter of recent geology.

Most of the naturalists arrived at Thorne N.E. Station at about 11 a.m., and immediately found interest in the large ponds that here occur on the sides of the line. These were bright with the flowers of both species of water lily, and many other less showy but interesting plants grew on the margins. The party walked through the town of Thorne to the Stanforth and Keadley Canal, where some followed one side of the water and some the other. Here were seen several plants of interest, a long stretch of bank being purple with the flowers of *Malva sylvestris;* while in the canal were *Potamogeton crispus*, *P. perfoliatus*, *P. lucens*, *P. pectinatus*, and *P. freisii*, all except *lucens* being in flower and fruit. A somewhat remarkable conchological feature was the great abundance of *Helix cantiana*, a species that seems to be extending its area of distribution.

Leaving the canal side and taking the field path towards 'the Waste,' *Iris pseudacorus* and *Nasturtium amphibium* became conspicuous in the ditches, while the bushy edges furnished only one species or variety of Rubus, namely, *R. dumetorum* v. *ferox* approaching *diversifolius*. This is eminently *the* bramble of the hedge rows and thickets of the cultivated fields near the waste, but almost disappears at the edge of the waste itself, where its place is taken by *diversifolius.* An oat field yielded some interesting weeds, e.g., *Valerianella olitorea, Ranunculus arvensis,* and *Scandix pecten-veneris.* A rough half re-claimed field showed the first real evidence of the true flora of the bog land, *Lysimachia vulgaris, Myrica gale,* and *Potentilla palustre* becoming plentiful, and just at the far side of this field, where the lane bordering the moor was crossed, was seen *Peucedanum palustre,* not yet in flower. Close by *Spiræa salicifolia* grows, and has grown for many years, though probably introduced. The peat itself proved disappointing to botanists. Turf cutting and felling of the birch trees has been going on so rapidly that the more interesting native plants had gone, and neither *Drosera* nor *Andromeda* were to be seen. In the lane leading to the moor *Rubus fissus* was in flower, and on again joining the canal several sedges were gathered, the most interesting being *Carex pseudo-cyperus.*

Medge Hall Station was reached nearly an hour before the time for the train to Thorne, so the peat bog was again visited, and here the mycologists found some interesting things. In a patch of fodder oats just reclaimed from the bog was abundance of *Erysium chæranthoides,* and by the ditch sides another rare crucifer *Barbarea stricta.*

At the station the entomological section joined the botanical, and reported poor results! the only good local insects seen being *Chortobius Davus* and *Chaerocampa elpenor* among the lepidoptera. One of them had, however, found an abundance of *Andromeda,* with which he supplied his botanical friends.

Returning to Thorne a very well served tea was taken at the Red Lion Hotel, and the sectional meetings were held, followed by the general meeting, under the Chairmanship of the President, Mr. C. Crossland. Votes of thanks were passed to the landowners for granting permission to visit their properties, and to the guides.

The following reports of sections and lists of species have been furnished by different members.

GEOLOGY.—Mr. Culpin writes :—The route taken was from Crowle by Godknow Bridge and Thorne Waste to Thorne, the sections examined on the way being :—

Gravels of Keuper marlstones on the south side of Crowle Hill.

Keuper marls, topped by blown sand, at the Crowle Brick and Tile Works.

Peat on Thorne Waste.

Gravels, being grits with some quartzites, on Tween Bridge Moors.

Bunter Sand at Thorne.

Gravels of Magnesian limestone at Thorne, mingled with which were some Carboniferous limestones and grits.

The VERTEBRATE SECTION was not officially represented at the meeting, but the Rev. F. H. Woods and Messrs. Butterfield and Corbett noted the following amongst numerous commoner birds :—Reed Warbler, Marsh Tit, Cuckoo (still singing), Reed Bunting, Lesser Redpole, Mallard, Snipe, Redshank, and Blackheaded Gull.

ENTOMOLOGY. Lepidoptera.—The following, among other commoner species, were found :—

Chortobius Davus.

Hesperia sylvanus.

Chaerocampa elpenor.

Arctia fuliginosa.

Coleoptera.—Mr. E. G. Bayford writes :—No doubt owing to long continued wet weather, beetles were not much in evidence, many of the species enumerated below having being seen in single specimens. Even the *Coccinellidae*, usually so ubiquitous at this time of the year, were by no means common, *C. 7-punctata* not being so frequent as on Good Friday, while the equally, if not more common, *C. 10-punctata* was not observed at all. The fine hot week which commenced with the visit of the Union to Thorne will no doubt have greatly increased the number of species, as also their comparative abundance. The following is a list of those met with :—

Carabus violaceus L.

Notiophilus palustris Duft.

Leistus ferrugineus L.

Bradycellus placidus Gyll.

Harpalus ruficornis F.

Pterostichus niger Schall.

    ,,   *strenuus* Panz.

Amara aulica Panz.

  ,, *plebeia* Gyll.

Amphigynus piceus Marsh.

Bembidium flammulatum Clairv.

Trechus minutus F.

Haliplus ruficollis De G.

Laccophilus variegatus Germ.

Hydroporus dorsalis F.

    ,,   *palustris* L.

Hydroporus pubescens Gyll.

Agabus bipustulatus L.

Gyrinus natator Scop.

Megasternum boletophagum Marsh.

Oligota inflata Mann.

Conosoma pubescens Grav.

Tachyporus chrysomelinus L.

    ,,   *hypnorum* F.

Tachinus marginellus F.

Quedius fulgidus F.

Creophilus maxillosus L.

Philonthus æneus Rossi.

    ,,   *politus* F.

    ,,   *fimetarius* Grav.

Xantholinus linearis Oliv,

Othius læviusculus Steph.

Oxyporus rufus L.
Oxytetus rugosus F.
Oxytetus sculpturatus Grav.
    ,,   tetracarinatus Block.
Homalium rivulare Payk.
Necrophorus mortuorum F.
Trichopteryx lata Mots.
Ptenidium nitidum Heer.
Adalia bipunctata L.
Coccinella 7-punctata L.
Halyzia 16-guttata L.
Rhizobius litura F.
Coccidula rufa Herbst.
Onthophilus striatus F.
Bracyhpterus pubescens Er.
    ,,   urticæ F.
Lathridius lardarius De G.
Coninomus nodifer West.

Enicmus minutus L.
Corticaria pubescens Gyll.
Elater balteatus L.
Athous hæmorrhoidalis F.
Cyphon variabilis Thunb,
Telephorus flavilabris Fall.
Strangalia armata Herbst.
Donacia simplex F.
Chrysomela polita L.
Anaspis frontalis L.
    ,,   maculata Fourc.
Phyllobius urticæ De G.
    ,,   argentatus L.
    ,,   viridi-aeris Laich.
Sitones hispidulus F.
Cionus blattariæ F.
Cœliodes quadrimaculatus L.

Of other orders of insects many species were taken, but these await examination and diagnosis.

For the CONCHOLOGICAL SECTION, Mr. J. E. Crowther reports :—Arriving at Thorne, somewhat earlier than the time stated on the circular, we made our way to a drain near the first turn-bridge, which had yielded very good results in the first week in May. On our way we found *Helix nemoralis* very sparingly on the road side, but fine in size and colour. The drain having been recently cleaned out, yielded nothing but *Limnæa pereger* and *L. stagnalis* of small size. *Helix cantiana* was very abundant on the sloping banks, along with *H. hispida* and *Succinea putris*. *S. elegans* was very plentiful close to the edge of the water. In the big ditch alongside the canal *Valvata piscinalis* was plentiful, with *Vivipara vivipara* and *Limnæa truncatula*, on the bricks of the bridge. *Sphaerium pallidum* occurred in the canal. In the road side ditch on the way to the moor *Bythinia leachii* and *B. tentaculata* were plentiful, while young *L. pereger* occured in great numbers, together with several species of *Planorbis*. In a drain near by *Physa hypnorum* and *Limnæa glabra* were found. The only species found on the moor itself was *Vitria alliaria*, obtained by Dr. Corbett a few days before the meeting.

The total number of species seen was thirty, composed of four slugs, seven terrestrial, and nineteen aquatic species. The complete list is as follows :—

Agriolimax agrestis.
Vitria alliaria.
Arion ater.
    ,,   hortensis.
    ,,   minimus.

Helix cantiana.
    ,,   hispida.
    ,,   aspersa.
Helix nemoralis.
Succinea putris.

Planorbis corneus.
,, albus.
,, carinatus.
,, vortex.
,, contortus.
Physa fontinalis.
,, hypnorum.
Bythinia tentaculata.
,, leachii.
Vivipara vivipara.

Valvata piscinalis.
Sphaerium corneum.
,, pallidum.
Pisidium fontinale.
,, elegans.
Limnæa pereger.
,, palustris.
,, truncatula.
,, stagnalis.
Limnæa glabra.

BOTANY. *Phanerogams.*—Mr. Bellerby and Mr. Corbett took note of the more interesting flowering plants. Altogether about one hundred and ninety species were seen either in flower or fruit. Of these the following list contains the more local and rare :—

Thalictrum flavum L.
Ranunculus arvensis L.
Nasturtium amphibium R. Br.
Barbarea stricta Andrz.
Erysimum cheiranthoides L.
Coronopus ruellii All.
Rubus fissus Lindb.
,, plicatus W. & N.
,, dumitorum W. & N.
Potentilla palustris Scop.

Peucedanum palustre Mœnch.
Andromeda Polifolia L.
Hottonia palustris L.
Lysimachia nummularia L.
Rumex hydrolapathum Huds.
Sagittaria sagittifolia L.
Potamogeton lucens L.
,, friesii Rupr.
Carex pseudo-cyperus L.
Calamagrostis lanceolata Roth.

ECOLOGICAL BOTANY.—The Rev. E. Adrian Woodruffe-Peacock writes :—The soils between Thorne and Crowle are most varied. Many of them are purely artificial or of true human origin. The rock bed consists of the water stones of the Keuper, but it is all buried by sand and gravel of river origin, blown sand and peat. *Phleum pratense maximum* was only met with once, on old canal dredgings. *Myosotis collina* only as a road-side casual. *Hypochœris radicata* was conspicuous on the peaty soils, but *Cnicus arvensis* was only found on dyke banks in shallow peat mixture. *Chrysanthemum Leucanthemum* was patchy in the same situation. An oat field illustrated what a combination of natives and aliens might be found on a rich, peaty, sand, alluvial soil :—*Viola arvensis, Spergula sativa, Anagallis arvensis, Lepidium campestre, Raphanus Raphanistrum,* .and *Myosotis versicolor, Matricaria chamomilla* were all found within a yard. A peaty sand pasture had the characteristic combination :—*Agrostis vulgaris, Festuca rubra, Galium palustre, typica,* though var: *witheringii* was the commonest form, *Polygonum amphibium, Myosotis palustris, Potentilla sylvestris, Hieracium Pilosella, Myosotis versi-*

color, Cnicus palustris, Veronica officinalis, etc., with *Carex ovalis* in one limited area of fairly short turf. The flora of the waste where I touched it first was characteristic of a desiccating quagmire. The peat was much higher than the surrounding cultivated soils, with *Pteris* as the predominating species, and with *Calluna* and *Eriophorum vaginatum* as first and second subsidiaries. The *Pteris* clearly overpowered everything but the *Betula verrucosa,* which had all been cut down. *Molinia varia* held the fourth place frequently in the varietal form *depauperata.* On picking up the tramway, which carries the dried turves to the manufactury, one of the prettiest ecological studies imaginable came under review. The line, without much attempt at levelling, had been laid east to west right over the original peat surface, and cinders from the engine-room fire had been used as a binder to compact the road for the horses' feet. These cinders provided sufficient mineral matter for the growth of common species on the rich, nitrogenous moor soil. I found no plant beyond six inches from the outside of the rails on either side, seldom as far, for not one of them can grow on pure peat, though almost all are more or less characteristic of peat mixed slightly with some introduced soil. Those I met with first are the most usual moor-side dwellers. The following is the order in which the species came into evidence :—*Festuca sciuroides, Bromus mollis, Cnicus arvensis, Poa annua, Holcus lanatus, Poa pratensis, Cerastium triviale, Dactylis, Plantago lanceolata, P. major, Veronica serpyllifolia, Veronica agrestis, Veronica arvensis, Polygonum Persicaria* was very sickly, *Tussilago* very fine, *Myosotis arvensis, Anthriscus sylvestris, Lolium perenne, Trifolium repens* poor, *Agrostis vulgaris* fine, *Urtica dioica* poor, *Bromus mollis,* var. *glabratus, Senicio sylvaticus* poor, *Rumex crispus, Geranium molle, Hypochœris radicata, Arrhenatherum, Urtica urens* very poor, *Poa trivialis, Bursa, Veronica chamœdrys, Senecio erucifolius, Polygonum aviculare,* var. *rurivagum, Bellis, Cnicus lanceolatus, Senecio vulgaris, Lychnis alba,* and *Festuca rubra.* The last five species were out of place altogether.

The road, as soon as the peat is left, up to Midge Hall railway station, is an estuarine alluvium or warp. All our best pasture grasses came into evidence at once, *Festuca elatior, typica,* being very characteristic. *Sonchus asper* and *Cnicus lanceolatus* were both five feet high. *Papaver rhœas* was mixed with the variety *Prioris, Geranium dissectum, Matricaria inodora, Reseda luteola, Linaria vulgaris,* and *Sisymbrium officinale,* illustrated the mixed nature of the road and ditch side vegetation. *Bromus mollis glabratus* was also just outside the station.

The above does not attempt to be an exhaustive list of all the notes made.

FUNGI.—Mr. C. Crossland writes :—The mycologists had a' pleasant and profitable time, especially in the birch, ling, and conifer portion of the strip of woodland bordering the Waste.

There were a good scattering of agarics. *Amanita rubescens* —'the blusher'—was very abundant. A nice little cluster of *Inocybe scabella* was found among short grass. *Mycena sanguinolenta* was very common among dead ling, and on some rotting birch logs. A noticable feature was the quantity of *Marasmius androsaceus* on damp decaying portions of prostrate ling. A bright, golden yellow *Flammula* was also common on the ground among ling, and on rotting birch stumps ; this form appears to have a preference for heathy ground : we find it on the moor edges about Halifax under similar conditions. Perhaps the best find was made by one of the entomologists while beating an *Epilobium hirsutum* bush, on the railway side near Medge Hall station ; this is a coral pink peziza which occurred in scores on the bare soil; it studded the ground for some yards near the *Epilobium ;* it appears to be nearest to *Peziza Adæ,* but is not that species so far as one can judge at present. *Echinella setulosa* Mass. & Crossl. was especially looked for on decaying branches of ling, and found in abundance ; this little discomycete was formerly confused with *Mollisia cinerea,* which, under a pocket lens, it much resembles ; the characters of the spores, however, are very different ; its appearance here and at other places, since it was properly diagnosed,* substantiates the remark made in the Yorks. Fung. Flo. that it would be 'certain to be found in additional localities if looked for.' There were few fungal leaf parasites.

Two species are new to Yorkshire ; these are marked with an asterisk.

All the following species were found on or near Thorné Waste :—

GASTROMYCETES.	Scleroderma verrucosum, On the
Crucibulum vulgare. On decaying twigs.	ground in plantation.
	HYMENOMYCETES.
Lycoperdon pyriforme. On the ground among rotting twigs, etc.	Agaricaceæ.
	Amanita rubescens. On the ground, margin of the wood.

---

* Mass. Brit. Fung. Flo. iv. p. 305.

Laccaria laccata. Among grass.

Collybia dryophila. Among decaying leaves.

Mycena vitilis. Among decaying twigs, etc.

M. sanguinolenta. Common on dead ling, and on rotting birch logs.

Omphalia umbellifera. On heathy ground.

O. fibula. Among moss.

* *Clitopilus popinalis.* Among grass, margin of the wood.

Nolanea pascua. Among grass.

* *Inocybe scabella.* Among grass in the wood. Quite a distinct species.

Flammula sapinea. Var. Among decaying ling, and on rotting birch stumps.

Naucoria semiorbicularis. Among grass in moist place.

Galera tenera. Among grass.

Agaricus arvensis. In pasture.

A. campestris. In pasture.

Stropharia semiglobata.

Panæolus fimicola.

Coprinus radiatus.

The last three on dung in pastures.

C. ephemerus. On rich soil.

Paxillus involutus. On woodland ground.

Hygrophorus ceraceus.

H. obrusseus. Both in pasture.

Lactarius turpis. On heavy soil, pathway near the wood.

L. quietus.

L. subdulcis.

Russula fragilis.

All three on soil in woodland.

Marasmius oreades. Among grass, but not in rings as one generally finds it.

M. androsaceus. Common among decaying ling.

POLYPORACEÆ.

Polystictus versicolor. On birch stump.

Poria vaporaria. On decaying trunk.

HYDNACEÆ.

Grandinia granulosa. On rotting wood.

TREMELLACEÆ..

Dacryomyces stillatus. On rotting wood rail.

Calocera viscosa. On birch stump.

UREDINACEÆ.

Coleosporium sonchi. On coltsfoot.

Puccinia poarum (*Æcidiospores*) on coltsfoot.

P. malvacearum. On *Malva rotundifolia.*

USTILAGINACEÆ.

Ustilago longissima. On leaves of *Glyceria aquatica.*

U. avenæ. In the flowers of *Arrhenatherum avenaceum.*

PYRENOMYCETES.

Hypocrea rufa (*Conidial condition*). On rotting birch log.

Xylaria hypoxylon. On birch stumps.

Daldinia concentrica. On birch logs.

PERISPORACEÆ.

Podosphæria oxyacanthæ. On living leaves of thorn.

DISCOMYCETES.

Peziza vesiculosa. On soil.

Dascypha hyalina. On dead birch stump.

Echinella setulosa. On prostrate ling branches.

Mollisia cinerea. On dead branches.

HYPHOMYCETES.

Penicillium glaucum. On rotting ling.

Isaria farinosa. On dead insect.

MOSSES and HEPATICS.—Mr. C. A. Cheetham writes :—The mosses at Thorne were few in number, two, however, were plentiful on the Waste ; *Campylopus pyriformis* on the bare walls of peat in sheets, fruiting abundantly, and *Webera nutans* amongst the ling and bracken ; one wet ditch at the edge of the moor was quite full of this latter, partly submerged in the water, no doubt due to the wet season.

These two mosses are typical of a peaty moorland ; both have vegetative methods of reproduction which serve the plants if the more ordinary method of spore formation is in any way checked. That of *Campylopus* is well known, leaves break off and often cover the barren plants, and such leaves will, under favourable circumstances, grow and produce new plants.

With *Webera nutans*, some plants are found with long innovations with adpressed leaves, these elongated stems easily break off and may then grow on in a fresh place.

The mosses seen on the roadsides and walls on the way were :—*Dicranella heteromalla, Ceratodon purpureas, Barbula muralis, B. recurvifolia,* and *Bryum argenteum.* Then on the clay banks of the canal *Dicranella varia.* Getting near to the Waste a good wet place was passed with *Sphagnum acutifolium, Hypnum cuspidatum, H. cordifolium.* This spot would probably have repaid a more careful search, but we were anxious to get to the Waste. Arriving there, a feeling of disappointment came over us, as the place is now far too well drained to harbour many mosses. The first two mentioned, *Campylopus pyriformis* and *Webera nutans,* were the principal, with small patches of *Polytrichum commune, Dicranella heteromalla, Dicranum scoparium, Brachythecium rutabulum, Eurhynchium myosuroides.* The only Sphagnum seen on the Waste was *S. acutifolium* in the wet bottom of the old duck decoy.

Mr. W. Bellerby adds :—Under a shady bank of one of the numerous channels cut through the peat, a large patch of a common hepatic, *Cephalozia bicuspidata* (L.), and *Odontoschisma Sphagni* (Dicks,), were the only hepaticæ seen.

A fine mass of *Sphagnum fimbriatum* (type) of bright green colour, was growing in a shady trench and also a slender form of the same peat moss, which I sent to Dr. Warnstorf for identification, he writes, ' The sphagnum sent for determination is *Sphagnum fimbriatum* Wils., var. *tenue* (Grav).' It is a very tall, slender and graceful form, bright green above, and with lateral gracefully arching branches curving downwards and distantly placed, a distinct plant much more slender than the type.

An interesting lichen growing among *Webera nutans* I sent to Mr. Wheldon. It is of white colour with crimson spherical fruit, *Cladonia Floerkeana* forma *trachypoda* (Nyl.). Mr. Wheldon writes, ' This Cladonia is one of the section of the Erythrocarpæ and is distinguished from *C. coccifera* and *C. digitata* by its chemical reaction K-C-.'

FIELD NOTES.

BIRDS.

Hawfinch Nesting near Sedbergh.—The Hawfinches have nested again this year in Ingmire Park, near Sedbergh.—Wm. Morris.

Red=backed Shrike nesting in Yorkshire.—A pair of Red-backed Shrikes nested and brought off their young this summer in the Pickering district.—Oxley Grabham, York.

Nightjar's Nest with unusual number of Eggs.—This year I was shown by Messrs. H. B. Booth and R. Butterfield a nest of the Nightjar near Shipley, containing the unusual number of four eggs. Apparently the eggs had been laid by two birds, as there were two distinct types. One bird was, however, sitting upon the eggs, covering them quite con-

Nightjar's nest with four eggs.

tentedly. There seemed to be no doubt that the eggs had been deposited *in situ*, and not introduced by human agency. One cannot but wonder at the cause which had induced two birds to deposit their eggs upon the same spot, especially as the whole district abounded in suitable nesting sites; indeed, I do not remember seeing so many Nightjar's nests in such a small district as I saw that day.—R. Fortune.

The Spotted Flycatcher in Yorkshire.—A note of mine appeared in the July 'Naturalist,' commenting upon the scarcity of the Spotted Flycatcher at Harrogate this year. When the note was written this was correct, but since then these birds have turned up, if anything in increased numbers. They were unusually late in arriving, and they are correspondingly late in nesting. To-day, August 11th, a pair is feeding their young in a rather unusual place ; the nest is situated on a string course just under the spout of a house in Harrogate, quite 30 feet from the ground, and another has a nest with young on a lamp-post. During the visit of the Yorkshire Naturalists' Union to Arncliffe a nest was found in the highest loop-hole window in the church belfrey. It contained newly hatched young. Many of the members were much interested in being able to obtain a very close view of the birds, which they were able to do from the steps inside the tower.—R. Fortune, Harrogate.

Late arrival of Migrants in Craven.—The migrants have been unusually late this season, owing to the continued spell of cold weather during the time when they generally reach their breeding haunts. We noticed that most birds arrived during a S.W. wind, which usually preceded a slight improvement in the weather. Below are the dates of their arrival in this district :—

March 21.—Ringousel.	May 4.—Treepipit.
,, 31.—Wheatear.	,, 5.—Spotted Flycatcher.
* April 8.—House Martin.	,, 6.—Greater Whitethroat.
,, 13.—Yellow Wagtail.	,, 6.—Swift.
,, 19.—Redstart.	,, 6.—Landrail.
,, 22.—Willow Warbler.	,, 10.—Blackcap.
,, 23.—Sand Martin.	,, 8.—Lesser Whitethroat.
,, 23.—Common Sandpiper.	,, 12.—Garden Warbler.
,, 24.—Swallow.	,, 16.—Pied Flycatcher.
,, 24.—Cuckoo.	,, 18.—Whinchat.
,, 24.—Chiffchaff.	

House Martin.—The first arrival, April 8th, was a very early date, but scarcity of insect food drove them away. They reappeared on April 15th, when they stayed a few hours, after which they were not seen again until April 22nd. Food then being more abundant they remained with us.

Spotted Flycatcher.—There is a great increase in the number of these birds this season.

Landrail.—The landrail has not been heard in many of its favourite haunts in this district ; no doubt the disastrous weather during migration time accounts for their scarcity.

Common Whitethroat.—Quite an increase in their numbers to this district.

Yellow Wagtail.—This bird is not as common as it was in this district in past years.—W. Wilson, Skipton-in-Craven, July 13th, 1907.

MOLLUSCA.

Arion ater var. castanea at Newsome.—On the 18th June last my sister drew my attention to a monstrous slug in our garden, which I boxed and sent to Mr. W. Denison Roebuck, for identification. Mr. Roebuck informs me that this slug was one of the finest *Arion ater* he had seen. It would stretch out to six inches and then not be fully extended. The variety is *castanea*, for though very dark, it is not black, but brown. This variety of *A. ater* does not appear to have been previously recorded for the Huddersfield district.—W. E. L. WATTAM.

—: o :—

FLOWERING PLANTS.

Potamogeton alpinus, Balb., near Doncaster.—This plant is abundant in a slow stream at Sandal Beat, near Doncaster, in Vice-County 63, at an altitude of about 35 ft. above O.D. As there is no record for this species in 'The Flora of the West Riding,' it is worth noting. The plant has been identified by Mr. Arthur Bennett.—H. H. CORBETT, Doncaster.

—: o :—

HEPATICS.

Ricciella fluitans (Braun.) at Mablethorpe.—While dredging for fresh water shells in the dykes on Poplar Farm, Mablethorpe, on Aug. 3rd, I brought up *Ricciella fluitans* in plenty from the dyke on the north-east side of the farm, and about 200 yards from the Theddlethorpe Road. I believe this interesting Hepatic is only recorded for one other locality in Lincs., in "Transactions of the Lincs. Nat. Union," 1906, in Miss S. C. Stows' List of Lincs. Liverworts on Scotton Common, 29th July, 1905, J. Reeves.'—F. RHODES, Bradford, 13th August, 1907.

———◇◆———

The Fungi of Ants Nests in Ceylon has been made the subject of a Memoir by the Government Mycologist, Mr. T. Petch. He considers, after careful examination, microscopic and otherwise, of fresh specimens, the typical species to be a Volvaria, so that its name will stand *Volvaria eurhiza* Berk. Originally described by Mr. Berkeley in 1847, from dried specimens, as an Armillaria, it has been by various authors regarded as a Lentinus, Collybia, Pluteus, Pholiota, and Flammula. The volva is very adnate to the base of the stem, resembling our *V. gliocephalus*. The fungus grows from the coomb, and comes up above the ground in rainy weather. Another species which grows on the nest is an edible one, *Entoloma microcarpum*. *Xyliaria nigripes* grows on deserted nests, while species of Mucor, Thamnidium, Cephalosporium, and Peziza are developed upon the coomb after it is removed from the nests. The paper is illustrated by 19 excellent plates from photographs, and is published in 'Annals of the Royal Botanic Gardens, Peradinyia,' November 1906, pp. 185-270.

NORTHERN NEWS.

'If a fly should get in your eye, keep your eye tightly closed for *two or three minutes*':—the prize 'Country-side hint' recently!

More *damages!* Mr. E. E. Austen describes 'A rare British Fungus-midge, re-discovered in London,' in the 'Entomologist's Monthly Magazine' for August.

In answer to an advertisement for a well-known geological memoir, a leading London bookseller sends us a quotation for 'The Yorkshire Liars, Tate & Blake.'

At the Arncliffe meeting of the Yorkshire Naturalists' Union the geological members were invited to examine a fine 'fossil trout,' which was so well preserved that even the spots were clearly shown. It proved to be a cast of a *Stigmaria!*

We must congratulate our Bradford scientific friends on their method of popularising their journal by the insertion of short scientific notes of a humorous nature. In a recent issue we learn that 'Gold has been discovered on the shores of Loch Fyne (*where the herrings come from*)'!

Mr. John Maclauchlan, the President of the Museums Association, thinks that had there been museums in Elizabeth's time, the greatest of all the poets would have written, 'All the world's a museum, and all the men and women merely specimens, who, in their time, play many parts.'

We quite agree with a recent writer, in referring to the courtship of grasshoppers, that 'one's fancy must stretch a good deal from the human point of view to realise that the suitor whispers his soft nothings with the back of his wings, and his sweetheart listens with her front legs.'

Early in August the local press recorded a 'pest of jelly fish' on the Yorkshire coast. 'Some of the pink variety were of immense bulk, measuring over three feet across, and from six to nine inches in thickness. The blue-tinted fish were also very numerous.' Salmon nets were quickly filled with them, and incautious bathers were severely stung.

We learn from a certain source, which can probably be guessed, that 'Insects have no conscious feeling of pain. Their knots of nerves are distributed down their body, and there is no central brain to enable the insect *to regard itself as an individual, and realise that it suffers.*' The information is so decisive that we presume the writer has had the information supplied direct by some 'insect.' Or was it the lyre-bird, which, by the way, is suggested as a badge for a certain natural history (*sic*) paper.

In connection with the recent meeting of Yorkshire Naturalists in Littondale, Mr. W. Morrison supplied the members with some interesting local information. The devil, locally known as Old Pam, takes the Thresh-field Grammar School for one night in the year, and teaches the little Wharfedale devils, who are 'that clever that they need nobbut yan nicht's schuling i' the year.' The devil also gives a supper at Kirkby Malham on a tombstone in the churchyard at midnight. As the dates on which these events occurred were not known, the members were not able to take part. Mr. Morrison also could not help, nor could he make enquiries, as, he said, he did not know the devil's address!

The editor of a certain 'natural history' newspaper, thinking that 'many may be glad to know' how he manages to be prepared for most eventualities, although, when starting for a walk, he may seem to be armed only with a walking stick [!], proceeds to the length of two columns to give his readers the necessary information. We have had the patience to peruse his article, and learn that he wears a *coat* with plenty of pockets, in which he puts his tobacco pouch, handkerchief, match-boxes, etc. He also carries a knife and a piece of string. We wonder if he wears a hat, and if so, what he puts in it. We wish the army of badge-wearers would see that their chief did not run short of matter. Have the 135 who have recently won sixpence each in a 'what is it' competition no gratitude?

505.42:

OCTOBER 1907.

No. 609
(No. 387 of current series)

THE NATURALIST.

A MONTHLY ILLUSTRATED JOURNAL OF
NATURAL HISTORY FOR THE NORTH OF ENGLAND.

EDITED BY

T. SHEPPARD, F.G.S.,
THE MUSEUM, HULL;

AND

T. W. WOODHEAD, Ph.D., F.L.S.,
TECHNICAL COLLEGE, HUDDERSFIELD.

WITH THE ASSISTANCE AS REFEREES IN SPECIAL DEPARTMENTS OF

J. GILBERT BAKER, F.R.S. F.L.S.,   GEO. T. PORRITT, F.L.S., F.E.S.,
Prof. P. F. KENDALL, M.Sc., F.G.S.,   JOHN W. TAYLOR,
T. H. NELSON, M.B.O.U.,   WILLIAM WEST, F.L.S.

Contents :—

LONDON:

A. BROWN & SONS, LIMITED, 5, FARRINGDON AVENUE, E.C.
And at HULL AND YORK.

Printers and Publishers to the Y.N.U.

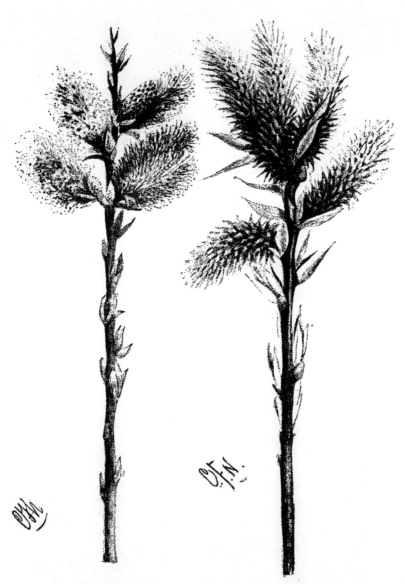

Sallow Catkins.
(Male (or Staminate) Blossoms).

Willow Catkins.
(Female (or Pistillate) Blossoms).

NOTES AND COMMENTS.

THE PACIFIC EIDER AGAIN.

We have more than once referred to the reported occurrence of the Pacific Eider (*Somateria v-nigrum*) in Britain, a record based upon a specimen shot in the Orkneys in 1904, which was sent to a Scarborough dealer, who sold it to a gentleman in Oldham. Mr. F. Smalley, in the August 'British Birds,' gives what is perhaps the last word on the subject, having examined the Oldham specimen, and also several others. He concludes that the Pacific Eider is still to be recorded for Britain, the supposed occurrences being merely examples of the Common Eider, which had more or less distinct V-marks under the throat. Mr. Smalley gives illustrations of the chins and throats of the Pacific Eider, and also of a variety of the Common Eider, from which the differences between the markings on the two species is clearly shown. It is considered that the occasional indistinct V-mark on the examples of the Common Eider is an instance of 'reversion.'

FOOD OF THE BLACK-HEADED GULL.

There has recently been much discussion on the usefulness or otherwise of the Black-headed Gull, but after the excellent and exhaustive report by Messrs. D. Losh Thorpe and L. E. Hope, recently issued for the Cumberland County Council, this discussion will surely cease. The report is really a fine piece of work, and reflects every credit upon the authors. Circulars were forwarded to leading naturalists, as well as to well-known farmers, gamekeepers, and anglers. These contained the questions :—(1) 'Do you consider the Black-headed Gull harmful to the fishing or farming industries? State reasons'; (2) 'Have you ever examined the gullet and stomach of this Gull? If so, what were the contents?' and (3) 'What in your opinion is the staple food of this Gull?' Sixty-two replies were given to these queries, and these are carefully tabulated and summarised. In addition, the authors have, during a period of thirteen months, examined the contents of the stomachs and gullets of 100 birds. The results of this examination are given in detail and tabulated—the material being preserved in spirits. The report is distinctly in favour of the birds, which are evidently not so black as they have been painted. We only regret our space prevents us from referring to this work in greater detail.

THE 'GREY WETHERS.

The 'Grey Wethers' of the Marlborough Downs are well known. They consist of hardened and solidified boulders of a stratum of Eocene, which formerly covered the Chalk. These 'Sarsen stones,' as they are locally called, are scattered about the surface of the ground, and vary in size from small boulders to vast masses weighing sixty or seventy tons. For many years the sarsens have been broken up for building purposes, etc., and these interesting geological relics are threatened with destruction. A fine collection of them occurs in Pickle Dean, near Marlborough (see Plate XXXV.). For a sum of £500 twenty acres can be secured and preserved for all time. An appeal for funds is being made by the Wiltshire Archæological Society, the Marlborough College Natural History Society, and the National Trust (25 Victoria Street, S.W.), and this we commend to the notice of our readers.

LAMARCK'S EVENING PRIMROSE.

The most conspicuous of the many alien plants which occur on the sandhills and sandy wastes of St. Anns-on-the-Sea is Lamarck's Evening Primrose (*Œnothera Lamarkiana*, Ser.). It is probably of North American origin and produces a daily succession of rich yellow flowers from July to November. This is the species which occurred at Hilversum near Amsterdam, and in 1886 attracted the attention of Hugo de Vries. He studied the variations long and minutely, and the result was his well-known 'Mutationstheorie.' Mr. Charles Bailey has examined the varieties at St. Anne's, and gave an account of his results in an address to the Manchester Field Club, which has been recently published.* This plant has been established at St. Anns upwards of thirty years, and occurs on both sides of the Ribble Estuary. Mr. Bailey gives a summary of De Vries' results, and reproduces in detail the characters of the principal mutations. He shewed that new elementary species attain their full constancy at once, no intermediates occurred amongst them, that they made their appearance suddenly, and there was no apparent evidence of a struggle for existence or anything that savoured of natural selection. The principal variations found at St. Anns consisted of slight modifications of the petals and styles. The striking modifications in height, branching,

---

* 'De Lamarck's Evening Primrose on the sandhills at St. Ann's-on-the Sea.' Chas. Bailey. Manchester: Hinchcliffe & Co., July, 1907.

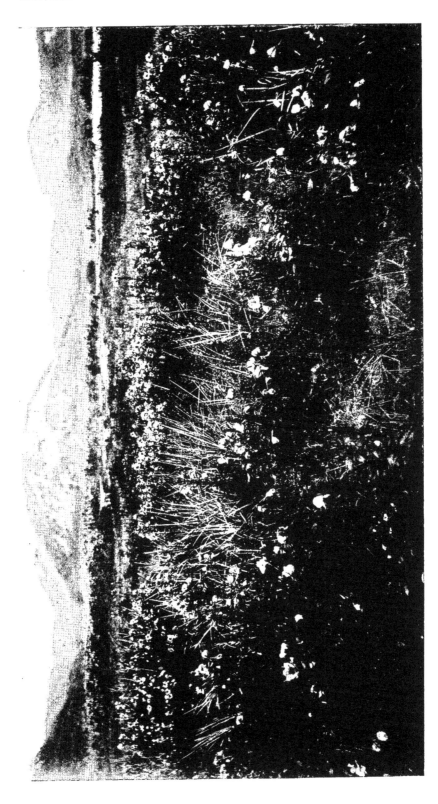

Œnothera Lamarkiana Ser. View looking across the Sandhills westward from the eastern end of Beach Road, St. Anns-on-the Sea.

PLATE XXVII.

leaves, fruits, and seeds, noticed by De Vries, do not seem to have been noticed at St. Anns, though one form has some of the characters of *rubrinervis*, and another, with sessile stigmas, agrees more closely with *brevistylis*. The paper is illustrated by six excellent photographs, one of which (Plate XXXVI.) we are permitted to reproduce by the kind permission of the author. We hope that these observations will be continued, and that a sharp look-out will be kept for forms agreeing still more closely with those described by De Vries. In a second paper* Mr. Bailey gives an account of upwards of forty species of aliens occurring on the sandhills of St. Anns. Besides plates of *Œno-thera* this is illustrated by two plates of *Ambrosia artemisifolia*.

FOOTPRINTS IN THE SANDS OF TIME.

Perhaps one of the most interesting contributions to the Geological Section at the British Association meeting at

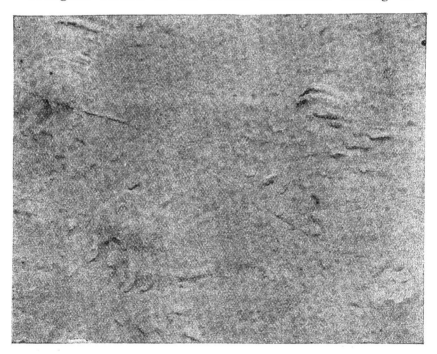

Leicester was the report of the Committee appointed for the investigation of the Fauna and Flora of the Trias. This included a paper by Dr. A. Smith Woodward 'On a mandible

---

* 'Further Notes on the Adventitious Vegetation of the Sandhills of St. Anns-on-the-Sea.' 'Memoirs and Proceedings of the Manchester Literary and Philosophical Society.' Vol. 51, Pt. III., 1907.

of *Labyrinthodon leptognathus* from the Keuper sandstone near Leamington.' An area of about 40 feet by 50 feet of footprint-bearing sandstone has been exposed in the quarries at Storeton, and a careful examination of these has resulted in some important additions to our knowledge of these footprints being obtained. Large slabs from this recent exposure are now preserved in the British Museum and the Museums at Birkenhead, Bolton, Hull, Leeds, Liverpool, and Manchester. The report also contains notes by Messrs. H. C. Beasley, J. Lomas, A. R. Horwood, and L. J. Wills. One of the illustrations accompanying the report we are kindly permitted to reproduce.

NORFOLK NATURALISTS.

'The Transactions of the Norfolk and Norwich Naturalists' Society' for 1906-7 * are to hand, and appear to contain a greater number and variety of valuable papers than usual, which is saying a good deal. To enumerate the titles of the papers even would be a lengthy matter, and their worth can be estimated from the following list of authors :—C. A. Hamond, J. T. Hotblack, A. Bennett, J. H. Gurney, H. Laver, H. M. Evans, J. O. Borley, T. Southwell, W. G. Clarke, R. Gurney, A. W. Preston, E. L. Turner, T. J. Wigg, F. Leney, A. H. Patterson, F. Balfour Browne, and Claude Morley. The subjects dealt with relate to Archæology, Botany, Birds, Conchology, Fishes, Biography, Crustacea, Meteorology, Entomology, etc., etc.

NESTS OF COOT AND CRESTED GREBE.

Perhaps one of the most interesting articles is by Miss E. L. Turner, who had the rare good fortune to watch and photograph at close quarters the nests of a Coot and Crested Grebe, which were built only eighteen inches apart, on Hickling Broad. Her description of the nesting habits of these two species, and of the tactics she made to get near to them and photograph them, are pleasant to peruse. Eight of her photographs accompany the notes, and by kind permission these are reproduced for the benefit of our readers (Plates XXXVII. and XXXVIII.). The photographs shew :—(No. 1) Coot on nest and Grebe uncovering her eggs ; (No. 2) the unusual height of the nests ; (No. 3) Coot on nest with two young ones, the

---

* Vol. VIII., Pt. 3. 1907, pp. 329-498, Plates. Fletcher & Son, Norwich, 5/-

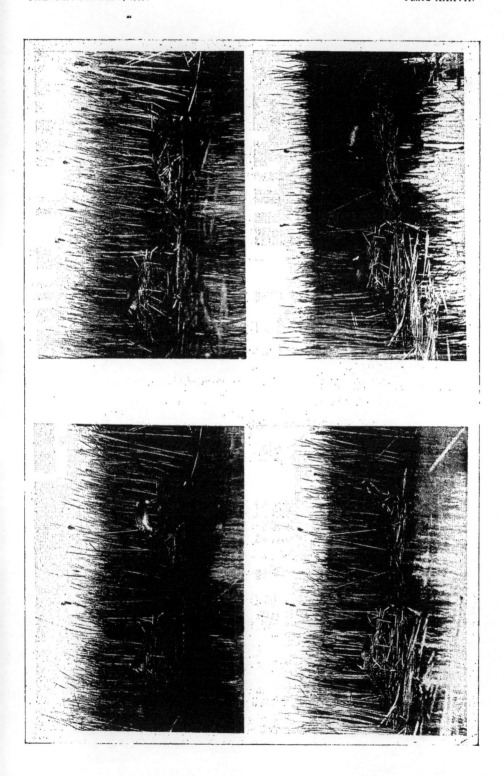

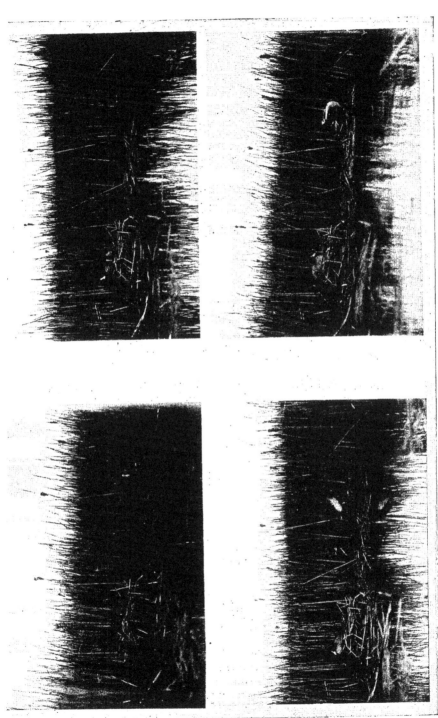

Nests of Coot and Crested Grebe.

female Grebe inspecting her newly hatched young one ; (No, 4) Female Grebe removing egg-shell from nest ; (No. 5) Coot visiting Grebe's nest during the latter's absence ; (No. 6) Both Coots bringing food to their young ones ; (No. 7) Coot carrying young, and male Grebe swimming away with chicks under his wing ; and (No. 8) Female Grebe feeding chick under her wing.

A NEW CRUSTACEAN.

Dr. Henry Woodward, F.R.S. in the Geological Magazine, has an interesting paper 'On *Pygocephalus,* a primitive Schizopod Crustacean from the Coal Measures.' The specimen was obtained by Mr. W. A. Parker from the well-known section at Sparth, Rochdale. At the York Meeting of the British Association Dr. Woodward described the specimen under the manuscript name of *Anthrapalæmon parkeri,* but now considers

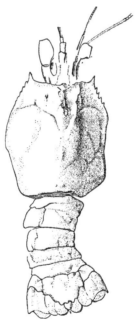

Pygocephalus parkeri,
From Sparth, Rochdale.

it should be placed in the genus *Pygocephalus,* the specific name, *parkeri,* being retained in honour of the discoverer of the specimen. Dr. Woodward's paper also contains a detailed description (with plate), of *P. cooperi,* Huxley, from the coal measures near Dudley. By the courtesy of Dr. Woodward we are able to give our readers an illustration of the Lancashire specimen.

NOTES OF CHANGES AMONG ANIMALS IN FYLING-DALES, NORTH-EAST YORKS., WITHIN THE LAST FIFTY YEARS.

J. W. BARRY.
Fylingdales.

(*Continued from page 309*).

BIRDS.

The Peregrine I have missed the last few years. I used to see it regularly, perhaps once a week, and, almost invariably, if the hour was advanced, making its way in the direction of the Peak, so there, I presume, it had its habitat among the high cliffs. It caused some little destruction among the grouse, and one of my workmen has seen it strike the wild duck close at hand to him. I believe that Sir C. Strickland's late keeper killed one or two Peregrines, but I observed that its disappearance was almost coincident with the conversion of the Peak Hall into the watering place of ' Ravenscar.'

The Buzzard I see, and have seen, at all seasons of the year, but not for long at a time. The last seen by me was just a month ago, and on two successive days. On the second occasion it flew in front of the house and almost within gun-shot.

The Raven was common a short time ago, but I have not seen it for the last four or five years. Increased use of the pole-trap, I think.*

The Jay I never saw or heard here until about fifteen years ago when some forty or fifty Jays suddenly appeared together. For many years afterwards they quite took possession of the woods, and became quite a nuisance from their discordant cries alone. The last year or so only one or two odd ones have been heard, and that not continuously. This year none at all. I cannot attribute the decline to their being shot at.

The nearest woods which they haunted before they appeared here were those of Harwood Dale about six or seven miles off in a straight line, but across the moor.

Woodpeckers have greatly increased in numbers of late years and also in boldness. Formerly, Woodpeckers used never

---

* The use of the Pole trap is now illegal, so that perhaps Ravens may fare rather better in the district for the future.—ED-

to be seen away from the main woods. Now, however, they come regularly round the house and their weird laugh is for a considerable part of the year one of the constant sounds in the garden. The green Woodpecker is, of course, the more common one ; but the Lesser Spotted Woodpecker is, or was, here. The Greater Spotted Woodpecker I remember to have seen only once. I should attribute the change in the matter of Woodpeckers generally to the planting done round the house.

Bullfinches.—To the same cause I should attribute the presence and increase of Bullfinches. Formerly, Bullfinches were quite unknown here. At least so I was told, and so I found, for I was always looking for them as a boy, and always being disappointed. Twenty years ago, however, when the plantations made in 1878 and 1879 on the north side of the house were beginning to get up, a pair made its appearance and nested there ; whilst in the course of a very few years they became so numerous that my gardener estimated that there could not be less than two hundred of them in that plantation and in the garden. At that time, the damage that they did in the way of picking out (for pure mishief) the young leaf-buds of the trees as they made their appearance began to be serious, the favourite objects of attack being the currants, thorns, and scyamores. Under the scyamores the ground was quite green every morniug with the buds that had been pulled out. This being so, there was a declaration of war. Sixty were shot one spring, and thirty-five the next. The Bullfinches then seemed to realise that they were being singled out for destruction, and, whilst no other birds left, or exhibited much shyness, the Bull-finches took their departure in a body, leaving not one of their number behind. A few seemed to take refuge in the main woods, particularly in the narrowest part, where there were young larch, and there they have remained ever since, but the quantity has remained small. This year, I understand, they have increased and have come to the plantation at the back of the house. The tradition, however, of their persecution seems to linger among them, for they do not venture into the plantation where they were so shot down or into the garden.

Since the above was submitted, I have seen one Bullfinch on the front lawn, but this was in the early morning, whilst I was dressing and I noticed that it was very different in its deportment from its ancestors twenty years ago, being suspicious as a corncrake and looking round every few seconds to see if anyone was about.

Hawfinches have, as elsewhere, made their appearance here within the last few years. As a choice has had to be made between letting the peas or the Hawfinches go, it has been decided, of course, adversely to the Finches. They have not, however, been detained like the Bullfinches from coming, but have developed a considerable amount of cunning in making their raids.

The Cuckoo is later in coming than it used to be. I used to hear it pretty punctually on April 27th. Now, however, it is generally May, and some way into May before I do so.

The House Martin.—The same is the case with the House Martins. This year, for instance, I saw them for the first time, after a daily look out, on May 10th, whilst in other recent years they have been even later. They have ceased to build under the cornice of this house (which is a very projecting cornice), and under the eaves of the houses in the neighbouring village of Thorpe. As regards this particular house the increase of sparrows in the ivy at the back might account for the desertion, though it must be observed that the front is quite clear of them.

The Swift which used to be common is now rare.

The Nightingale, on the other hand, has visited us. The first time that it was heard was about Whitsuntide in 1902. It was heard both at this house and at Thorpe. In the former case, it was recognised by a servant who came from a part of the country where Nightingales were abundant. In the latter case, it was identified by a family who had recently come to settle there, and by others as well. Up here it sang only in the early morning. At Thorpe it sang almost always in the evening and until past eleven o'clock at night, keeping the family already mentioned awake, since it always chose a certain tree opposite their house and bordering the public road. It invariably sang its loudest, they noticed, the moment the lamp was lighted, so that they thought it took it for the moon. The singing was continued every night for three weeks, then for some weeks there was a cessation, and then, for a short time, a resumption, suggesting that in the interval the bird had been nesting.

On March 22nd, of the present year, I was amazed to hear, when walking in the garden at dusk, what seemed to be exactly the same notes as those heard in 1902. I approached the tree where the bird was and stood for a long time listening to it, but could not attribute the strong plaintive lament with its rich *contralto* of ' heu heu ' to any other bird but the Nightingale. I have not, heard it again. On mentioning the matter to the lady at Thorpe who was favoured in 1902, I was told by her that she, too, had, about the same date, heard what certainly seemed to her, to be the Nightingale, but that she had not liked to say anything about it on account of the time of the year and of her being thought to be mistaken. The weather, however, it may be observed, was more summerlike than any we have had since.

The Curlew, which used to be so shy, has became wonderfully tame of late years. It has nested regularly at the moor edge on one of the farms of this estate (though owing to health I cannot speak as to last year), and became quite a companion in these parts, flying round and round close at hand uttering its cry.

A Semi-feral Parakeet.—A female Parakeet belonging to my gardener has been living out in the woods and shrubberies for three years. It comes to his cottage regularly to be fed, except during the nesting time, when it is absent for many weeks together. In appears to make its nest somewhere in the heart of the oak woods, not less than half-a-mile away, but exactly where no one has been able to discover. To reach these woods it crosses about three-eighths of a mile of open ground. There was also a male bird with it, but this perished in the first winter that it was out.

----◆◆----

'Grouse Disease,' what it is and how it spreads, with suggestions for stamping out disease, and the gradual improvement of moors, by the Rev. E. A. Woodruffe-Peacock. Louth : Goulding & Son, 1907. 111 pp. Price 5/. In this, No. 10 of his 'Rural Studies Series,' the Vicar of Cadney brings together all that he knows of the Grouse Disease. In his 'Foreword' the author states, 'So far as they go, these notes are my own, *i.e.*, no other person is in any way responsible for them . . . I have never done any field observation for the Grouse Commission, or had a specimen of theirs through my hands . . . who their advisers are I do not even know, as I have never been before them, or received any communication about their methods. . . . I always understood I was appointed field observer to them as an honour for work done in the past; but to show that this study has no connection whatever with the Commission's special line of enquiry, I have resigned the honorary post they conferred upon me.' Having thus washed his hands of the Grouse Commission, and notwithstanding his opening remarks, Mr. Peacock declines to accept credit for any originality in his notes, and believes that if a file of ' The Field' ' were carefully sought through, the notes I have brought together here from the lips and letters of many men, and ephemeral literature, would be found to have been put on record over and over again during the past fifty years by practical sportsmen and field observers.' Having thus a clear statement of the nature of the present contribution to this subject, and having noticed that from the quotation on the title page, ' The worst enemies of the human race—ignorance and superstition—can only be vanquished by truth and reason,' we can only hope that we have truth and reason from the soil, grass, and game specialist of Cadney, Lincs.

THE ANCESTORS OF THE ANGIOSPERMS.

MARY A. JOHNSTONE, B.Sc., F.L.S.

AT the present time, when the attention of every prominent worker in botanical research is turned to the last magnificent contribution towards the solution of the problem of the phylogeny of the angiosperms, it may not be inappropriate to set forth in a journal of special interest to Yorkshire a few brief statements concerning the most important recent additions to our knowledge of the subject. Condensed within the space of a few pages, such a résumé must necessarily be the merest outline, and can make no pretentions to attempting a critical survey of the various hypotheses which have been formulated, much less to add to their number.

The question involves not one problem, but a whole plexus of problems, most important of which may be mentioned the following :—the line of evolution along which the typical flower and fruit of the angiosperm have progressed ; the sudden appearance in Lower Cretaceous times of the angiosperm type of foliage ; which of the modern angiosperms are the more primitive—the simple, unisexual, often apetalous forms, or the complete, bisporangiate ; is the group monophyletic, or have monocotyledons been derived from dicotyledons, or *vice versâ ?*

All recent palæontological evidence tends to confirm the view that modern flowering plants are descendants of a long line of fern-like, Cycado-filicean, and Cycadean ancestors. As this evidence has accumulated, the necessity for continual revision and rapid re-adjustment of our principles of classification has become more and more apparent. For instance, secondary growth can no longer be regarded as a mark of Phanerogamic affinity, since it has been shown to be the rule amongst the dominant Palæozoic Cryptogams ; from amongst the Palæozoic 'Ferns' many types had to be transferred to a new group of *Cycado-filices*, combining the characters of both ferns and cycads, and from these again we get the *Pterido-sperms*. These last furnish a strong example of that law of evolution which states that the various organs of a plant need not necessarily have an ancestral history marked by contemporaneous development. If we take *Lyginodendron* as a typical Pteridosperm we find that it has retained its fern-like foliage as well as certain fern-like anatomical characters ; that its microsporangia or pollen-bearing organs are so fern-like in appearance that, until their connection with *Lyginodendron* was recently established by Mr. Kidston,[*] they were regarded as the sporangia of a true fern ; but that, in strong contrast to all these, it bore remarkably complex *seeds*, evolved apparently long before anything approximating to a *flower* was in existence in this group.[†]

It is, however, from Mesozoic times that we get types more closely allied to our existing forms. As a result of the study of these, a new family has been constituted—the *Bennettiteæ*—which includes, besides the true Cycadales, those species which combine the habit of growth and the foliage of Cycads with a fructification differing absolutely from that of any known plant. The type genus was founded in 1868 by Mr. Carruthers[‡] to include species from the Middle Oolites to the Lower Greensand. In their general appearance, in the mode of attachment of the remarkable fruits to the plant axis, and in their detailed structure the English specimens agree so closely with American species (to be referred to below) that the same description may suffice for both.

Another allied form was obtained from the Lower Oolite of Yorkshire ; this was the *Zamia gigas*, of which an account was presented by Williamson in 1868, and which was placed by Mr. Carruthers in the new genus *Williamsonia*. The fructifications found with the *Zamia*-like foliage of this plant, were very puzzling, exhibiting externally a globular form clothed with bracts, and internally a disk-like structure bearing a pyramidal upgrowth. Dr. Wieland's recent work strengthens the view expressed by Mr. Seward that these plants were allied to the *Bennettites*.

But by far the most important of recent discoveries are those of Dr. Wieland.[§] These, the results of eight years' labour, are embodied in a magnificent volume published last year on 'American Fossil Cycads.' The wealth of material which has been at Dr. Wieland's disposal has been obtained from the Mesozoic beds of America, ranging from Upper Trias to Lower Cretaceous horizons : its classification has been the work of Professor Lester Ward.

It is these *Bennettiteæ* (*Cycadoideæ* of Wieland) which seem to furnish the key to the evolutionary problem of the angio-

---

[*] Kidston, Trans. Royal Soc. (1905).
[†] Oliver and Scott, Trans. Roy. Soc. (1904).
[‡] Carruthers, Trans. Linn. Soc. xxvi. (1868).
[§] Wieland, American Fossil Cycads (1906).

spermic flower, and that being so, a short summary of their structure may be of interest.

The habit of growth and stem-form agree very closely with those of the typical modern cycads; but the first great difference is apparent in the mode of attachment of the fructification, which consists of a strobiloid axis borne on the main stem and wedged in amongst the persistent leaf-bases. This short axis terminates in a convex receptacle bearing (*a*) at its lowest level a series of spirally arranged bracts, (*b*) above these a whorl of frond-like sporophylls, fused proximally, bipinnate, and having numerous pollen sacs of a synangial type, (*c*) a central conical region, arranged on which are numerous stalked ovules interspersed with elongated scales partially united at their distal ends to form a kind of ovary wall, but still allowing the micropyles of the ovules to project a little beyond them.

The marked general agreement in the arrangement of the various parts of the flower, the perianth-like lower members, the hypogynous stamens, and the central ovary are all most suggestive of the angiospermous flower. This very close approximation was at once recognised by Dr. Wieland, and has met with very general acceptance. Minute examination of the structure strengthens the case by showing that not only were the seeds almost exalbuminous, but the embryo was in all respects that of a highly organised dicotyledon.

The *Bennettites* of Mr. Carruthers and the earliest specimens examined by Dr. Wieland showed only the female organs.; this probably simply meant that the stamens had withered, and was quite in accord with the fact that the seeds were in a fully ripe condition as shown by the stage of development of the embryo. The basal portion left after the stamens had dropped probably represented the ' disc ' of *Williamsonia gigas*.

Such a striking assemblage of widely contrasting characters as we have in these surprising fossils has of course given rise to much fresh speculation. The fern-like sporophylls with their synangia of a Marattiaceous type indicate a connection with the oldest types of ferns. Side by side with these we find the highly evolved seeds and the ovary wall suggestive of the angiosperms, though the plant is yet essentially a Gymnosperm, inasmuch as the pollen is collected not by a modified carpel, but by the ovule itself. The spiral of bracts, which possibly may foreshadow the modern perianth, and the dicotyledonous character of the seed complete a most astonishing combination of great type features.

From a consideration of such fossil evidence as the above, combined with a comparison of existing forms of flowers, as well as evidence from other sources, which must be passed over at present, Messrs. Newell Arber and Parkin* have formulated a set of conclusions, which they present as a 'Working Hypothesis' for future guidance. They maintain that the simple apetalous types of modern angiosperms are *not* primitive forms, but have been derived by reduction from hermaphrodite forms with a perianth; that the unknown ancestor of the angiosperms was built, generally speaking, on the plan of the Mesozoic *Bennettites*; that the angiosperms are a monophyletic group, the monocotyledons having been derived from the dicotyledons; and that entomophily was a primitive feature of the race.

It will be seen from the above outline of new evidence that no light has yet been shed on the dark period of the emergence of the typical foliage of the angiosperm, and it is in connection with this that the practical suggestion of this paper may be made. The estuarine beds of Yorkshire have unfortunately not yet provided us with any evidence of a microscopic character bearing on these problems; but it is, of course, well known that they are extremely rich in well-preserved impressions of leaf form, and to a less extent of flower form. Might not the Research Committees of the Yorkshire Naturalists' Union take this into consideration as a probably fruitful field on which to expend some of their energies? The exposures of fossiliferous beds along the coast have been well investigated, but there would seem to be many promising sections lying further inland which have been neglected, and which would repay investigation. Systematic collecting from these, with especial care expended on localising the exact horizon of all specimens, might possibly do something towards elucidating a sequence from the cycadean or filicinean type of leaf to the angiospermic —hereby contributing in some degree towards the solution of one of the greatest botanical problems of the day.

———◆◆———

We have received the August number of the **Nature Reader Monthly** —a small magazine printed in large type—apparently for the use of school children. It deals with a variety of subjects—Poppies, Butterflies and Moths, Protective Colouration, the Sea Shore, Rocks and Scenery, etc. The language used is generally clear. The pamphlet is edited by Mr. F. H. Shoosmith, is illustrated, contains 32 pp., and is sold at one penny by Charles and Dibble, London.

* Journal of Linn. Soc. London, xxxviii. (1907).

THE NATURAL HISTORY OF LITTONDALE, YORKS.

THE numerous members of the Yorkshire Naturalists' Union who reached Arncliffe for August Bank Holiday week-end were well rewarded for their pains. To some of them the recollection of the journey from Grassington, notwithstanding the glorious view of the surroundings, whilst the daylight lasted, will be anything but pleasant, though these will doubtless remind them of methods of crossing the country in pre-railway days. Still, even the out-of-the-way headquarters of the meeting had many advantages, and although it was a holiday season, the visitors were quite comfortable, albeit that an influx of some

Photo by   *R. Fortune*

Arncliffe Village.

forty Yorkshire Naturalists was a serious tax on the sleeping accommodation of this small and old-world village.

But what a glorious district! Could anything be more charming to the eye accustomed to crowded thoroughfares and smoky chimneys than those far-reaching dales and cloud-capped fells. To see them alone was well worth the longest journey—to investigate them, to unravel the riddles they held, was paradise. One can quite understand this district being the retreat of many busy men. Kingsley stayed at Arncliffe—it is the Vendale of his 'Water Babies.' Part of that well-known book was written here. Malham Cove, close by, is the Low-thwaite Crag of that same work, and still shows the black mark made by Little Tom, the sweep, when he slipped down the Scar!

Local tradition hath it that there are ghosts and barguests in the neighbourhood, and that Old Pam (the devil) pays periodical visits, gives suppers, and on one night each year teaches the little Wharfedale devils in Threshfield Grammar School. But we don't believe it. No sane person would leave Upper Wharfedale for any place Old Pam could offer. And insane persons he doesn't cater for.

With more or less poetic feelings, therefore, most of the members arrived at Arncliffe on Friday evening, but lost their sentiment in the scramble for beds. On the morrow, however, the high moorlands were ascended, and from their tops the

Photo by   *R. Fortune.*

River Skirfare below Arncliffe.

valleys and nestling stone-built villages appeared typical of peace and rest :

> A region of repose it seems.
> A place of slumber and of dreams,
> Remote among the wooded hills,
> For there no noisy railway speeds,

and we hope one never will.

But the Yorkshire Naturalists were bent on work. Many were astir long before breakfast, and most remained out as long as the light allowed ; a few stayed longer.. The evenings were profitably spent. On Saturday the Vicar, the Rev. W. A. Shuffrey, read some valuable notes on the Botany of Littondale (see pp. 354-356), and Mr. W. Denison Roebuck gave an account of 'Yorkshire Hemiptera.' He also noted some recent additions to the list of Hymenoptera of the county.

At this excursion, too, the sectional recorders were well in evidence, and each took charge of his department. Littondale had not previously been visited by the Union, and consequently the district was of more than usual interest, and in view of an excursion to be presently held in an adjoining dale, particular note of the fauna and flora was taken, and on this account details of the records made in the various departments will be given more fully than usual.

As might be expected from the geological features of the area, the hammermen were much in evidence, and their President, Mr. Cosmo Johns, by example and precept, caused his party to work hard. The result was a substantial addition to our knowledge of the fauna and stratigraphy of the Carboniferous Limestone of the area.

Of perhaps more general interest was the visit paid by the more venturous of the party to Skotska cave. This runs underground, tunnel-like, for a great distance, and is remarkable for its regular course and perfectly flat roof. A good coat and a pair of trousers, far into the cavern, were 'finds' of a somewhat unexpected nature !

On Monday evening the usual general meeting was held, under the presidency of Mr. C. Crossland. At this, votes of thanks were passed to the landowners, and to the Rev. W. A. Shuffrey for allowing the party to examine his fine botanical garden, and some new members were elected. Brief reports on the work of the sections were presented by the officers. These are given at greater length in these pages.

VERTEBRATE ZOOLOGY.—This section was officially represented by its President, Mr. Riley Fortune, and one of its Secretaries, Mr. H. B. Booth, who report :—Several working members being present, this Section accomplished good work during the limited time at its disposal.

Mammals.—Eleven species of Mammals were indentified, viz. :—Mole, Common Shrew, Water Shrew, Stoat, Weasel, Squirrel, Brown Rat, Red-backed or Bank Vole, Long-tailed Field Mouse, Hare and Rabbit. Of these, the Hare was decidedly scarce, but the Rabbit was extremely abundant throughout the valley. Several black, and a few white, varieties were noticed, probably the descendants of tame rabbits turned down. 'Rabbits everywhere' was a common remark, and yet their greatest enemy, the Stoat, was very numerous. Several of the latter were seen, and a local gamekeeper had the tails of fifty-six Stoats and ten Weasels strung up at his house, and all killed since February. Judging by the results of a dozen small traps continually set, the Common Shrew must be very numerous in the valley, and one specimen each of the Water Shrew, the Long-tailed Field Mouse, and the Red-backed or

Bank Vole, were captured. A few Bats, about the size of the Pipistrelle, were noted in the evenings, but no means of securing one for identification were available.

Birds.—Unfortunately this is one of the worst seasons of the year for observing the avifauna of an inland district. Most of the birds keep quiet, and well under cover with their young

The work of Hawfinches.

broods ; and more particularly this is the case with the woodland species. The Common Buzzard was seen on both days, and on Monday the members of this Section were entertained by an exciting chase of a large Rabbit by a Buzzard, in which the Rabbit eventually made good his escape underground. Three Stockdoves were observed mobbing the Buzzard for some time. The Hawfinch was detected by his onslaught on some rows of peas in the vicarage garden, and the bird was afterwards seen. It would appear that the Hawfinch first made his advent into Littondale in 1906, as the Rev. W. A. Shuffrey informed us that although he has grown peas for many years, yet they were attacked last year for the first time.

The following fifty species of birds were identified* :—

Missel Thrush (in flocks).	Redstart (only one seen).
Song Thrush (common).	Redbreast.
Blackbird (common).	Willow Warbler (very common).
Ring Ouzel (common).	Dunnock (common).
Wheatear (very common).	Dipper (very common).
Whinchat.	

---

* To this list might be added the following four species, which were seen n Littondale this year, at Easter, but which were not observed during the Union's visit, viz. :—Peregrine Falcon, Merlin, Mallard, and Redshank.— H.B.B.

1907 October 1.

Great Titmouse (one or two pairs only seen).

Blue Titmouse (one or two pairs were seen).

Wren (very common).

Pied Wagtail.

Grey Wagtail.

Yellow Wagtail (abundant).

Meadow Pipit (abundant).

Spotted Flycatcher (common around Arncliffe).

Swallow (abundant).

House Martin (common).

Sand Martin (fairly common).

Greenfinch (only heard once).

Hawfinch.

House Sparrow (common at Hawkswick and Arncliffe, uncommon higher up the valley at Litton, Halton, etc.

Chaffinch (common).

Starling (fairly common).

Magpie.

Jackdaw (common).

Carrion Crow.

Rook.

Swift (common).

Tawny Owl (young heard calling in the woods).

Common Buzzard.

Sparrow Hawk.

Kestrel (fairly common).

Common Heron (fairly common).

Wood Pigeon (fairly common).

Stock Dove.

Red Grouse.

Pheasant (a few near Arncliffe).

Partridge (very few seen).

Waterhen.

Golden Plover.

Lapwing (in flocks now).

Common Snipe (rather scarce).

Common Sandpiper (apparently the bulk of this species had left the dale.)

Curlew (common).

Black-headed Gull (stray birds).

Herring Gull (stray birds).

In investigating the fauna of a district it is interesting to note what expected species do not occur, or are very rare, and Littondale gave plenty of scope for this class of investigation. Search as we would we could not discover a single Skylark ! Neither could we detect the Corncrake, nor any of the Warblers with the exception of the Willow Warbler, which was plentiful. Not a single Bunting of any kind was noted, and with the exceptions of the Chaffinch, Hawfinch, Greenfinch, and Sparrow, none of the finch family was seen. Titmice were not by any means so common, neither in numbers nor in species, as one would have expected. It would be useful if future observers in this dale would take note of the apparent absence, or rarity, of these otherwise common birds.

Pisces.—The Trout was plentiful, and the Loach, Minnow, and River Bullhead were also noted in the Skirfare.

In *Reptilia* and *Amphibia*, the only species seen was the Common Frog, and it was not by any means numerous.

ENTOMOLOGY.—Mr. G. T. Porritt writes :—But little wa attempted entomologically. Mr. J. Beanland found specimens of the local *Coremia munitata* on the high hills, and *Nudaria mundana* was plentiful on old walls. *Prays curtisellus* occurred among ash. As showing the extraordinary lateness of the season too, it may be mentioned that *Acronycta ligustri* and *Abrostola urticæ* were taken as imagines.

Among Neuroptera I was very pleased to find the pretty and rather scarce *Hemerobius marginatus*, common in the wood opposite Arncliffe village, and near where also a very sparsely spotted form of *Panorpa germanica* occurred. *Chloroperla grammatica*, *Rhyacophila dosalis*, etc., were about the river.

Mr. J. Beanland gives the following list of MACRO-LEPIDOP-TERA for Littondale :—

Chortobius pamphilus.
Lycæna alexis.
Acronycta psi.
   ,,   *megacephala.*
   ,,   *ligustri.*
Nudaria mundana.
Miana fasciuncula.
Caradrina cubicularis.

Noctua festiva.
Abraxas grossulariata.
Larentia cæsiata.
   ,,   *didymata.*
   ,,   *olivata.*
   ,,   *pectinitaria.*
Melanippe montanata.
Coremia munitata.

I took one *C. munitata* at Airton (Malham) on the 20th July, and two captured in Littondale were taken to Hull by Mr. Wilford.

Mr. E. P. Butterfield adds :—

Metrocampa margaritata.
Hypsipetes elutata.
Melanthia ocellata.
Cidaria corylata.
Eubolia mensuraria.
Hepialus hectus.
Scoparia muralis, abundant
   ,,   *mercuralis.*
   ,,   *truncicolalis.*
   ,,   *ambigualis.*
Crambus culmellus.

Crambus tristellus.
Totrix ribeana.
   ,,   *forsterana.*
Penthina corticana.
   ,,   *cynosbana.*
Sericoris lacunana.
Sciaphila subjectana.
Grapholita penkleriana.
Pædisca corticana.
   ,,   *solandriana.*
Pepilla curtisella.

Mr. H. V. Corbett sends the following list of beetles, all common species :—*Harpalus latus* (L.), *Pterostichus strenuus* (Panz.), *Brachypterus urticæ* (F.), *Serica brunnea* (L.), *Agriotes obscurus* (L.), *Rhagonycha fulva* (Scop.), *Crepidodera ferruginea* (Scop.), *Otiorrhynchus picipes* (F.), *Phyllobius oblongus* (L.), *Phyllobius urticæ* (De. G.), *Hypera nigrirostris* (F.).

Mr. J. Beanland adds that three Glow-worms were obtained at Hawkswick.

CONCHOLOGY.—The conchologists, under the leadership of Mr. W. Denison Roebuck, were particularly pleased with the excursion, as they were able to report the confirmation of every previous record of the molluscan fauna of Littondale. They were exceptionally gratified to find *Pupa secale.*

BOTANY.—Dr. T. W. Woodhead writes :—Botanists in every branch found here ample material to engage their close attention.   Excursions were made to Foxup, Heselden Gill, and Penyghent : also over Old Cote Moor to Kettlewell and to Great Whernside.   This area is included in the district surveyed by Smith and Rankin,* and an excellent opportunity was afforded to examine some of the chief plant associations there dealt with.

In these notes, however, mention is only made of species observed in Littondale and the drainage area of the Skirfare. We were fortunate in having the assistance of the Rev. W. A. Shuffrey, who, on the Saturday evening, gave an interesting account of the ' Botany of Littondale.'   This was illustrated by herbarium specimens (including the Lady's Slipper Orchid), and he afterwards invited the members to visit his delightful garden, which it was a pleasure to see.

My own observations were directed mainly to plant associations and their distribution in relation to soils, the results of which will be published in a future paper.   Lists of species have also been kindly supplied by Mr. J. Beanland (85 species) and by Mr. C. Waterfall (74 species).

In the neighbourhood of the village striking plants by the river side were *Myrrhis odorata* and *Senecio saracenicus*, the latter a garden outcast by Cowside Beck.   The fields and lanes were purple with *Geranium pratense*, while other conspicuous plants were *Cnicus heterophyllus* and *Campanula latifolia*. *Geranium lucidum* decorated the walls, and in a rough pasture on the drift was an abundance of *Colchicum autumnale*.

Among the rarer plants met with were *Saxifraga umbrosa*, *S. aizoides*, and *Polemonium cæruleum*.

In the woods on the steep drift-covered slopes below the Scars, Ash (*Fraxinus excelsior*) was the dominant tree.   Hazel is also abundant, and here were noted Mountain Ash, Hawthorn, Whitebeam, Larch, *Euonymus europæus* and *Viburnum opulus*. The mixed soils here, together with the shelter afforded by the trees, give support to a rich and varied undergrowth.   The following are among the species noted :—

Anemone nemorosa.	*Geranium sanguineum.*
Aquilegia vulgaris.	,, *sylvaticum.*
Actæa spicata.	,, *Robertianum.*
Hypericum hirsutum,	*Spiræa Ulmaria.*

* ' Geological Distribution of Vegetation in Yorkshire ; Harrogate and Skipton District, 1903. Smith and Rankin.'

Rubus saxatilis.
Geum urbanum.
    ,, rivale.
Agrimonia Eupatoria.
Poterium officinale.
Fragaria vesca.
Circæa lutetiana.
Sanicula europæa.
Cnicus heterophyllus.
Primula aculis.
Galium sylvestre.
Scrophularia nodosa.
Veronica chamædrys.
Melampyrum pratense.

Origanum vulgare.
Prunella vulgaris.
Lamium Galeobdolon.
Teucrium Scorodonia.
Mercurialis perennis.
Convallaria majalis.
Allium ursinum.
Scilla festalis.
Luzula maxima.
Carex pallescens.
    ,, sylvatica.
Pteris aquilina.
Polystichum lobatam.
Lastræa Filix-mas.

Above the woods are the Scars of the Great Scar Limestone and the Limestone Pavements. The following were among the species noted here :—

Thalictrum minus.
    ,,   collinum.
Arabis hirsuta.
Draba incana.
Helianthemum chamæcistus.
Arenaria verna.
Dryas octopetala.
Ribes alpinum.
Saxifraga tridactylites.

Oxalis acetosella.
Listera ovata.
Sesleria cærulea.
Asplenium viride.
    ,,   trichomanes.
    ,,   Ruta-muraria.
Scolopendrium vulgare.
Cystopteris fragilis.
Phegopteris calcarea.

Higher still, extensive tracts are occupied by ' Hill pasture ' dominated by *Festuca ovina.* There are many local variations, however ; ericaceous plants have wandered down from the heather moors, and are growing alongside species characteristic of limestone soils. These inclnded :—

Trollius europæus.
Cochlearia danica.
Viola lutea.
Polygala vulgaris.
Cerastium triviale.
Sagina nodosa.
Arenaria verna.
Linum catharticum.
Geranium sylvaticum.
Anthyllis Vulneraria.
Potentilla sylvestris.
Poterium Sanguisorba.
Saxifraga hypnoides.
Parnassia palustris.
Galium verum.
Scabiosa Columbaria.
Carlina vulgaris.

Vaccinum Myrtillus.
Calluna Erica.
Primula acaulis.
    ,, farinosa.
Gentiana Amarella.
Pinguicula vulgaris.
Pedicularis palustris.
Thymus Serpyllum.
Plantago media.
Habenaria conopsea.
    ,,   viridis.
Carex dioica.
Scripus caricis.
Sesleria cærulea.
Koeleria cristata.
Ophioglossum vulgatum.
Selaginella selaginelloides.

The upper slopes of the Yoredales are peat covered, and dominated by heather moors broken into considerably in places by grassy and rushy tracts. The chief species noted here were:—

Calluna Erica.	*Carex dioica.*
Vaccinium Myrtillus.	,, *pilulifera.*
,, *vitis-idæa.*	,, *pulicaris.*
Erica tetralix.	,, *fulva.*
Rumex acetosella.	*Agrostis vulgaris.*
Juncus squarrosus.	*Nardus stricta.*
,, *glaucus.*	*Deschampsia cæspitosa.*
Scirpus cæspitosus.	*Molinia varia.*

The top of Old Cote Moor is a plateau capped with Mill-stone Grit and covered by peat 1 to 3 feet deep. The species met with here are in order of dominance :—

Calluna Erica.	*Eriophorum angustifolium.*
Vaccinium Myrtillus.	*Juncus squarrosus.*
Rubus Chamæmorus.	*Nardus stricta.*
Empetrium nigrum.	*Festuca ovina.*
Eriophorum vaginatum.	

FUNGI.—Mr. C. Crossland writes :—The attention of the mycologists was given to the immediate neighbourhood of Arncliffe and up Littondale as far as Foxup. There were only two or three small woods, and these lacked the conditions —moisture and plenty of rotting branches—favourable to the growth of larger fungi. Very little variety in the larger agarics was seen. Having gone with the express purpose of finding something, we turned our investigations in other directions. Nettle beds by the way-side, meadow-sweet beds in moist corners, and cattle dung in pastures and about farms found us plenty to do, in addition to odd places here and there looked into. Decaying stems of last year's nettles proved most prolific in micro-species, some of the stems having five or six different species, in as many inches, growing upon them. Among the meadow-sweet was found *Belonidium deparculum ;* which had been sought for by the writer on decaying Spiræa for some years, and here it was in abundance ; it is an aberrant Discomycete having only four spores in each ascus, and was particularly wanted for the purpose of checking a previous examination of the same species found at Hornsea Mere in 1900. No less than one fourth of the 107 species found were coprophilous fungi, mostly common habitues of cow, horse, sheep, and rabbit dung. *Coprinus Gibbsii*, a tiny but very distinct species first discovered by Mr. Gibbs near Sheffield in 1903, being amongst them ; they sprung up in quantity on a piece of cow dung kept a few weeks under a bell glass for observation. Several others, included in the following list, came forward. One of the least expected to put in an appearance on this matrix was a Trichia or Hemitrichia—*T. Karstenii* Mass. Mon. p. 168 = *H. Karstenii* (Rost.) Lister. There was a small colony of four or five sporangia. There also came up a very small agaric, less than a line across the pileus, with pale brown spores ; this appears to be an undescribed species.

There was an almost entire absence of the colt's foot and butterbur Uredines, though both plants were plentiful.

The absence of decaying wood and shaded moist corners accounts for the comparative absence of Myxogastres : they were sought for.

We have to thank several members of other sections for collecting specimens, two or three of the more important ones being brought in by them.

The following is a summary of the work done :—107 species found, 2 New to Yorkshire, marked * in the list ; 21 New to Mid W. Div., marked † ; 8 Confirmations of hitherto single county records.

The initals added to the records are :—A. = Arncliffe ; A. C. = Arncliffe Clouder ; L. = Litton ; F. = Foxup.

Collybia dryophila. In plantation, A.

Mycena galericulata. On old stump, A.

M. acicula. On soil among butter-bur, A.

† *Omphalia griseo-pallida,* A.

GASTROMYCETES.

Lycoperdon bovista. Remains of a last year's plant, A. C., at 1700 ft.

HYMENOMYCETES.

Agaricaceæ.

Lacaria laccata. In plantation, L.

† *Pleurotus applicatus.* On rotting wood on the Scar behind Hawkswick.

P. septicus. On decaying meadow-sweet, F.

Entoloma sericeum. In pasture, A. ; L.

Nolanea pascua. In pastures, A. ; L.

† *Inocybe incarnata.* On soil near the stream side, A.

I. rimosa. On the ground in a small wood, L.

Crepidotus mollis. On decaying gate post, L.

Bolbitius tener. Among grass in pasture, A.

Agaricus campestris. In pasture, A. C.

Stropharia stercoraria. On cow dung in pasture, A,

S. semiglobata. Common on cow dung all the way up the dale.

Panæolus campanulatus.

P. papilionaceus.

P. fimicola.

Anellaria separata.

All on cow dung about Arncliffe and on the hill sides.

Psilocybe semilanceata. Among grass in hill side pasture at 1700 ft.

Psathyrella atomata. Among grass, road side, A.

P. disseminata. On rotting, moss-covered trunk, A.

Coprinus niveus.

C. radiatus.
    Both on horse dung in pastures A.

† *C. Gibbsii.* On cow dung, A.

C. plicatilis. Among grass in pastures, and on lawn, A.

Hygrophorus chlorophanus, A.

Marasmius rotula. On dead twigs L.

POLYPORACEÆ.

Boletus flavus. Among grass, A.

B. laricinus. Among fallen larch leaves, A.

Polyporus squamosus. On ash stump, A. L.

Polyticstus versicolor. On gate posts, A. L.

† *Fomes connatus.* On stump, A.

F. igniarius. A portion of last year's growth picked up from the ground on hill side in plantation, A.

F. annosus. On pine stump, A.

HYDNACEÆ.

Grandinia granulosa. On rotten wood, L.

THELEPHORACEÆ.

Corticium calceum. On dead branch, L.

Cyphella capula. On dead stems of butter-bur, A.

TREMELLACEÆ.

Tremella mesenterica. On fallen trunk, A.

Dacryomyces deliquescens.

D. stillatus.
    Both on dead wood, A. L.

UREDINACEÆ.

Uromyces alchemillæ. On ladies' mantle, A.

Puccinia pulverulenta. On *Epilobium montanum,* L.

P. pimpinella. Uredospores on sweet cicely, L.

P. poarum. Æcidiospores on colt's foot, F.

P. caricis. Æcidiospores on the common nettle, F.

P. centaureæ. On Knapweed, A.

Phragmidium sanguisorbæ. Æcidiospores on *Poterium sanguisorba,* A. L. ; F.

Triphragmium ulmariæ. Uredospores on meadow sweet, A.

USTILAGINACEÆ.

Ustilago violacea. On the anthers of rose campion, L.

PYRENOMYCETES.

Nectria cinnabarina. On dead ash branches, A.

Hypomyces rosellus. On *Polystictus versicolor,* A.

Diatrype disciformis. On dead beach branch, A.

Melanomma pulvis-pyrius. On dead decorticated branch, A.

† *Sordaria curvula.* On rabbit dung, A.

† *S. minuta.* On rabbit dung, A.

† *Sporormia intermedia.* On rabbit dung, A.

Rhaphidospora rubella, F.

R. urticæ, A.

Heptameria doliolum, A. F.

† *H. acuta.* A. F.

Pleospora herbarum, A.
    The last five all on dead nettle stems.

Sphærella rumicis. Common on *Rumex obtusifolia,* A. L. ; F.

Sphærotheca pannosa. The Oidium stage on wild rose, A. L. ; F.

S. castagnei (S. humuli). On living leaves of lady's mantle near the Hotel door, A.

DISCOMYCETES.

Acetabula vulgaris. On sandy ground near stream, A.

Humaria granulata. A. L.

Lachnea stercorea, A.

† *L. ascoboloides,* A.
    All three on cow dung.

† *L. fimbriata.* On soil in butter-bur bed, L.

Dasyscypha virginea, F. On dead stems of meadow sweet, F.

D. nivea. On dead wood, F.

† *Erinella Nylanderi.* On dead nettle stems, A.

Chlorosplenium æruginosum. On dead ash branch in the wood on the hill side behind the church, A.

Helotium claro-flavum. On rotting, partly moss-covered branch, L.

H. pallescens. On rotting twig, F.

H. cyathoideum. Extremely common on decaying herbaceous stems, especially nettle.

H. scutula, F.

† *Belonidium deparculum,* A. Both on decaying stems and leaf stalks of meadow sweet.

Mollisia cinerea. On dead twigs, L.

M. atrata. On decaying stems of meadow sweet, A.; F.

M. urticicola. On dead nettle stems, F.

† *Pseudopeziza trifolii.* On living leaves of white clover in meadow, A.; on road side, L.

† *Ryparobius dubius.* On rabbit dung, A.

* *Ascophanus cinereus.* On rabbit dung, A.

† *A. minutissimus.* On cow dung, F.

A. carneus. On rabbit dung, A.

A. equinus. On dung of cow and horse, A.; L.

Ascobolus furfuraceus. On cow dung, A.; L.; F.

† *A. glaber.* On rabbit dung, A.

† *A. immersus.* On dung of cow and rabbit, A.

Calloria fusarioides. Conidial stage, on dead nettle stems, A.

PHYCOMYCETES.

Pilobolus crystallinus. Common on cow dung, A.; L.

Mucor mucedo. On rabbit dung, A.; L.

DEUTEROMYCETES.

† *Phoma nebulosum.* On dead nettle stems, A.

P. longissima. On dead stems of some umbelliferous plant, A.

† *Vermicularia dematium.* On dead nettle stems, A.

HYPHOMYCETES.

Botryosporium pulchrum. On decaying nettle stems, A.

† *Periconia podospora.* On dead herbaceous stems, F.

Stilbum tomentosum. On *Hemitrichia Karstenii,* A.

Cladosporium herbarum, A.; F.

Arthrobotryum atrum, A.

† *Fusarium roseum,* F.

The last three on dead nettle stems.

MYXOGASTRES.

Reticularia lycoperdon. On old ornamental stump at the Hotel door, A.

* *Trichia Karstenii* (= *Hemitrichia Karstenii* (Rost.) Lister). On cow dung, A.

T. S.

COLEOPTERA.

Uncommon Beetles near Barnsley.—Late at night, on the 3rd of August, I took a specimen of *Deleaster dichrous* Grav. at rest on an electric lamp post in the town. The specimen is the var. *leachii* Curt. The species has not hitherto been recorded from the West Riding. On the afternoon of the 15th, under a small fungus at Darfield, I found several specimens of *Cis bidentatus* L. and one *Mycetophagus quadripustulatus* L. The only record for the West Riding hitherto for the former species is Studley, where it was taken many years ago by E. A. Waterhouse.—E. G. BAYFORD, Barnsley.

THE BOTANICAL FEATURES OF LITTONDALE.*

Rev. W. A. SHUFFREY, M.A.

IF we would realise how different the flora in this valley is from that of the east side, for instance, of Yorkshire, we must bear in mind that we are here at an elevation of 750 feet above the sea, that the highest land in the Dale rises to 2270 feet, and that the flora exists in an average rainfall of sixty inches, or about three times that of the east side of England. This means that rain falls on two days out of three, taking the year through. We may wonder that there are any flowering plants at all. One would imagine that instead of phanerogams the whole country would be filled with mosses and lichens. But if we have a heavy rainfall we have a compensation in a very light soil, and one particularly favourable to plant life, viz., the limestone, and there is very little clay in the valley. It is the limestone that makes our flora so rich and extensive. As soon as the grit is reached, to the south—south east of Burnsall and Rylstone—many of our characteristic plants disappear, and the few that are peculiar to the grit and are not found on the limestone are unimportant and not rare. The first flower which blooms when Spring comes in after snows disappear, is not the "celandine" (*Ranunculus ficaria*) which Wordsworth noticed and sang of, in the Lake Country. It is the Tussilago Farfara. I have made a note for the last twenty years or more of the date of its first appearance in each year, and I find that the dates range from February 21st, 1884, to April 7th, 1904. Our rarest plants are, first the lady's slipper orchid (*Cypripedium Calceolus*). I don't like to omit this, though I am sorry to say that the plants which I had in my garden for many years, and which were taken by my predecessor from the slope of the hill, have from some cause or other disappeared. But I understand that a single specimen was found in the Dale not many years ago, and I have a dried specimen of a flower of it which was found not far outside the parish, not long ago. And I think that it may yet be found in the Dale. Then there is the *Polemoneum cæruleum*, which grows plentifully not more than half a mile from the village; and *Sedum Telephium* grows not far from it. We can also boast of *Saxifraga oppositifolia* and *S. aizoides* on the slopes of Penyghent. Of course the *Dryas*

---

* Read at Arncliffe, to the Yorkshire Naturalists' Union, on August 3rd, 1907.

octopetala is well known as growing on the Clouders. This was recorded as far back as 1782 (Curtis). I am sorry to say that it is not as plentiful as it was, owing, I am afraid, to the number of people who gather it and sometimes try, but of course unsuccessfully, to make it grow at a low elevation. It is said to be one of the few survivors of the Arctic flora in those regions and prefers a northern aspect. *Rubus Chamæmorus* grows on the moors. The berries of this are sometimes gathered by careful housewives and made into a preserve, which is not unpleasant to the taste. The *Actæa spicata* flourishes only in one area. I am glad to say that the *Daphne mezereum* has been found in bloom this year by Mr. Booth. I saw one in Hesleden Gill years ago. I had twelve in my garden, but strange to say, during this year they have all, with the exception of a seedling, died out. The root seemed to dwindle away, and then the stems lost their vigour and became rotten. I have found *Sedum anglicum* near the river, it is probably an escape. *Plantago maritima* is plentiful on the road side between here and Kilnsey. Another feature is the wonderful growth and variety of the Geranium family, and the way in which several varieties seem to have their fixed abode, and will go no further. The prevailing type is *Geranium sanguinium* from Hawkswick to Arncliffe. One pasture near Hawkswick is full of it, and the crimson and russet colours of the decaying leaves and branches in the Autumn are quite a pretty picture. Past Arncliffe the *G. sanguineum* ceases, and makes way for *G. pratense*, which, in its turn, finds its extreme limit near to Heather Gill; and about a quarter of a mile south of this village *G. sylvaticum* is as luxuriant as *G. sanguinum* is at the bottom of the valley. All these specimens flourish in my garden. *Geranium phœum* is known to exist in a wild state in one locality in the Dale.

Another feature in our plant life is that of the movement and the disappearance of plants. Some plants seem to partake of the restlessness of the age, and desire a change of residence. I have noticed since I have been here that the *Inula Pulicaria* has shifted its quarters, travelling about a mile in twenty-five years. The *Tragopogon pratensis* has followed suit. It occurs only sparingly in one or two localities; each year it seems to change its place. It is making up the valley now. This may be accounted for from the fact that the hairs of the pappus are long and very feathery, and like the *Leontodon*, are carried along easily by the wind. But then the prevailing wind is

Western, and yet the plant seems to be travelling East. I may also notice here that *Epilobium hirsutum,* which is not known in this valley, though it occurs eleven miles away, has recently unexpectedly appeared in considerable quantities in my garden.

With regard to the disappearance of plants, I may say that the *Menyanthes trifoliata* once grew in a wet pasture not more than half a mile from this village. It is now extinct, probably from the draining of the pasture. My predecessor, Archdeacon Boyd, who was here for fifty-eight years, took a great interest in the flora. He told me that when he came here the Holly Fern grew on the Clouders; he remembered seeing forty or fifty plants. It is now not to be found there. This he ascribed to the depredation of the collectors of ferns. I am sorry to say that we lose a great many every year. Another scarce plant is *Allium Scorodoprasum.* It grows only in one place, and during the last twenty-five years I have watched it I have seen no tendency for it to diminish or increase, nor does it move its position, probably the bulbous nature of the plant prevents migration. I found three or four plants of the Fly orchis on a bank in a wood years ago, but I have not been able to find any plants in the same spot since. The *Lamium album* is scarce *Parnassia palustris* is our last flower.

FLOWERING PLANTS, Etc.

Plants on Allerthorpe Common.—During a visit to Allerthorpe Common on August Bank Holiday a small party noted the following plants :—

Teesdalia nudicaulis.
Drosera rotundifolia.
Gentiana Pneumonanthe.
Gnaphalium sylvaticum.
Filago minima.
Carduus pratensis.
Epilobium angustifolium var. *alba,*
Ornithopus perpusillus.
Hypericum humifusum.
,, *pulchrum.*
,, *perforatum.*
Radiola millegrana.
Teucrium Scorodonia.
Geranium pusillum.
Silene noctiflora.
Urtica urens.
Erodium cicutarium.
Galium saxatile.

Galium palustre.
Senecio sylvaticus.
Calluna vulgaris.
Erica Tetralix. •
,, *cinerea.*
Œnanthe fistulosa.
Polygonum amphibium.
Comarum palustre.
Lycopsis arvensis.
Malva rotundifolia.
Genista anglica.
Spergula arvensis.
Buda rubra.
Juncus articulatus.
Aulacomnium palustre, (Moss.)
Marchantia polymorpha. (Hepatic), in splendid condition.

—J. MARSHALL, Beverley, August 17th, 1907.

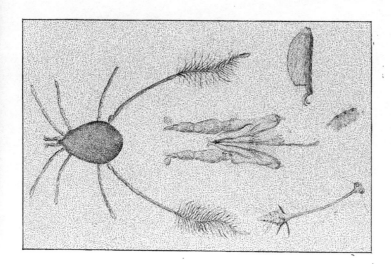

Fig. 9.　*Eatoniana plumifer.*
Fig. 10.　Palpi and rostrum.
Fig. 11.　Anterior tarsal joint.
Fig. 12.　Chitinous rod with capitulum.
Fig. 13.　Skin of body with scales attached.

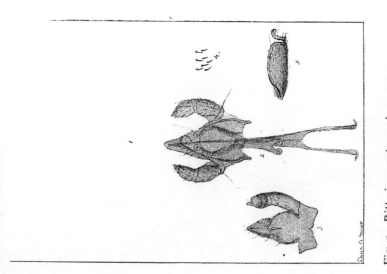

Fig. 1.　*Ritteria mantonensis,* n. sp.
Fig. 2.　Palpi and rostrum, much magnified.
Fig. 3.　Palpi, under side.
Fig. 4.　Body hairs.
Fig. 5.　Anterior tarsal joint.

LINCOLNSHIRE MITES.

RHYNCHOLOPHIDÆ.—(*Concluded*).

(PLATE XXXIX.)

C. F. GEORGE, M.R.C.S.
Kirton-in-Lindsey.

RITTERIA. This genus, or as Professor Sig Thor calls it, sub-genus, was first differentiated from *Rhyncholophus* by Kramer in 1877. It will be seen to differ in many respects from *Rhyncholophus.* The legs are not nearly so long, nor is the penultimate internode of the last pair of legs markedly longer than the others. The eyes, one on each side, consist of a single ocellus. The central part of the body in front projects in a snout-like manner, on which are a few hairs and two stigmatic openings, each provided with a rather fine and long tactile hair, well shown in Mr. Soar's figures. There is no ball-like capitulum as in *Rhyncholophus,* the modification of the palpus is perhaps the most important difference here; instead of a prominent pear-shaped fifth joint, there is only a slightly projecting hair pad, close to the base of the claw which terminates the fourth joint. The hairs or spines which cover the body are smooth, and appear to arise from the centre of a chitinous plate, they are also somewhat curved (see Plate XXXIX., fig. 4). This mite somewhat resembles *Erythræus nemorum,* but is easily differentiated from that creature by the structure of the palpi and body hairs. It is of a fine red colour, and is not unlike Thor's *Ritteria norvegica,* yet I am inclined to consider it to be a different species, and will therefore give it the provisional name of *mantonensis,* from its being found in the Parish of Manton, near to Kirton-in-Lindsey. I have also found several small specimens of a beautiful yellow colour belonging to this genus, whose specific names I am not able to give, each having the peculiar structure of the palpus. This structure was described and figured by Brady 'On *Rhyncholophus hispidus,*' in the Proceedings of the Zoological Society for 1877, pl. iv., and doubtless there are many more species of *Rhyncholophus* in England which require searching for, describing and recording. Many species, some of which, judging from the figures, appear to be very handsome creatures, have been described by foreign writers. One genus, which, so far as I know, has not yet been found in England, is remarkable for having a number of long hairs on the two last internodes of the hind legs, set in whorls and resembling very much a bottle-brush. The first

specimen I ever saw was sent to me in 1895 by Mr. Luff, of
Guernsey. It was a carded specimen which I had to return. I
recorded it in "Science Gossip" for November, 1896, under the
name of *Rhyncholophus plumipes.* Six years afterwards, in
August, 1902, I received from Mr. Luff a few living specimens,
some of which I dissected and mounted for the microscope.
Mr. Soar has kindly drawn a figure of this mite from a mounted
specimen of his own, assisted by mounted dissections made by
myself. Fig. 9 is a drawing of the dorsal aspect, showing
the general appearance of the mite ; the curious formation of
the hind leg is well shown. It must be very curious to see it
running on the sand in bright sunlight, with these legs carried
perpendicularly above its body. When alive, even when rather
feeble, if stimulated by a touch, the legs are immediately
elevated. Their utility to the creature, so far as I know, has
not yet been made out.

Fig. 10 shews the mouth parts much enlarged. The palpi
are seen to be distinctly five-jointed, the fifth joint pear-shaped
and arising near the middle of the fourth. The circlet of fine
hairs round the end of the proboscis is also indicated. In the
centre only one mandible is shewn, the other having been withdrawn during dissection. Fig. 11 shews the tarsus of the first
leg, which is compressed sideways, with claws and hair pad.
Fig. 12, the chitinous rod, which lies in the dorsal groove, with
its capitulum, which, in this case, is an inverted cone and not
ball-shaped as in *Rhyncholophus.* Fig. 13 shews the peculiar
leaf-like hairs or scales which cover the body.

In the Proceedings of the Zoological Society of London for
December 14th, 1897, will be found another species of this
genus figured and described by the Rev. O. Pickard Cambridge,
under the name of *Eatonia scopulifera* (the name *Eatonia* being
preoccupied was changed to *Eatoniana, loc. cit.* May 3rd, 1898).
The mite which was found in Algeria will be seen to differ in
several important particulars from the one here illustrated. In
the "Canadian Entomologist" for February, 1900, Vol. xxxii.,
No. 2, page 32, is a paper on these mites by Nathan Banks of
Washington, who states that Cambridge's mite is the same as
one described by Lucas in 1864 as *R. plumipes;* that Birula
described one in 1893, from Russian Armenia, as *R. plumifer,*
and this appears to be the same as the one figured by me from
Jersey. He also mentions another species from Switzerland,
described by Haller, which he would therefore call *halleri.* I
have not seen Haller's paper, and therefore do not know wherein

it differs from the othes two—at all events, it appears that three species are already known, viz :—

1. *Eatoniana plumipes*—Lucas.
2.   ,,   *plumifer*—Birula.
3.   ,,   *halleri*—Banks.

Possibly other species may eventually be discovered. It does not seem certain what the sex of the mite is. Cambridge considers his to be an adult female; the situation of the genital aperture, however, appears to me to be peculiar, certainly different from the female *Rhyncholophus* described in my previous notes. See 'The Naturalist' for July 1907, p. 260.

We regret to record the death of Mr. J. Romilly Allen. F.S.A., for many years editor of the ' Reliquary.'

A portrait of the late J. F. Walker of York, together with a lengthy memoir, appears in the August ' Geological Magazine.'

Mr. Alexander Ramsay continues to publish his remarkable ' Scientific Roll and Magazine of Systematised Notes,' Bacteria, Vol. II., No. 19, dealing with Vital Chemistry :—General, Acetates, Acids, has recently been received.

Mr. E. A. Martin, F.G.S., favours us with a reprint of his interesting article on ' Dewponds '; and from Mr. Joseph Kenworthy we have received a reprint of his notes on ' Antiquities of Bolderstone and Neighbourhood,' in which some British Cinerary urns, etc., are figured and described.

The late Mark Stirrup has bequeathed to the Museum of the Manchester University specimens of volcanic rocks and fossils; £1000 for the maintenance of a geological and palæontological collection; and £1500 for the foundation of a palæontological scholarship, tenable for two years by anyone who has studied in the university.

Referring to the note on the Bramble Finch near Halifax (' Naturalist, August, p. 291), Mr. Fred Stubbs of Oldham informs us that in May last he liberated a male brambling, which remained in the district two or three days, and was singing in an adjacent garden. He suggests that the Halifax bird may have been an escape.

In the ' Reliquary ' for July, Mr. E. Howarth describes and figures a pre-Norman cross-shaft at Sheffield. It appears to have been made from the local Carboniferous sandstone, and at one time did duty as a hardening trough in a cutler's shop. The representation of an archer on the Sheffield cross is ' not devoid of natural anatomy,' as are so many figures on crosses of the period,

The Council of the Leeds University has appointed Dr. Walter Garstang to the Professorship of Zoology, and Mr. V. H. Blackman to the Professorship of Botany, the two chairs which are to take the place of the Professorship of Biology hitherto held by Professor Miall. The Council has also appointed Miss Alice M. Cooke lecturer in history, in association with Professor Grant.

Sir Joseph Hooker, G.C.S.I., C.B., F.R.S., who celebrated his ninetieth birthday on June 30th, has been appointed by the King to the Order of Merit. Sir Joseph first gained fame by his work in the Antarctic, whither he accompanied Sir James Ross as botanist. Later he became the pioneer of Himalayan exploration, and was rewarded with a knighthood. He succeeded his father as director of Kew Gardens in 1865, and held that post for twenty years. He was an intimate friend of Darwin.

TORTRIX SEMIALBANA AT DONCASTER:
A Lepidopteron new to the County.

L. S. BRADY,
Sheffield.

ON Augnst 4th, whilst collecting in one of the large woods in the Doncaster neighbourhood, I beat out of a hedge, along with other Tortricina, a fine specimen of *Tortrix semialbana.* Barrett writes as follows of this species :—'It was to be found near Darenth, Greenhythe, and Dartford in Kent, and Mickleham in Surrey before 1860, but then became scarce, the latest capture I know in these localities being in 1873. Twenty years later it was discovered at Folkestone, and still occurs there.' I believe there has been no additional locality recorded since, so that its occurrence so far north as Yorkshire is interesting and unexpected.

[This is a very gratifying record, because, as Mr. Brady says, the species was totally unexpected to occur so far north. Mr. Brady sent me the moth for examination, and it is a very fine and well-marked specimen.—G. T. P.].

Proceedings of the Cleveland Naturalists' Field Club.

Edited by the Rev. J. Cowley Fowler, F.G.S. 1905-6, Vol. II., Pt. 2. Middlesborough. pp. 85-142, 2/-

In this volume our Cleveland friends have gathered together much useful information. Whilst by far the greater proportion of it had been previously printed elsewhere, it is none the less welcome in the present form, and as apparently the society has funds at its disposal for the purpose, we must not grumble. By far the largest article is devoted to a series of useful notes, on all manner of subjects, contributed by the late Rev. J. Hawell to his Parish Magazine. These, perhaps, were not generally accessible in their original form. From 'Country Life' Mr. Nelson's paper on 'The Ruff in the North of England' [Durham] is reprinted 'by the permission of the Editor'; a lengthy paper on ' The River Tees : its Marshes and their Fauna ' by the late R. Lofthouse is reprinted from this journal for 1887, without obtaining or asking for permission. Mr. T. A. Lofthouse gives an account of 'Cleveland Lepidoptera in 1905,' and Mr. M. L. Thompson writes a 'Report on the Coleoptera observed in Cleveland' in 1905. The publication is cheap at two shillings. There is a fair sprinkling of misprints, the assistant secretary's name is spelt wrongly more than once, and we don't like the word ' Mammalogy.'

GEOLOGY AT THE BRITISH ASSOCIATION.

J. LOMAS, F.G.S.,
Birkenhead.

THE shadow of the forthcoming Centenary of the Geological Society was cast over the proceedings of Section C at Leicester. Many distinguished foreigners, and not a few British geologists, were prevented from attending the meeting, owing to their inability to spare time for both functions. Nevertheless, the proceedings were full and interesting. Local papers were both numerous and important, reflecting the vigour and enthusiasm of the local workers.

Mr. Fox Strangways and Prof. Watts described the country about Leicester. Drs. Bennett and Stracey, who followed with papers on the Charnwood Rocks, did not see eye to eye with the Professor, and the conflicting views were discussed both in the meetings and in the field, during the admirable series of excursions which were held. Although the combatants remained unconvinced, a happy ending is promised by the appointment of a Committee to conduct analyses of the rocks and report at the next meeting.

Such is the interest now shown in Triassic problems, that a whole day proved insufficient to discuss the papers offered on this formation. Mr. H. T. Ferrar opened most appropriately with an account of the features shown in existing deserts. Mr. T. O. Bosworth applied these with great force and insight to explain the puzzling characters of the local Trias. At Croft, during one of the excursions led by Mr. Fox Strangways, we were enabled to see for ourselves the igneous rocks denuded of Marl, with wind etching, desert screes, desert crusts, and many other characteristic desert features. Dr. Cullis announced the discovery of dolomite crystals in the Keuper Marls of the South West of England, and he attributed their origin to precipitation from the waters of an inland sea. That similar crystals may be formed under other conditions was shown by their occurrence in marine sands.

Messrs. Bolton and Waterfall gave a description of the great masses of Strontia found at Abbot's Leigh, near Bristol.

In presenting the fifth report on the Fauna and Flora of the Trias, the Secretary communicated an important paper by Dr. A. Smith Woodward on 'A Mandible of *Labyrinthodon leptognathus*' recently obtained from Cubbington Heath, near

Leamington,' which showed in its structure an approach towards the Palæozoic Crossopterygian fishes.   Mr. H. C. Beasley, after a careful study of the wonderful find of footprints at Storeton last year, was able to give further information regarding some of the forms he has described, and Mr. Lomas gave an account of a large slab containing fifteen impressions made by the same animal.   So perfect were the prints that the full structure of the foot, including the skin, claws, and muscular pads could be ascertained,   It was suggested that *Cheirotherium* walked erect on the pes and only used the manus for support when bending down.   The Triassic fauna and flora of Leicestershire were well described by Mr. A. R. Horwood.

The set discussion of the Section on 'Iron Ore supplies'* was opened by Mr. Bennett Brough and Prof. Sjögren.   The former took a pessimistic view of the future of the British Iron supply, but we were relieved to hear from Prof. Sjögren that vast stores of ore are available from Scandinavia.   Prof. Lapworth pointed out the changes in the centre of gravity of the iron industry in Britain.   When the native forests were used to smelt the ore, the Weald was the great centre.   As coal came to be used, the centre moved to the coal-fields, and now the Lias and Oolite rocks supply the greater part of the iron supply of England.   Mr. Lamplugh pointed out that when the richer ores were exhausted the leaner kinds would be used, and of these there is still a great supply.   The President thought that Australia, with its vast coal supplies and good means of transport, would eventually be the great centre of iron supply.

Palæontology was not much in evidence, but two papers by Mr. F. Raw on the 'Trilobite fauna of the Shineton Shales' and 'The development of *Olenus salteri,*' were most admirable.   A marine remarkable peat was described by the present writer.

The report of the Kirmington Committee dealt very fully with the mammaliferous gravels at Bielsbeck, in the Vale of York.   The age of the deposit is still uncertain, as no evidence of ice action was observed in the material associated with

---

* By some mischance Mr. Brough, in the abstract of his valuable paper, is made to attribute the whole of the 'brown iron-ore beds' of Lincolnshire, Leicestershire, and Northamptonshire to the Inferior Oolite.   We are sure Mr. Brough would not desire so misleading a statement to remain long uncorrected.   By far the greater part of the Lincolnshire ore is derived from the Lower and Middle Lias, the Northampton ironstone being worked only at two places, and in Leicestershire very large quantities of ore are obtained from the Middle Lias.—P. F. K.

the fossils, and the accompanying flora does not suggest any special degree of cold. One of the speakers who had seen the excavations made laid great stress on the smell of the bones, and this reminds us of an incident told by a geologist whose veracity has never been questioned. He was at Ludlow, and wishing to locate the well-known bed containing fish remains, went to the Market Place. He sniffed the air all round, and at last detecting a fishy odour, he followed the scent, and was led straight to the exposure !

Prof. J. Joly found the rocks of the Simplon Tunnel to contain radium in unsuspected quantities. He is led to enquire whether this may not account for the high temperatures experienced in making the tunnel, and the thermal convection caused by the removal and deposition of radium bearing sediments may be a factor in mountain building.

Prof. J. Milne delighted the large audience which came to hear him with a racy account of his recent researches on Earthquakes. He finds that photographic plates exposed in dark caves and mines are affected by a mysterious light emanating from the rocks, and some of these coincide with the times of recorded earthquakes. His catalogue of important earthquakes shows that a maximum occurred between the years 1150 and 1250 A.D., and another increase commenced about the year 1650, and is still in progress.

The 1907 meeting, although not a large one, was most enjoyable, and Section C will never forget the warm welcome and the kindness extended to its members by the local geologists.

———◆◆———

Familiar Indian Birds, by Gordon Dalgleish. West Newman, London, 1907. 70 pp. Price 2/6 net. Apparently this book is for the benefit of those people in India who are likely to take an interest in the more common birds met with in that Empire. Several of the notes have previously appeared in English and Indian journals. In twenty-nine short articles the author describes the more common representatives of the Indian avi-fauna. The names of some of them sound odd to English ears :—the Amethyst-rumped Sunbird would doubtless have a different name in England, and several are 'Crimson-breasted,' 'Rose-ringed,' 'Indian spotted,' Bengal green,' etc. The author describes how the ' Paddy-bird' threw itself into a fighting attitude, though this may have nothing to do with its name. But the illustrations, by two artists, are very poor. If the 'Blue-faced Barbet' is anything like the drawing on p. 15, we are sorry for it, although its bright colours may be 'blended together with exquisite harmony ;' and the ' Purple Sunbird' on p. 13 is surely drawn from a model carved in wood. The frontispiece is called ' Bird-scaring in Bengal.' Apparently in India deformed, long-necked, hump-backed and macrocephalous humans are selected for the purpose of frightening the birds, or are the four bipeds perched in a tree rather out of drawing ? In view of its size, etc., the price of the book is too high.

BOTANY AT THE BRITISH ASSOCIATION.

DR. C. E. MOSS.

THE address of the President of the Botanical Section was a clear, clever, and incisive statement of some of the more abstruse problems of physiological botany, and Professor Farmer holds that it is only by the help of the elder sciences of chemistry and physics that the accurate formulation, to say nothing of the final solution, of the problems will be achieved. The majority of the papers were of a very technical nature; and close students of botany should make themselves acquainted with the details of the communications of Professor Bower on the embryology of the Pteridophytes; of Dr. H. C. J. Fraser, on nuclear fusions and reductions in the Ascomycetes; of Professor and Dr. Armstrong on enzymes; and of Mr. Gwynne-Vaughan on the real nature of the so-called tracheids of ferns. We understand that Professor Bower's communication will be embodied in the final chapter of his forthcoming book. The real nature of anything is always of interest to students of any branch of science, and it has to be confessed that the results embodied in Mr. Gwynne-Vaughan's work on the real nature of the so-called tracheids of ferns justified his rather ambitious title. It was more than interesting to see a young man pointing out to the veterans of botany their shortcomings in this particular matter, and more than pleasing to note the favourable reception given by them to his observations and conclusions. Mr. W. Bell, the local secretary of the botanical section, gave a highly interesting account of Charnwood Forest, and his remarks were illuminating in connection with two excursions of the section to that interesting region. The forest is partly under cultivation now; but extensive and primitive oak woods, and bracken-clothed hills still exist. The oak woods have a carpet of *Holcus mollis* and *Pteris aquilina*, varied in one case by extensive plant societies of *Luzula maxima*. Professor Weiss gave an excellent semi-popular lecture on " Some recent advances in our knowledge of pollination of flowers." Mendelism was very much to the fore, both in individual communications, and in the joint discussions with the zoological section on the physical basis of hereditary transmission. One of the most profitable excursions was to Burbage, to see Mr. C. C. Hurst's experiments in heredity on sweet peas, rabbits, and school children. Particulars of these experiments were supplied on a special circular. There was also a joint discussion with the zoological and educational sections on the teaching of biology in schools. Mr. Hugh Richardson, of York, was one of the official speakers in this discussion. Mr. D. M. S. Watson, a student of the University of Manchester, gave an account of the cone of *Bothrodendron mundum*, and Professor Oliver spoke on the structure and affinities of a fossil seed from the coal measures. Professor Bottomley brought up to date our knowledge of the root-tubercles in Leguminous and allied plants ; and his recent experiments in this regard on wheat are likely to revolutionise things. Mr. Bentley, of the University of Sheffield, essayed to speak on that thorny topic, the nuclear divisions of the Cyanophyceae. Mr. Wager was inclined to be sceptical, and pointed out the number of papers on this subject in recent years—almost every paper contradicting the rest. Mr. Bentley's results, however, are remarkable ; and thoroughly justify him in proceeding further with a very difficult investigation. One of the most interesting lectures was given by Professor Conwenz, of Danzig, State Commissioner in Prussia for the Preservation of Natural Monuments. His theme was the care of natural monuments, and his remarks were listened to with rapt attention by the members of the geological, geographical, and botanical sections. Professor Conwentz alluded to the 'Central Committee for the Study and Survey of British Vegetation' (secretary, Dr. W. G. Smith, of the University of Leeds), and thought that, of all bodies in Britain, this was the most suitable for taking up and propagating the work which he (Professor Conwentz) had at heart. Only one ecological paper was read, and that was by Professor Yapp on the 'Hairiness of certain Marsh Plants.' In particular, Professor Yapp pointed out many interesting peculiarities of the comparative hairiness of the leaves of *Spiraea Ulmaria*, the common meadow-sweet.

The following gentlemen attended the meetings of the botanical section :—Mr. W. N. Cheeseman, of Selby ; Dr. C. E. Moss, of Manchester ; Mr. M. B. Slater, of Malton ; Dr. W. G. Smith, of Leeds ; Mr. W. West, of Bradford ; and Dr. T. W. Woodhead, of Huddersfield.

———◆◆———

The attendance at the Leicester meeting of the British Association was over 300 less than at York the previous year. £1288 was appropriated for scientific purposes during the coming year.

A young seal was caught at the foot of Speeton Cliffs early in July, and for some time kept alive at Filey. The occurrence of seals on the Yorkshire coast is much more frequent than is usually supposed.

ANTHROPOLOGY AT THE BRITISH ASSOCIATION.

G. A. AUDEN, M.D.

THE increasing recognition of Archæology as a subject worthy to hold a place by the side of the 'more exact Sciences,' was amply exemplified in the programme of Section 'H' at Leicester ; for an actual majority of the papers contributed to the Anthropological Section dealt with Archæology in one aspect or another. The strong classical tone which pervaded the meeting is a point of some importance ; and in these days when the comparative merits of classical and scientific education are so frequently discussed, it is worthy of remark that in Section H, representatives of both educational traditions can meet upon common ground. Here, for example, a criticism of Usener's Theories upon the 'Augenblick-gotter' of the Greek religious cults may divide the day with 'Notes upon the Maories,' 'The Tribes of Perek,' or on 'The Souterrains of Ulster.'

The choice of a scholar so well-known as Mr. D. G. Hogarth (the author of 'A Wandering Scholar in the Levant'), to fill the Presidential Chair, gave promise of a memorable address, a promise which was more than amply fulfilled. Those who heard his 'Religious Survivals' will not readily forget the masterly analysis and development of his thesis or his delightful grace of style and diction.*

The discussion of greatest general interest was that initiated by Professor Ridgeway upon 'The Beginnings of the Iron Age,' wherein he showed by a process of exclusion that it was in the highest degree improbable that a knowledge of the properties and use of iron was brought from India, China, Egypt, South Africa, or Babylonia, and that, *per contra*, there was some presumptive evidence that this knowledge was diffused from the region of Noricum (within a few miles of Hallstatt), the iron mines of which have always been famous, and which is still the centre of the great Central European cattle routes. He believed that this knowledge was carried into the Aegean during the Dorian immigration period, together with the use of cremation—a custom adopted from the tribes of the great Germanic forest.

---

* As the address has been published in full by the British Association ('Presidential Addresses'), and in 'Nature' (Aug. 15th), no abstract has here been given ; any such attempt would fail to give an adequate summary of the argument.

Professor Flinders Petrie referred to the marked distinction to be drawn between the sporadic and general use of iron, and pointed out that Egyptian History showed 4000 years of the sporadic use of iron, obtained, as a rule, from Haematite, the so-called 'Stone of Heaven'; statuettes of which material are not uncommon. The earliest dated tools of iron are a saw and centre bits referred to 680 B.C. He alluded to the analogous case of flint, which was used in Egypt right down to Roman times, although bronze had been known for 800 years. Prof. Edouard Naville, of Geneva, argued that iron was but little used even in the new Empire, and referred to the use of iron battle axes as a tribute. Prof. J. L. Myres (who has recently been appointed to the Chair of Greek and Ancient Geography at Liverpool) drew attention to the difference in the blast furnaces used north of the Alps and the open hearth furnaces which had existed around the shores of the Mediterranean and in Egypt as far back as the 18th Dynasty, and to the resulting difference in the quantity of metal produced by the two processes. Mr. Arthur Evans found his chief difficulty in accepting Prof. Ridgeway's views in the fact that an earlier phase of iron age culture is found further south than Hallstatt in the Cemeteries of Bosnia, and that iron was found in the Palace at Knossos in undisturbed earth even of the 12th century B.C.; whereas the generally accepted date of the Hallstatt period was from 1000 to 800 B.C. The discussion was in the highest degree valuable, and to some of those present recalled the discussion on the same subject at Liverpool eleven years ago, when a good deal of dialectic heat was evolved.

At onother meeting Professor Flinders Petrie gave a description of the pottery Soul-houses disclosed by the last winter's work of the British School of Archæology at Rifeh. The object of these models was to provide shelter and provision for the soul, to keep it satisfied, and thus to prevent it from returning to the village. He proved that the increasing complexity of the models was a reflex of the evolution of the dwellings which they represented. In some of the more complex models not only was a stairway provided whereby the soul might mount to the upper storey, couches and chairs upon which it might rest, fire-places, water jugs, and a little model of a woman making bread under the stairway, but even a manger was added for the donkey, and a pond from which it might drink.

Professor Naville contributed an important paper upon 'The Beginnings of Egyptian Civilisation,' and described the discovery at Deir-el-bahari of the shrine of Hathor—the Goddess of the Mountain of the West—here, as usual, represented in the form of a heifer, the modelling of which shows such exquisite workmanship and power.

Professor Bosanquet described the continued excavations at Sparta, on the site of the Temple of Artemis Orthia, and the wonderful richness of the find of votive offerings in the neighbourhood of the Altar. In connection with this he gave an interesting explanation of the 'Scourging Festival' at Sparta, the cruelty of which is described by Cicero and Plutarch. The Scourging ordeal, the victor which was known as the ' *Bomonikos,*' or 'Victor of the Altar,' and which not infrequently resulted in death from the injuries inflicted, has hitherto been assumed to be a survival of an older Spartan test of endurance, comparable to the initiation rites of many primitive peoples into the privilege of manhood. Professor Bosanquet, however, traced the evolution of the Festival from a rough game in the 4th century B.C., in which the young Spartans had to snatch cheeses from the altar while others, armed with whips, tried to beat them off. Under the Roman rule there seems to have been an artificial revival under a mistaken idea as to the origin and meaning of the traditional usage, perverted, however, from being a mere game into being a regular competitive examination in the power of endurance, conducted before crowds of spectators, who flocked to the theatre built round the altar about A.D. 200 for that purpose. The game itself may have had its origin in a custom of the lads striking one another for luck with boughs from the *Agnus Castus,* which grew in the river bed, and was sacred to the goddess.

Dr. Ashby, who has charge of the excavations on the Roman site at Caerwent ('Venta Silurum'), presented a report of the excavations of the newly discovered Basilica and Forum, and of the large building with two Hypocausts, known as 'Building No. 7,' which was excavated last year. Caerwent, some five miles from Chepstow, is well worth a visit by those interested in the subject, and it is to be hoped that ample funds will be forthcoming to explore the large area which the generosity of Lord Tredegar has secured for excavation. The south wall of the fortress is in part in a very good state of preservation, and in its course has two multangular towers, which have a close likeness to the well-known Multangular Tower at York.

Turning to other themes the Section joined with that of

Education to discuss the question of Anthropometrics in schools. The subject, the importance of which can hardly be over-estimated, was introduced by Sir Victor Horsley, who was supported by Mr. Ramsay Macdonald, M.P., and others. The following Resolution, communicated to persons interested in the matter, gives to the discussion a concrete form :—

'Resolved that, in view of the national importance of obtaining data on the question of physical deterioration, this Association urges upon the Government the pressing necessity of instituting in connection with the Medical Inspection of school children, a system of periodic measurements which will provide definite information on their physical condition and development.'

A novel method of illustrating a paper was that of Dr. Seligmann, who, by means of a cinematograph, gave a graphic description of some of the dances of New Guinea. Dr. Seligmann also read a highly important paper upon pre-historic Stone Weapons, engraved Shells, and Potsherds from various sites on the Coast of New Guinea. The present inhabitants do not recognise the use or nature of these objects, while the depth at which many of them were found attests their great antiquity.

Other papers of interest were Mr. G. L. Gomme's 'The Origin of Totemism'; that of Mr. Crowfoot upon the 'Anthropological Field in the Anglo-Egyptian Soudan'; and the Report of the University of Wales upon the 'Ethnological Survey of Wales.' But where all were excellent it is perhaps invidious to call attention to individual papers.

Finally, allusion should be made to the valuable Reports of the Sectional Committees, the work of which goes on from year to year, *e.g.*, that for exploring the lake village of Glastonbury, a work begun in 1892 now nearing its completion, and that for estimating the age of Stone Circles, which reported that a work of the utmost importance is about to be begun in the examination of the Avebury Stone Circle. The Committee for determining the best method of registration and cataloguing the Megalithic remains of Great Britain has not been idle, and presented a useful Interim Report.

On the Saturday during the Meeting, a large number of those who attended the Section availed themselves of the Geological excursion through Charnwood Forest, arranged by Professor Watts, whose experience of this area as a Member of the Geological Survey, gave exceptional interest and value to the expedition.

1907 October 1.

In Memoriam.

JOHN WILLIAM FARRAH, 1884-1907.

OUR readers will greatly regret to hear of the untimely death of Mr. John William Farrah, which took place on the 7th of September. He was the son of Mr. John Farrah, of Harrogate, and frequently accompanied his father on the excursions of the Yorkshire Naturalists' Union. On these occasions his kindly disposition secured him many friends. He was very successful with his camera, and the pages of this journal have frequently been enriched by reproductions of his negatives. Like his father, he had a distinct taste for Natural History, and his death at so young an age as twenty-three is a distinct loss to our county society. In recent years father and son were rarely separated. We can all extend to Mr. John Farrah our real sympathy in a loss, the nature and magnitude of which very few of us can form any idea.

━━━◆◆━━━

We have received Charter's **Bridlington and District Guide** and Apartments' Directory (36 pp., with numerous illustrations and good maps). Hull : Harland & Sons. Price 6d. It has been carefully drawn up, and the principal attractions of the place are set forth in a clear manner. The maps are particularly useful.

Hull Museum Publications Nos. 43 and 44 have been issued. The first named is the twenty-first quarterly record of additions, and contains notes on ' The Chariot-Burial recently found at Hunmanby,' ' The History and Evolution of Coins and Medals,' ' Old Hull Pottery,' etc. The second contains reprints of Mr. Sheppard's papers on ' A Deformed Antler of a Red-Deer ' and ' Recent Geological Discoveries at Speeton,' which recently appeared in the ' Naturalist.' The pamphlets are on sale at Messrs. A. Brown & Sons, Hull, at one penny each.

The recently issued **Report and Transactions of the Manchester Microscopical Society** bears unmistakable evidence of the enthusiasm of its members. In addition to the various reports on the year's working, the volume contains the President's address on ' The Differentiation of Species of Coelenterata in the Shallow Water Seas,' by Prof. S. J. Hickson ; ' Snakes,' by R. Howarth ; ' An Introduction to the British Hepaticæ,' by W. H. Pearson ; ' Notes and Criticisms on Microscopical work,' by A. Flatters ; ' Practical Bacteriology,' [by Dr. A. Sellars] ; ' Spring Notes on Natural History,' by W. H. Pepworth ; ' Notes on Scolytidæ or Bark Beetles,' by A. T. Gillanders ; ' British Forest Trees,' by Dr. F. E. Weiss ; and ' The Internal Structure of some Insect's Heads, as revealed by the Microscope,' by W. Hart. Altogether the Report is a very creditable production, and worthy of the city which does so much in furtherance of natural science. In our copy, however, the pages containing the papers dealing with ' Snakes ' and ' Coelenterata ' are mixed up in such a way that we can only assume the binder was very busy, and working late, very late, on a Saturday night !

MANX CROSSES.*

WE can imagine no greater pleasure to a conscientious worker than to find the results of his labour given to the world in a volume which does credit alike to himself, to the publishers, and to the subject with which he deals. For many years our contributor, Mr. P. M. C. Kermode, has devoted his energies to the geology, natural history, and antiquities of his island home. To him is in a large measure due our present extensive knowledge of the scientific history of Manxland. To him future generations will pay tribute for the care he has taken in rescuing and preserving the various evidences of human handiwork, which have a peculiar interest to the archæological world, from the geographical portion of the well-defined area with which he deals. The present writer is fairly familiar with the literature dealing with the archæological and natural history of Manxland, and for some time has been impressed with the great good that has been done by Mr. Kermode. His present work, however, eclipses everything else he has accomplished. In it he describes a collection of undescribed and sculptured monuments, which are as extraordinary in their variety as they are quaint in their design. They cover the period from the end of the fifth to the beginning of the thirteenth century, and upon them is a series of records dating from the Introduction of Christianity in the Isle of Man, and 'form a connecting link between the early sepulchral stones of Wales, the inscribed slabs of Ireland, the cross-slabs of Scotland, and the Celtic, Anglian, and Scandinavian stones of the North of England.'

There are no fewer than one hundred and seventeen monuments of this class in the island—of which number several have been found during the past few years, and about seventy are figured and fully described for the first time in the excellent volume before us. Some of the sculptures depict characters and illustrate stories in the Norse Mythology, in connection with which Mr. Kermode has made a careful study of the Saga literature. It is also remarkable that, so far as is known, every Manx Cross still remains on the island—one or two examples which had been taken away having been returned.

In a scholarly Introduction Mr. Kermode discusses the period of the monuments; the arrival of the Scandinavians as Heathens; the dates, materials, nature, purpose, evolution, and distribution of the crosses; their art and development, the

---

* By P. M. C. Kermode. London: Bemrose & Sons, Ltd., 1907. 222 pp., plates. Price £3 3s. net.

inscriptions, etc. The volume is then divided—the first portion dealing with the art of the crosses, and the second with detailed descriptions of them.

For many years to come 'Manx Crosses' will be a text book, not only so far as it relates to the small island with which it deals, but also to students interested in the early history of north-western Europe. It is marvellous that so important a chapter in our history should have emanated from so small an area. But the Isle of Man has yielded glorious opportunities, and of these every possible advantage has been taken by Mr. Kermode. Quite apart from its historical worth, all book-lovers will be thankful to Messrs. Bemrose for the excellent way in which they have done their share of the work. They have intended 'Manx Crosses' to be a standard work for all time, and neither trouble nor expense has been spared towards this end. The paper and illustrations are such that they recall the old days when printing was a pleasure, and sixpence-halfpenny cloth bound volumes were unknown. We can only hope that the work has the large sale that it certainly deserves.

Fifty-two Nature Rambles, by **W. Percival Westell.** The Religious Tract Society, 1907. 237 pp., price 3/6. Uncle W. Percival Westell has written another book. His previous books have had introductions by Mr. Aflalo, the Rt. Hon. Sir Herbert Maxwell, and Lord Avebury respectively. When noticing his last book (see ' Nat.,' March, 1907, p. 115) we wondered who would write the introduction to Mr. (beg pardon, Uncle) Westell's next. Uncle Westell has scored; it is by W. Percival Westell! Having now reached the top of the tree we cannot even hope to guess who will introduce the next work from Uncle Westell's pen—or is it a type-writing machine that makes multiplex copies? We know the author is an Uncle, he tells us so something like a hundred and seven times, and he dedicates the book to his nephew. If we mistake not there is a photo of Uncle Percival (or Percy, or William, or is it plain Uncle Bill?). He is 'listening to the chiff-chaff.' And there is a photo of his nephew, who has also listened to the chiff-chaff for a year, and we hope he has picked up a few grains of knowledge from it. He is an appreciative young man, this nephew; and says 'Oh, Uncle' on nearly every page, and ' Uncle, please tell me. It is all so very interesting,' 'Oh, Uncle, how quickly you found one,' 'Uncle, I am so enjoying this *dandelion* talk,' and ' Uncle, a fond nephew offers you a thousand thanks.' In his opening remarks to his 'Dear boys and girls' Uncle Westell truly says they may not all have Uncles willing *or able* to take them into the fields and lanes, etc. Nephew Stanley is careful to ask suitable questions on each ramble, and unlike most boys, he does not ask any awkward ones. He likes poetry, too, or at least his Uncle says he does, and this affords opportunity for giving quotations from the various poets, followed by ' Isn't that nice'? etc. Uncle Westell himself grows poetic, and wants to sow 'nature study seeds' in the gardens of his readers' minds. 'Fifty-two Nature Rambles' tells us much that we have been told before. It is largely ornithological, as usual, but on account of its style and particularly by the reason of the many beautiful illustrations, will be suitable as a prize-book for young children. Some of the illustrations are unusually fine—that of the Iris (fig. 59) being perfect. There are a few coloured plates, one of which, shewing Sallow and Willow Catkin, the publishers kindly permit us to reproduce (Plate XXXIV.).

LINCOLNSHIRE NATURALISTS.*

WE have here another number of the 'Transactions of the Lincolnshire Naturalists' Union,' which keeps up the interest, the varied nature, and the excellence done by the active workers in that county.

The frontispiece, a good likeness of Mr. F. M. Burton, F.G.S., F.L.S., who was the Union's second president, is accompanied by a sketch of the career of one who has done sterling work in various departments of natural history, more especially geology.

Mr. G. W. Mason has a good list, with remarks on localities and a summary of the districts range of each, of the Lincolnshire butterflies, of which there are no fewer than fifty-six, including such good things as *Lycæna semiargus, Polyommatus dispar, Aporia cratægi, Papilio machaon, Apatura iris, Limenitis sibylla*, etc. Mr. Mason has drawn upon the work and experience of numerous observers, to whom due credit is given.

A similar excellent list of the Lincolnshire Liverworts, forty-three in number, by Miss S. C. Stow, follows. To this paper are prefixed some general remarks on the group, by Mr. J. Reeves, F.L.S.

The presidential address, from which the President's name is conspicuous by its absence, no doubt merely an inadvertence, deals with 'Natural Habitats and Nativeness,' in which the Rev. E. Adrian Woodruffe-Peacock, F.L.S., F.G.S., discusses an interesting botanical subject in his own inimitable style, and we note that there is scarcely such a thing as 'a natural habitat' in the second largest English county, so great has been the influence of man in altering the surface. Mr. C. S. Carter follows with a list, with habitats, of 'Additions to Lincolnshire Non-Marine Mollusca'—the additions being varieties and fresh localities for thirty-eight species. 'Notes' on Local Occurrence of *Neritina fluvaitilis*,' by Mr. John F. Musham, a new and welcome writer in these Transactions, discusses local distribution near the city of Lincoln. The Rev. E. Adrian Woodruffe-Peacock has a page on 'Rare Lincolnshire Plants,' really *a* plant, *Cyclamen hederæfolium* (Ait.) The County Museum is the subject of a page, and also a plate of

* Lincolnshire Naturalists' Union Transactions, 1906. Edited by Arthur Smith, F.L.S., F.E.S. Printed by Wiggen Bros.; Louth, Lincs. (8vo., pp. 73-128, and two plates).

its Lower Storey, which by permission we here reproduce. We have to congratulate the City of Lincoln on establishing a genuinely county museum, and trust under the able curatorship

of Mr. Arthur Smith it will flourish and develop. An interesting and valuable paper, 'Notes on the Birds which inhabit Scotton Common,' is from the pen of the Rev. F. L. Blathwayt, M.A., M.B.O.U. There are various short notes and records of field work, an illustrated paper, by Mr. C S. Carter, on the 'Pairing of *Limax maximus*,' based on original observation, a good account of the field meetings for 1906 and the excellent work done at them—and there are also lists of officers, new members, and a balance sheet.

We observe a note of a resolution to 'ignore' the first part of the Transactions, the reason assigned being the erratic pagination, and a suggestion to reprint two of the papers in it. Might we suggest that from a bibliographical point of view there can be no ignoring by a Society of a portion of its own publications, especially so excellent a part as the one referred to, and from a practical point of view, that with so great a wealth of material in a county like Lincolnshire, it would be wisest to utilise funds in printing fresh papers rather than reprinting former ones?   R.

FIELD NOTES.

SHELLS.

Paludestrina confusa at **Saltfleetby.**—We have recently found *Paludestrina confusa* in abundance in drains at Saltfleetby. It has been identified by Mr. E. A. Smith.—C. S. CARTER, Louth, September 9th, 1907.

—: o :—

LEPIDOPTERA.

Acherontia atropos at **Paddock, Huddersfield.**—I had a perfect specimen of the Death's Head Hawk Moth (*A. atropos*) brought to me on the evening of the 3rd September. It had been captured very early the same morning in the bedroom of a house in Brow Row, Paddock.—W. E. L. WATTAM, Newsome.

—: o :—

FLOWERING PLANTS.

Potamogeton alpinus, a Correction.—Since recording the above species as new to V. C. 63 in last month's 'Naturalist,' I have received letters from six different botanists calling my attention to the fact that *P. alpinus = P. rufescens,* and has been found in several stations in 63. It is very pleasing to see how well the Yorkshire records are watched.—H. H. CORBETT.

—: o :—

FISHES.

Large Sunfish at Whitby.—Yesterday, 16th inst., a large Sunfish (*Orthagoriscus mola*), measuring from tip of dorsal fin to tip of ventral fin 5 feet $6\frac{1}{2}$ inches, and from snout to outer edge of caudal fin 4 feet 3 inches in length, was caught in the herring nets by the crew of the fishing boat 'Mary,' 35 B H, about five miles off Saltburn, and brought into Whitby, where it was sold and exhibited.—THOS. STEPHENSON, Whitby, September 17th, 1907.

NORTHERN NEWS.

An interesting note on 'The Manx Slates' appears in 'The Quarry' for August.

Mr. F. W. Sowerby records *Papilio machaon* near the shore at Tetney, North Lincs., in July, 1906 ('The Entomologist,' August, 1907).

We take this opportunity of congratulating our contributor, Mr. G. Grace, B.Sc., of Doncaster, on his appointment as principal of the Technical School, Barrow.

We are pleased to notice that the Ilkley Urban District Council has declined to allow an extension of the quarries on Ilkley Moors, on account of the spoilation of the beauties of the moorlands which would result.

In a paper on 'Nestling Birds and some of the problems they present,' Mr. W. P. Pycraft describes 'the active, down-clad type, and the type which *leaves the egg perfectly naked, and with sealed eyelids.*' Fancy leaving an egg like that!

In 'British Birds' for August, reference is made to the breeding of Ruffs in Yorkshire recently. As this has been copied in more than one scientific journal it is perhaps as well to correct it. The record refers to *Durham*, as the nests, which were figured and described in three or four journals at the time, occurred on the Durham side of the Tees.

There has been some correspondence recently in reference to the provision of a Municipal Museum for Leeds. A 'prominent member of the Corporation,' however, who was interviewed, whilst admitting the idea as 'worthy of consideration,' said that it 'must not be talked of when the city was being committed to such vast expenditure as was involved in the sewage scheme, etc.'

Mr. E. A. Martin, in 'Knowledge and Scientific News' says, 'In speaking of times intervening between one [geological] formation and another, we have no titles which in a single word would explain these possible breaks.' He considers such titles would be useful, and makes a number of suggestions, such as Marrian, Binneyan, Harmerian, Juddian, Whitakerian, Seeleyan, and Sillimanian. 'Sillimanian' is good; we'll stop there!

The Manchester Microscopical Society has sent us a copy of their syllabus of lectures given by its 'Extension Section.' In this there are titles of no fewer than forty-four lectures by well-known students. These lectures are given gratuitously by the members of the society, but actual out-of-pocket expenses are to be paid. We feel sure that the societies in the district will avail themselves of this offer, and that good will result from the scheme.

In a recent number we gave some examples of Newspaper Natural History. A correspondent sends us the following examples of legal botany and zoology culled from official papers :—' Any bud, blossom, flower, or leaf of any tree, sapling, shrub, underwood, gorse, furze, fern, herb, or plant.' ' "Animal" means any beast or other animal.' ' The carcase of any head of cattle. The expression " cattle " includes horses, mules, asses, sheep, goats, and swine.' ' It might be argued that rabbit skins are "animal matter," but we very much doubt it.'

By a new rule Junior members have been admitted to the Chester Society of Natural Science, Literature, and Art. In this way fifty-two additions have been made to the membership, which now stands at the satisfactory figure of 1072. As well as a brief account of the year's work of the Society, the Thirty-Sixth Annual Report contains a list of additions to the Grosvenor Museum, and a meteorological report for the year. As Part 6, No 1 of its 'Proceedings' the same Society has issued a paper 'On some rare Arachnids captured during 1906 by Dr. A. R. Jackson. These are from various localities, but are largely from Cheshire.

A MONTHLY ILLUSTRATED JOURNAL OF
NATURAL HISTORY FOR THE NORTH OF ENGLAND.

EDITED BY

T. SHEPPARD, F.G.S.,

THE MUSEUM, HULL;

AND

T. W. WOODHEAD, Ph.D., F.L.S.,

TECHNICAL COLLEGE, HUDDERSFIELD.

WITH THE ASSISTANCE AS REFEREES IN SPECIAL DEPARTMENTS OF

J. GILBERT BAKER, F.R.S. F.L.S.,   GEO. T. PORRITT, F.L.S., F.E.S.,
Prof. P. F. KENDALL, M.Sc., F.G.S.,   JOHN W. TAYLOR,
T. H. NELSON, M.B.O.U.,   WILLIAM WEST, F.L.S.

Contents :—

LONDON :

A. BROWN & SONS, LIMITED, 5, FARRINGDON AVENUE, E.C.

And at HULL AND YORK.

Printers and Publi: rs to the Y.N.U.

PRICE 6d. NET. BY POST 7d. NET.

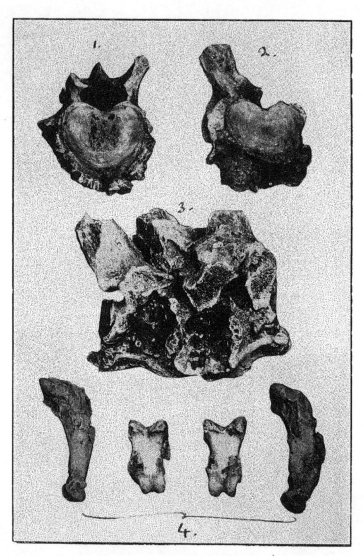

Diseased Bones of Bear.

NOTES AND COMMENTS.

THE GEOLOGICAL SOCIETY'S CENTENARY.

Probably at no previous period has there been gathered together in one place so many prominent geologists as assembled in London on the occasion of the recent Centenary Celebration of the Geological Society, and it is very unlikely that there will ever be such a gathering again.   There were delegates and representatives from Austria-Hungary, the Argentine Confederation, Belgium, Denmark, Egypt, France, Germany, Greece, Holland, Italy, Japan, Mexico, Norway, Portugal, Russia, Sweden, Switzerland, the United States, Canada, India, South Africa, Australia, and New Zealand.   Most of these countries were represented by several delegates.   From Great Britain were representatives of all the Universities, Colleges, Museums, and important Societies, and from most, congratulatory addresses were handed in to Sir Archibald Geikie, who presided.

DISTINGUISHED VISITORS.

Amongst those present were many whose names are world-wide.   Perhaps one of the most remarkable features in connection with the gathering was the great number of geologists present who have been well known for their work during the last fifty years.   And though many were bordering on four score years, their years seemed to sit lightly on them.   Every facility was given for making new and renewing old acquaintances, and in this way the successful conversazione held in the Natural History Museum at South Kensington did much in the interests of geological science.   One man, a Yorkshireman, was much missed—an accident which he had some little time ago depriving him of a pleasure which would only have been surpassed to the many savants who would gladly have seen him present; but Dr. Sorby was not forgotten, and several sent him some token of their regard.

THE HISTORY OF THE GEOLOGICAL SOCIETY.

In connection with the Celebration, Mr. H. B. Woodward has prepared a charming History of the Society, which is a mine of useful information, pleasantly written.   In this are some curious pieces of information relative to the beginnings of the Society.   It was at first an offshoot from two older bodies of limited membership, the Askesian and British Mineralogical

Societies. In the diary of William Allen, Quaker and analytical chemist, is the following entry :—' On the 13th of the eleventh month, 1807, dined at the Freemasons' Tavern, about five o'clock, with Davy, Dr. Babington, etc., etc., about eleven in all. Instituted a Geological Society.' On the same date, too, Humphry Davy had written to a friend, ' We are forming a little Geological Dining Club, of which I hope you will be a member. I shall propose you to-day.

'THE FATHER OF ENGLISH GEOLOGY.'

Mr. Woodward's book contains portraits of the leaders of geological science. As a frontispiece is a coloured plate of William Buckland, with his quaint costume, top hat, gloves, umbrella, and green bag. That of John Phillips, at the age of sixty, represents him in a waistcoat and trousers of a pattern which even geologists would hardly dare to wear to-day. Perhaps the most interesting of all is that of William Smith,

William Smith.

who over a century ago had outlined the general stratigraphy of Britain; and was the first, in this country, to determine the succession of strata by means of the fossils they contained. Sedgwick, in 1831, conferred upon William Smith the proud title of Father of English Geology.

TOADS EMBEDDED IN ROCKS.

The question of toads being found alive in cavities in rocks was formerly, as now, a frequent theme for discussion. The Geological Society Club, founded in 1824, conducted its researches somewhat after the following manner :—

November 19th, 1824.—Mr. Lyell having stated that an experiment had been instituted of enclosing toads in several cavities in rock in the month of September last, with the view of opening the cavities in a succession of years, one in each succeeding year :

The President (Buckland) bets Mr. Warburton two bottles of champagne to one that at the end of one year from the time of closing one toad will be found alive.

Mr. Warburton also bets the President a bottle of champagne that no toad will be found alive at the end of the second year.

Mr. Taylor bets Mr. Stokes a bottle of champagne that at the end of one year one of the toads will be found alive ; also another bottle that one will be found alive at the end of two years ; and another bottle that one will be found alive at the end of three years.

December 2nd, 1825.—Mr. Lyell stated that the cavities enclosing the toads had been opened on November 15, 1825, and that two toads in them had been found alive.

Resolved that the bets between Dr. Buckland and Mr. Warburton and Mr. Taylor and Mr. Stokes, referring to the period of one year, are decided by this evidence to be lost by Mr. Warburton and Mr. Stokes respectively, one bottle of champagne each.

THE LEICESTERSHIRE COALFIELD.*

Mr. C. Fox-Strangways in 1893 commenced a re-survey of the ' Leicestershire Coalfield,' by which is usually meant the joint coalfields of Leicestershire and South Derbyshire. The survey was completed in 1898 ; maps and explanatory memoirs were issued in 1899 and 1905. A memoir dealing with the coalfield generally, prepared by Mr. Fox-Strangways, has now been published. Besides an interesting introductory chapter, this contains an account of the History of the Development of the Coalfield; Pre-Cambrian Rocks ; Carboniferous Limestones and Shales ; Millstone Grit, Coal-Measures, etc. ; Physical History ; Extension of the Coalfield beyond Present Workings ; and Economic Geology. There is also a chapter on the Palæontology of the Coalfied, by Mr. A. R. Horwood. There are three useful appendices, (1) Glossary of Technical Terms ; (2) Bibliography ; and (3) Pit-sections, bore-holes, etc., the last of which occupies more than half the volume. Like all Mr. Fox-Strangways' work, this memoir is a thorough and conscientious production, and for a Government publication it is cheap, fairly well printed and bound, though the illustrations are poor, as usual in these Memoirs. And where else, but in a Government publication,

---

* Memoirs of the Geological Survey : The Geology of the Leicestershire and South Derbyshire Coalfield, by C. Fox-Strangways, 1907. E. Stanford, 373 pp., plates. Price 6/-

would one expect to find, at the foot of the Preface, anything like this :—'8556. 750.—Wt. 20003, 8/07, Wy. & S. 3577r. a.'?

SYCAMORE LEAF BLOTCH.

Perhaps one of the most disfiguring and certainly one of the most common of the diseases of trees in different parts of the country is the Sycamore Leaf Blotch. This has recently formed the subject of one of the useful leaflets issued by the Board of Agriculture and Fisheries. From it we learn that the blotches are due to a fungus, *Rhytisma acerinum* Fries., and when once infected the trees get worse year by year, until eventually the tree dies, as the fungus prevents the leaf from doing its work, enfeebles the tree, and thus exposes it to even more deadly

Sycamore Leaf Blotch, *Rhytisma acerinum* Fries.

parasitic fungi, such as the Coral Spot fungus. 'The method for preventing a continuance of this disease is both simple and effective. . . The young leaves are infected in spring by floating spores which escape at that season from dead leaves which have been lying on the ground during the winter. If all such dead leaves are collected and burned directly they fall in the autumn, or at latest before the young leaves unfold in the spring, the disease will be arrested.' We are indebted to the Board of Agriculture and Fisheries for permission to reproduce the illustration.

WALTON BONE CAVE, NEAR CLEVEDON.

In the 'Proceedings of the Bristol Naturalists Society' (Vol. I., Pt. 3), recently to hand, Prof. S. H. Reynolds, F.G.S., has an interesting note on a Bone Cave at Walton, near Clevedon. In addition to the mammalian remains (horse, bear, wolf, vole, rabbit, etc.) were bones of an exceptionally large number of birds, viz., eagle, buzzard, wheatear, skylark, robin, redwing, thrush, blackbird, raven, greenfinch, swift, ringed plover, golden plover, turnstone, dunlin or sandpiper (?), godwit or greenshank (?), whimbrel (?), heron, common gull, cormorant, wild duck, wigeon (?), pintail (?), goose. With regard to the bones of the bear, many of them exhibited a markedly diseased character. The vertebræ and phalangeal bones particularly showed a pronounced form of osteo-arthritis. An excellent illustration is given showing the nature of the bone disease. This we are kindly permitted to reproduce (Plate XL.).

THE LION AND THE MOUSE.

We are anxiously awaiting the reports from our agents in various parts of the world, in order to see to what an extent our sales have recently increased. For Mr. E. Kay Robinson has mentioned 'The Naturalist' in his paper. Truly, the reference to our journal is not a long one, perhaps not a very flattering one—thank heaven! but it *is* mentioned. Mr. Robinson has not seen this journal, and does not know whether it is a monthly or quarterly, but from letters he has received from the members of the Yorkshire Naturalists' Union he learns there is such a paper. We are quite prepared to believe what Mr. Robinson says, but there's a lot wouldn't. It is a pity the various marked copies that have been sent to him have all gone astray. We presume also that the copy of our journal containing remarks (which were made at Mr. E. K. Robinson's request) on 'Mammal *v.* Animal,' sent to him by registered post, has also gone astray. We understand from the Post Office that it was duly delivered. But we can't believe it. We have more faith in Mr. E. K. Robinson's word. We understand the members of the Yorkshire Naturalists' Union have complained to him about the remarks appearing in this journal. Might not these members have allowed us one more chance to live by warning us before writing to Mr. Robinson? Fortunately, however, generosity and greatness often go together, and that gifted naturalist has dealt leniently with us. How awkward it would have been for the publishers of 'The Naturalist' if, with a few strokes of his pen, Mr. E. K. Robinson had 'proposed to drop "The Naturalist,"' just as he did the word mammals.* We notice, however, that we are warned as to our conduct, and if we 'really wish to serve the interests of the Union,' we must keep our pen under better control. We shall certainly pay due regard to that warning. We were in the Natural History Museum at South Kensington a few days ago, and were horrified to find the word 'Mammal' still in general use. In fact, there was a 'Guide to the Mammals' offered for sale—though in fairness to Mr. Robinson we ought to say that we did not see any of the visitors (presumably mostly B.E.N.A.'s, though they didn't wear their badges) buy it. *But,* we hear on good authority that the Director, Sir E. Ray Lankester, is 'retiring' at the end of the present year. *Ah!*

COUNTRY SIDE 'NATURAL HISTORY.'

According to the 'Country Side,' the recent Limerick, etc., competitions in the newspapers has 'sent up the social value of brains.' Mr. E. Kay Robinson's finger-nails (and presumably his toe-nails too, though his observations have apparently not extended thus far) clearly shew that he has 'narrowly escaped death.' For the second time 'E.K.R.' has ridden 'Home' in a Motor Car, and for the second time has given his readers an account of his achievement. By a curious coincidence the 'species' of the car is the same as that 'puffed up,' with fair regularity, in the 'Motor Notes,' and in the advertisement column of the same paper. With the view of furthering the study of Natural History, presumably, 'Country Side' has started a 'fine contest of skill and brains'—a Limerick competition, and the B.E.N.A.'s can 'cudgel their brains' to produce the 'last lines.' From the following brilliant example it will be seen that there is a distinct 'natural history' flavour about the com-

---

* Nothwithstanding Mr. George Washington Robinson's assurance that he would drop the word 'mammal' and use the word 'animal' instead, we notice that in his paper recently, in asking for information about '*Creatures and Plants,*' he puts the Mole, Mouse, Rat, and Vole under the heading 'Mammals.' Possibly this is a misprint, and the unfortunate 'comp' is seeking a new situation. In the same list, under 'Weeds and Fungi,' is included 'Finger and toe in turnips.' Is this a weed or a fungus? or should it come under 'animals.'

petition; it is also remarkable that the letters 'e. k. r-o-b-i-n-s-o-n' occur in the first four lines :—

> Said the humming-bird to the shrew
> ' If I were as dowdy as you
> I'd keep to the *house.*'
> ' Fiddle '! answered the mouse
> .   .   .   .   .
> I'm a *Country Side* piffler, arn't you?

We have filled in the last line, in case any of our readers care to compete for a badge—or whatever the prize may be. There need be no fear of the source being detected, as Mr. Alfred Austin Robinson never sees 'The Naturalist.' *

SAGINA REUTERI BOISS. :

A NATIVE OF THE BRITISH FLORA.

W. INGHAM, B.A.
York.

ON the 11th August, 1906, I found this small flowering plant on the bed of a very old pool, almost dried up. This was on Skipwith Common, near Selby, in V.C. 61, in a truly wild habitat, with no sign of alien plants of any kind associated with it.

It is known as an alien plant from Worcestershire, Cheshire, and Lancashire, but there is the strongest evidence that it is native on Skipwith Common. Its associates there were *Mentha pulegium, Apium nodiflorum* var. *repens, Limosella aquatica, Veronica scutellata* var. *hirsuta,* and the Hepatic *Riccia crystallina.* Additional evidence given by Dr. F. N. Williams, our authority on the *Caryophyllaceous* order of plants, who has examined the specimen sent to him by Mr. Wheldon from me, is as follows :—'Although differing in appearance from the very glandular forms of this plant you (Mr. Wheldon) have previously sent me (*Vide* Ex. Club Rep., 1902), they belong to *S. Reuteri,* and are indeed more like the original Portuguese specimens than the Lancashire plants.' The Skipwith Common plant is quite eglandular, and a few plants only were gathered, as it was first thought to be *S. apetala.*

It is clear that *Sagina Reuteri* is a *native* addition to the British Flowering Plants.

---

* This 'fine contest of skill and brains' is apparently not suitable for the readers of the ' Country Side,' and has been discontinued !

THE HAIRY-ARMED BAT.
(*Vesperugo leisleri*).

(PLATE XLI.).

ARTHUR WHITAKER,
Worsbrough Bridge.

THIS bat appears to be the least common of the eight species which occur in Yorkshire, but it is more than probable that its apparent rarity is due to some extent to the species being so easily mistaken for the common Noctule Bat (*Pterygister noctula*), a species which abounds, and to which it bears so close a resemblance as to render it impossible to differentiate between them, with certainty, without actually handling them.

In the British Isles the Hairy-Armed, or Leisler's Bat has been found plentifully in the North East and Eastern Counties of Ireland, but is unknown in Scotland, and rare and local in England.

Mr. H. Charbonnier recorded the capture of several bats of this species at Mexbrough in May and June, 1890 (Zoologist 1892, page 329), and Messrs. Clarke and Roebuck mention three specimens taken from an old factory chimney at Hunslet, near Leeds, 'about forty years ago,'* and these appear to be the only records of the occurrence of this species in our county, except for the specimens taken in the neighbourhood of Barnsley by Mr. Armitage and myself.

On May 13th, 1904, Mr. W. Broadhead, Woodman, of Stainbrough, heard some bats squeaking in a slit in the trunk of a beech tree which had just been felled. He chopped open the hole, but in so doing accidently killed one of the occupants. Two others he removed without injury, and the following day gave them to me.

Although from the first I had doubts as to the identity of these specimens, I regarded them for some time as being unusually small and darkly coloured Noctules; eventually, however, I became convinced that they were Leisler's Bats, and upon one of the specimens being submitted to Mr. Oldfield Thomas, I was delighted to find that this was actually the case. Shortly afterwards Mr. Armitage discovered that he had another specimen of this bat, in spirits, which had also been taken at Stainbrough in March, 1905.

---

* Vertebrate Fauna of Yorkshire, 1881, page 4.

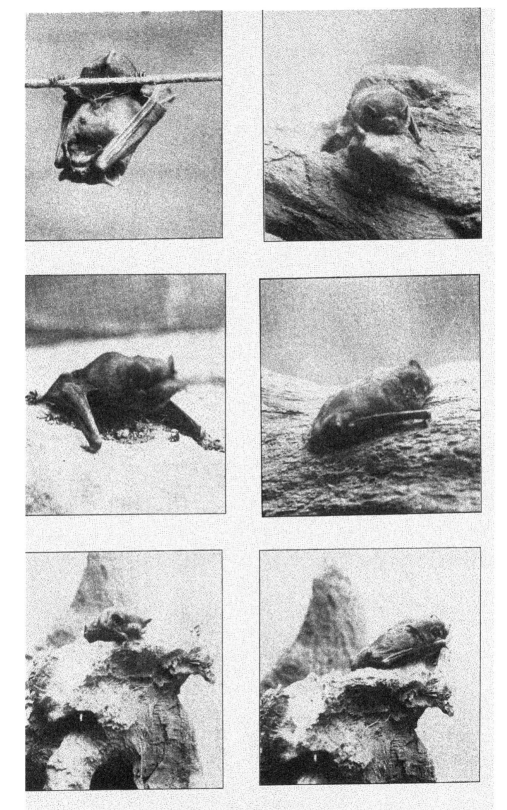

The Hairy-armed or Leisler's Bat.
(*Vesperugo leisleri*).

It was satisfactory to know that the species occurred in the district, but a serious obstacle to learning much of its habits lay in its very close resemblance to the Noctule.

In August, 1906, Mr. Sykes, of Old Mill, near Barnsley, mentioned to me that he had frequently observed a number of large bats flying about the vicinity of his house at dusk, and, at his suggestion, Mr. Wakefield and I met him there on the evening of August 18th.

We walked along a pasture which skirts the upper side of a long, narrow plantation covering a steep bank lying between this field and the river Dearne. The plantation is composed almost entirely of oak trees of fair growth; these we examined carefully, and I climbed up to several likely holes without finding anything.

By this time it was getting dusk, and several fairly large bats made their appearance and commenced to hawk up and down the top edge of the plantation, at about the elevation of the tree tops. They resembled the Noctule in flight so closely, that at the time I had no idea but that they were bats of this species.

We failed to see any of them leave their den, and, as it quickly grew darker, we began to dispair of being able to locate their retreat that night. Just as we were about to leave, however, Mr. Wakefield noticed one of them disappear somewhere in the dead top of one of the oak trees. Upon ascending the tree I soon found their hiding-place, which was at a greater elevation than any Noctules' den I have yet examined. The top portion of the tree was dead and without twigs, but a few of the larger branches remained, and it was through a hole in a short stump of one of these, which projected at about right angles from the trunk, that the bats found an entrance to a hollow in the trunk itself.

The height from the ground to the hole was about forty feet. I could hear the bats squeaking and shuffling about inside, but the opening was only about two inches diameter, and the surrounding wood otherwise comparatively sound, so that it was impossible for me to get my hand in or break into the den without tools. Moreover, it would have been foolish to have made any attempt to examine the colony at the time, as some of the bats were still out feeding in the vicinity, and it was also getting quite dark. So we left them undisturbed for the time being.

On the 22nd of August I revisited the tree, in company

with Mr. Armitage, and this time we took care to arm ourselves with a good mallet and wood chisel, and with the aid
of these I soon. cut a hole straight into the den in the main
trunk, having previously blocked up the exit hole through the
stump of the side branch. I found, on cutting the retreat open,
that it commenced, as before stated, with an entrance hole
some two inches in diameter, running from the end of the
broken side stump straight to the centre of the main trunk, a
distance of about a foot. The trunk itself, at this point, was
about ten inches in diameter, and in the heartwood was a
circular hole about four inches in diameter, extending upwards
from the point where the other hole joined it, for about fourteen
inches. It also went downwards for a short distance, but the
bottom part was inclined to be damp. In the upper portion I
found six bats, which all proved to be female Hairy-Armed
Bats, a fact which seems to indicate that, as in the case of the
Noctule, the sexes have a strong inclination to form separate
colonies during the summer months.

Another interesting feature about this little colony of
Leisler's bats was that it quite lacked the strong and offensive
odour so characteristic of a Noctules' den. In fact, the
absence of this unpleasant smell struck us even before we
commenced to cut into the hole, and was the first thing which
led us to suspect that the occupants were Leisler's, and not
Noctule bats.

The average wing expanse of these six Leisler's bats was
304 m.m., decidedly less than that of the Noctule, which
averages about 340 m.m. The colour of the fur, however,
appears to give by far the most ready method of distinguishing
between these two species. The Noctule is glossy (almost
oily-looking) golden brown on the back and only very slightly
paler and less golden underneath. Leisler's bat is of a glossy
clove brown above and paler, with quite a grey tinge, beneath.
This, however, is a method of identification which can only
be used when specimens of both species are available for
comparison.

It will be found upon a more careful examination, however,
that the colouration does afford a ready method of distinguishing
these two species ; and the distinction is one that can easily be
borne in mind, and used to identify even a single specimen
which may be obtained when one is in the country, with no
other specimens or books to refer to, and when general
impressions as to whether it looked small and darkly

coloured or the reverse, might be most unreliable. Take a Noctule in the hand and blow back the fur anywhere, above or below, and the place thus disturbed will show lighter in colour than the surrounding fur, the base of every hair being paler than the tip. Do the same with a Leisler's bat, and exactly the opposite will be the case, especially on the under surface of the creature, where the pelage becomes almost black towards the roots. Of course there are several other differences between the two species, as in the relative size of the feet, which are smaller and weaker in Leisler's bat, and in the dentition ; but as these points would be unintelligible to some readers and un-interesting to most, it is not necessary to go into them here, especially as the above method of recognition is the easiest to remember and to apply.

Of the six Hairy-Armed bats we obtained, I decided to try to keep a pair, in captivity, in order to learn a little more of their habits. I got them to take food from my hand the evening following their capture without any difficulty. Like most bats, they seemed to show no particular fear of human beings, only a certain amount of anger and irritation at being handled, and this quickly disappeared when they began to associate the event with 'feeding time.' During the first week I got a few pretty severe nips from them, though they never succeeded in drawing blood, as a Noctule often does. Their method of feeding was identical with that of the Noctule ; they would snatch the meal-worm or other insect from my forceps or fingers if it were held within three-quarters-of-an-inch from their nose, but if it were more than an inch away they did not seem to realise its vicinity at all. They made no attempt to 'pouch' their food, but ate it quite openly, with the head slightly raised, moving the jaws very rapidly, and making a decided 'champing' noise. Sometimes the whole of the insect would be consumed, and at other times the head and legs would be rejected. If one of the bats got hold of a mealworm crosswise, it would usually jerk it along until it got the end. It would then masticate the part in its mouth rapidly, and, with a slight jerk of its head, secure a further portion, repeating this process until the whole had disappeared ; occasionally it would commence to eat in the centre, but this generally resulted in one or both ends being lost. I always noted that the more nearly its appetite was satiated the more careless it became in consuming its food, larger and larger portions of the insects given to it being dropped. It would keep on snatching fresh ones, however, even though it only took a bite or two out of them. It would often refuse obstinately to take back a piece it had dropped ; in fact, it was seldom that I could induce it to do so. I used to take this extravance as an indication that it had had sufficient, and when it began to drop portions of its food I stopped the supply. I found that an average of about five dozen mealworms each per day seemed to meet the requirements of these bats fairly well ; at any rate they had to subsist on that allowance, for mealworms cost money.*

(To be continued).

ARACHNIDA.

A Phalangid new to Yorkshire.—*Oligolophus alpinus* Herbst., now recorded for our county for the first time, is an addition to the list of Yorkshire Harvest Spiders which appeared in the 'Naturalist,' November, 1906. The total list is thus increased to sixteen species. I secured specimens in Butternab Wood, Huddersfield, in July and near the summit of Ingleborough in September, 1907.—WM. FALCONER, Slaithwaite, 28th September, 1907.

A Pseudo-scorpion new to Northumberland.—Pseudo-scorpions are apparently, with one exception, very scarce in the North of England, only eight of the twenty-three British species having occurred in the six northern counties (*vide* 'Naturalist,' August, 1903 and June 1907). Only the commonest one, *Obisium muscorum* Leach., has up to the present been met with in Northumberland, but on August 16th, while sifting the accumulated refuse in an obscure and neglected corner of a barn at Manor Mill, Haltwhistle, on the South Tyne, I came across one example of the eyeless *Chernes rufeolus* Sim. This false-scorpion has only recently been added to the British list ('Trans. of Dorset Field Club,' 1905). It was first discovered in a London granary, but has since been taken in Kent, Essex, Wilts., Derby, Cumberland, and Cheshire.—WM. FALCONER, Slaithwaite, 28th September, 1907.

---

* This a certain gamekeeper discovered, to whom I was talking on the subject. He was enquiring after the welfare of a couple of Noctules he had given me some little time previously, and asking me upon what I fed them. I told him that the joint efforts of the bats accounted for some hundred and fifty mealworms per day. Upon this he enquired what I paid for my mealworms, and after I had told him, relapsed into profound silence, broken by fitful mutterings, which warned me that he was engaged in the mental solution of some problem in advanced practical mathematics. Just when I was beginning to think that he had either forgotten the matter or become hopelessly muddled, he burst out triumphantly—'I'll tell you what, Mister, I'm feeding a couple of pigs for three ha'pence a week less than them there bats is costing you'! I could quite believe him, too.

A NET-BUILDING CHIRONOMUS LARVA.

A. T. MUNDY.

In the summer of 1904, Prof. Miall and Mr. Taylor discovered a tube-building larva of special interest on account of its strange habits. The fly was hatched out, and has been identified as *Chironomus pusio*. The larvæ were found in great abundance at Toutech, Windermere. In 1905 I found that the same insect was quite common in some of the South Devon streams on the borders of Dartmoor. In the Yealm, which is only a few minutes' walk from my home, they are very plentiful, and so I have had exceptional facilities for studying their anatomy and life history. At first I had the greatest difficulty in keeping them alive and active, for they delight in the swiftest running water, while I had to keep them in dishes of still water. However, this difficulty was finally overcome by using a clock-work arrangement, which set in motion a revolving paddle, thus stirring the water round and round. And so by this means I have been able during the last two years to study their habits and metamorphoses very closely.

Some of the *Chironomids* make free cases just as do the caddisworms (*Trichoptera*). Reaumur (1743) was the first naturalist to describe one of these; Lyonet described another in 1832, which built a gelatinous case. This larva has been quite recently found in England. Dr. Lauterborn has now added three more to the number of case-builders. One larva makes a brown opaque case, rather flat with rounded ends, opening by a slit, thus resembling in a remarkable degree those old-fashioned spectacle cases which open at either end by gentle squeezing. The second makes a straight reed-like case covered with scales; the posterior end is closed by a tough membrane with a central pore. And the third makes a case somewhat the shape of a cigar. None of these cases are more than a few millimetres in length, and I am not aware that any of them have been as yet found in England, but that is probably only on account of their smallness. On the other hand the tube-building larvæ, among which we place our *Chironomus*, are found in great variety in all parts of England. Some make simple mud tubes buried in slime; others build tubes three or four inches long rising perpendicularly from the bottom of stagnant pools; while still another group constructs elaborate nets, supported by a strong framework of arms, to be used for catching food.

To this last group belongs *C. pusio*, to which I must now return.   The case (Fig. 1) is found in swift running moorland streams, anywhere except in a waterfall, attached to moss or rock.   It consists of a very compact tube one centimetre long ; the free end is surrounded by from four to seven long thin arms sticking out like spokes.   The whole case is strangely suggestive of the common brown *Hydra* of our ponds.   With

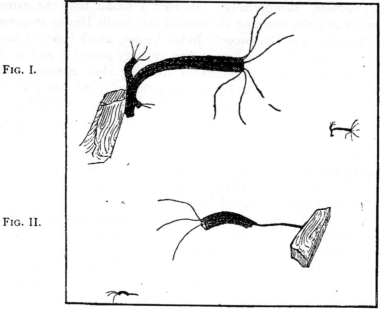

FIG. I.

FIG. II.

FIG. I.—*C. pusio*, larval case of a full-grown specimen, with two immature cases attached at the base.   Mag. ⅚, and natural size.
FIG. II.—The case of another larva, a closely related species.   Mag. ⅚, and natural size.

head projecting from the tube, lies the larva ; and if watched closely, it will be seen to dart up one of the arms, carry silken threads between this and the next until there is a strong network formed.   This is done between each pair of arms.   The larva will then retire into its tube once more, and wait while the net gets loaded with particles of mud containing desmids, diatoms, and other organic matter.   Then out it comes again, pulls in the web and all it contains, and either devours it, or, binding it into a firm mass, adds it to the wall of its house.   When full grown the larva enlarges the end of its case, and after removing the arms, covers it in with a circular disc containing a small central hole.   Then after a rest of twenty hours or so the larva changes into a pupa, and out through the central pore of the disc are pushed the ends of the respiratory organs. What a wise provision it seems, to have left that central hole, and yet after all, if I mistake not, the breathing organs have lost their function! Three days' rest, and the pupa pushes off the disc, swims to the surface, where in a moment out bursts the perfect fly to join his fellows in the air. The female fly lays about ninety eggs in a jelly mass, 1 m.m. in diameter, attached to moss just below the water. The development of the larva can be watched through the transparent egg shell, and in six days the larva hatches out.

The newly hatched larvæ are but specks, and quite transparent. They differ from the older ones in not having five pairs of plumed bifurcating hairs arising from the sides of the body. Yet these minute creatures, which I have myself hatched out from egg masses placed in moving water, as described above, build little mud tubes with arms and nets just as their elders dö. They are, however, so small as to be scarcely visible without a magnifying glass.

Another species, very closely related to *C. pusio*, makes a still more graceful dwelling. The tube of this larva is shown in Fig. II., and has lately been described and figured by Dr. Lauterborn, who found it very plentiful in the Speirbach and Helmbach, above Lambrecht (Palatinate). The two larvæ resemble one another so closely, that when taken from their cases, it is almost impossible to distinguish the species. I mention this larva because I have found it moderately plentiful in the same streams where *Chironomus pusio* is so abundant. And as the two larvæ live together in Devon, there seems no reason why this latter should not also be found in the same locality where *C. pusio* was originally discovered, or in other places in the north where moorland streams exist.

—◆◆—

Handbook to the Roman Wall, by the late **J. C. Bruce,** edited by R. Blair. Andrew Reid & Co., Newcastle, 1907. 284 pp., 2/6 net. In view of the extraordinary interest attached to the Roman wall, the glorious country it traverses, and the unique character of the discoveries that have been made at different points along its course, it is surprising that it is not visited by a far greater number of people. Bruce's Handbook has done much to popularise this magnificent relic of our early rulers, and the fact that it has recently reached a fifth edition alone speaks for its value. Seeing that the original author is no longer with us, perhaps no better qualified person could be found to revise and bring the handbook up to date than the energetic editor of the Society of Antiquaries of Newcastle, Mr. R. Blair. Since the fourth edition was issued in 1895 many important excavations and discoveries have been made along the wall—notably the great camp at Housesteads and the discovery in Cumberland of a fragment of a wall of turves ; of these Mr. Blair gives good account. The guide is well illustrated, and has an excellent folding map.

NOTES ON THE VARIATION OF *ABRAXAS ULMATA* AT SKELMANTHORPE.

B. MORLEY,
Skelmanthorpe.

LAST June, when so many familiar and common species of lepidoptera were failing to appear, and others only in few numbers, it was an agreeable surprise to find *Abraxas ulmata* in a countless swarm in its haunt in this neighbourhood. Such a host I had never seen before. To walk about in the wood meant killing them at every step. The herbage and bushes were simply alive with them, everything was spangled all over with their white wings. But evidences of tragedy abounded everywhere, bodiless wings littered the ground; thousands were drowned in the stream; hopelessly deformed examples were crawling about on every hand, crippled in every conceivable way. Their wretched plight was almost depressing. The satisfaction was the perfect ones, of course; how pretty they looked sitting with their wings streched out and laid flat on every leaf, and yet how very much alike they seemed. Having recently had the pleasure of seeing some of the remarkable forms found at Sledmere some years ago, it occurred to me that probably something out of the ordinary might be found amongst this multitude. Previously I had not found any great variation, or even found a good aberration. This time, however, I was more fortunate, for what I obtained surpassed anything I had ever dreamed of. Strange forms were really not uncommon.

. Roughly, the variation can be separated into two classes, one lighter than the figure given by Newman and the other darker. Newman's figure seems to exactly represent what I found to be the predominant type, so that it will be convenient as a base to work from.

Dealing first with the varieties that are lighter than Newman's type, it was very easy to find all the grades of variation to an almost white form. The brown spots at the bases of the wings and those on the inner margins gradually fading into a light ochreous shade, the cloudy blotches becoming smaller and fainter. In a few examples of the extreme light form the brown in the spots is replaced by an almost bright yellow colour, and all that remains of the transverse series of cloudy blotches beyond the middle of the wings is a short faint streak on each wing ray, the ground colour being pure white. On some of the very lightest specimens another very curious feature of variation begins to appear, and leads on to a very distinct variety. Some specimens have the yellow spots slightly blurred, the colour being faintly suffused on the wings outside the ordinary limits of the spots. The suffusion becomes more pronounced in other specimens until the ground colour becomes rust coloured, this colour being most pronounced in the region of the spots. But I regard these as only approaches to what I found of the extreme yellow suffused forms, which have the yellow spots very blurred and undefined, and the ground colour very dirty and greasy-looking. Perhaps the strangest feature about these yellow suffused forms is the fact that not one of them has all the wings equally affected. In one or two specimens an almost complete suffusion obtains, but generally the fore-wings are the most affected, and both sexes are subject to it.

Another very plentiful form, darker than Newman's figure, seems to make another distinct variation. The brown spots are very much darkened; indeed, in some cases they are almost obliterated by a thick lead colour. The transverse sets of blotches are united and form broad deep leaden bands across the wings. The central blotches become enlarged also, and are usually confluent. It is worthy of note, however, that this form has the spots well defined, and even in the most heavily marked examples what remains of the ground colour is white and not suffused.

A few of the deep leaden varieties were obtained, with pretty examples intermediate between them and the type. The first has only the outer half of the left forewing affected, the narrow smoky line on the outer margin and the leaden blotches being almost washed out, the area between them lead colour. It is worthy of note that although the brown spot on the inner margin is enveloped in the smoky cloud it is well defined and full coloured. The next specimen has all the wings equally affected; all the spots and blotches show out plainly, the ground colour being pale leaden grey more or less shadowy, but with no white showing. The remaining specimens of the leaden type seem to vary only in intensity of colour, with the cloudy blotches less apparent as the colour becomes deeper.

Judging from this series it would seem that two quite distinct forms of variation are developing in this locality, and although it may be premature to come to that conclusion, the fact remains that where a yellow suffusion obtains it is at the expense of the brown spots; and in the leaden forms the cloudy blotches disappear, while the brown spots remain complete and well defined. Probably the banded forms may be only heavily marked examples of the ordinary type, and without direct relation to either of the suffused varieties, but an opinion in this case is scarcely warranted when the real causes of these extreme variations are so little understood.

There are three specimens in the series which deserve special mention and a separate description. The first has the right forewing a dirty brown colour, all over the wing, the markings and the white ground being all obliterated; all the other wings, however, are of the usual typical colour in every respect, a strange looking specimen indeed.

The second is a really pretty aberration, the basal two-thirds of the wings being of a deep bronze colour flushed with blue, and divided in the centre by a narrow white band reaching from the inner margin almost to the costal margin, the outer portion of the wings being a broad belt of pure white. The extraordinary appearance of this specimen was noticed when I was ten yards away from it, although it was surrounded by scores of others of its species.

The third has all the wings a pale lead colour, but instead of the six brown spots being complete and well defined, as in the other leaden forms, they are replaced by round leaden spots similar to the ground colour, surrounded with brown lines, the wings being simply ornated with brown rings. It is worthy of note that the head, thorax, and body are all normally coloured in all the varieties except the lead coloured forms, in all of which they are invariably black.

———◆◆———

Wild Life on a Norfolk Estuary, by A. H. Patterson. London: Methuen & Co. 352 pp., 10/6 net. We have previously noticed Mr. Patterson's work in these columns. The present substantial book teems with equally interesting matter, and can be recommended for its racy style and for the valuable natural history observations it records. It deals largely with Breydon—so well known to East Coast naturalists. The second part of the present volume is really a continuation of the author's 'Nature in East Norfolk.' There is also added a series of interesting local notes, which owe their existence to the author's system of 'when found make a note of.' Some of these are exceptionally good—the 'rabbit yarn' we should have reprinted for the benefit of our readers had it not been quite so long! Throughout the work there are numerous anecdotes of mammal, bird, and fish. To one of them, however, we observe the footnote, 'This is a fact'! The book is illustrated by several of the author's own sketches, some of which are rather crude, and a reason for their insertion is given in the Preface. Altogether it is a most entertaining volume, and we only wish that the dozens of recent writers on birds had as much first-hand information to give as Mr. A. H. Patterson, and could put it in such a pleasant form.

FUNGI AT HORTON-IN-RIBBLESDALE.

T. GIBBS.

THE district of Horton-in-Ribblesdale is rather wanting in those rich moist woodlands, with deeply bedded leaf mould, which mycologists find such prolific hunting grounds. On the occasion of the September excursion of the Yorkshire Naturalists' Union, therefore, fungi were hardly so abundant as we expected to see them on the last excursion of the year. In addition to this, although it is a curious complaint to make after the summer of 1907, a pleasant little spell of delayed summer had somewhat checked the development of the moisture-loving pasture species. Notwithstanding these drawbacks, the writer, who was spending some days in the district, and consequently was able to extend the area of his operations beyond the strict bounds of the excursion programme, with the assistance, on the Saturday, of Messrs J W. Tindle and M. Malone, of Bradford, noted the occurrence of about eighty species. Some of these were of considerable interest, and one had not been previously certainly recorded for the county.

The most prolific locality visited was Douk Gill, a narrow wooded ravine where the beck, which sinks underground at Hell Pot, on the side of Penyghent, resumes its above ground course ; here large Agarics were fairly plentiful, the plantation yielding six of the nine *Russulæ* seen, also the beautiful edible *Clitopilus orcella*, and a fine large Discomycete *Otidea cochleata*.

The bare moorlands yielded few species, but a pretty white or primrose yellow form of *Omphalia umbellifera* was frequent on peaty banks high up Penyghent and Ingleborough, and *Naucoria semiorbicularis* abounded in the Sphagnum swamps.

The Limestone Scar area proved the most barren, as was to be expected, nevertheless it provided *the* find of the excursion, a solitary specimen, in perfect condition, of the beautiful and uncommon puffball *Lycoperdon velatum* Vitt., a species only recently added to the British Flora.

The following is a complete list of the species found :—

GASTROMYCETES.

Lycoperdon velatum. Among grass on rocky, bushy bank, Oxenber Wood.

L. echinatum. Among grass, Oxenber Wood.

L. pyriforme var. *excipuliforme* Desm. Several specimens among grass at foot of Ash tree, Oxenber Wood.

Bovista nigrescens. Among grass.

HYMENOMYCETES.

Agaricaceæ.

Amanita rubescens. Bransgill.
Lepiota cristata.
L. granulosa.
Armillaria mellea. One only seen.
Tricholoma terreum. Douk Gill.
Clitocybe infundibuliformis.
Laccaria laccata.
Collybia radicata. Douk Gill.
Mycena rugosa.
M. galericulata. Both on dead wood.
M. epipterygia.
M. galopoda.
Omphalia umbellifera. On peaty banks high up Penyghent and Ingleborough.
O. muralis. Among grass and moss, Gillet Brae.
Clitopilas orcella. Douk Gill.
Nolanea pascua. In pastures, abundant.
Inocybe asterospora.
I. rimosa.
I. scaber.
I. geophylla. Douk Gill. All under trees.
Naucoria semiorbicularis. Common among Sphagnum.
Galera tenera.
G. hypnorum. Both common.
Tubaria paludosa. Among Sphagnum.
Crepidotus mollis. On dead trunk.
Agaricus arvensis. In pasture.
Stropharia aeruginosa.
S. semiglobata. On cow and horse dung.
Hypholoma fasciculare. On stumps.
Panæolus retirugis.
P. campanulatus.
P. papilionaceus. In pastures.
Anellaria separata. On dung in pastures.
Psilocybe semilanceata. In pastures.
Psathyrella atomata.
Coprinas atramentarius.
C. tomentosus. In hilly pasture, Wharfe.
C. plicatilis.
Hygrophorus pratensis.
H. psittacinus.
H. coccineus.
H. miniatus.
H. conicus.
H. unguinosus.
H. chlorophanus.
The last eight species in pastures.
Lactarius pyrogalus.
L. quietus.

L. turpis.
Russula densifolia.
R. cutefracta.
R. azurea.
R. cyanoxantha.
R. emetica.
R. ochroleuca.
The six last named in Douk Gill.
R. alutacea. Bransgill.
R. foetens. Among grass, Gillet Brae.
R. vesca. Wood side near the Lime Works.
Marasmius androsaceus.

POLYPORACEÆ.

Boletus flavus.
B. viscidus. Wood side near the Lime Works.
Polyporus adustus.
Polystictus versicolor.

THELEPHORACEÆ.

Stereum hirsutum. On dead wood, and on *Polyporns adustus.*

CLAVARIACEÆ.

Clavaria cristata.
C. rugosa.
Both in Douk Gill.

UREDINACEÆ.

Puccinia violae. On Wood Violet.
P. poarum. *Æcidium* stage on Coltsfoot.
P. oblongata. On *Luzula sylvatica.*

PYRENOMYCETES.

Rhytisma acerinum. On leaves of Sycamore.
Sphaerella rumicis. On leaves of Dock.

DISCOMYCETES.

Otidea cochleata. On the ground in Douk Gill.
Humaria granulata. On cow dung.
Dasyscypha calycina. On living twigs of Larch.
D. virginea. On dead wood.

PHYCOMYCETES.

Mucoraceæ.

Pilobolus crystallinus. On cow dung.

FUNGUS FORAY AT GRASSINGTON, BOLTON WOODS, AND BUCKDEN.

C. CROSSLAND, F.L.S.
Halifax.

THE seventeenth Annual Fungus Foray was held Sep. 21st-26th, the head-quarters being at the Wilson Arms Hotel, Grassington, a good centre for the three localities selected. Permission to explore the woodlands near Grassington and Bolton Abbey had been unhesitatingly granted by His Grace the Duke of Devonshire ; and for Buckden Woods by Miss Stansfield, of Buckden Hall. There was an unusually large muster in addition to the Mycological Committee. Affiliated Societies from Halifax, Bradford, Crosshills, Hull, Rotherham, Huddersfield, and York were represented. There were also Mr. Thomas Smith, Alderley Edge, a member of the Manchester Microscopical Society ; and Mr. Thomas Hey, of the Midland Railway Nat. Hist. Soc., Derby. Consignments of Fungi, including many uncommon species, were sent from the Isle of Wight and the New Forest by Mr. J. F. Raynor, Southampton. Mr. Smith brought a fine collection of interesting species from Alderley Edge, among which was a *Boletus*, at present undeterminable.

After all had secured their berths, a short run out was made through the pastures by the river side as far as Grass Woods. A few of the members who arrived in the forenoon had got together a fair collection ready to hand.

Monday was set apart for Bolton Woods, and Tuesday for Buckden ; the remainder of the time being given to Grass Woods and the pastures. The unusually cold summer, with the ground almost constantly much below the normal summer temperature, had been unfavourable to the development of fungi, especially of the larger, more fleshy kinds. The ordinary field mushroom has been similarly affected. This fact led us not to expect too much, and it was fortunate that all three localities were included in the programme. One of the members, a few weeks previously, expressed a doubt as to whether there would be anything worth coming for. The reply to this was, 'If we don't find more stuff than we are able to determine it will be the first time.' We did, and all came right at the end of the Foray, everyone being well satisfied with the results. There now remains no doubt as to the prolific nature of the grounds selected had the season been ordinarily good. On the way to Bolton a promising fir-wood, between Burnsall and Barden Tower, was looked into ; this was a typical place for many fir-loving species, but few were seen. There is one fact which has been remarked upon several times in years past, that is—when a season, or a district, is less bountiful than usual, some uncommon or rare species are sure to turn up. It has proved so this time.

The exploration of Bolton Woods was worked in two companies ; one keeping the right bank of the river, the other the left, each under the guidance of a woodman. The investigation began at Barden Bridge, the conveyance being sent forward to Bolton Abbey to await us there for return. The first place looked into—a bit of moist, bracken-free woodland with plenty of rotting branches, herbaceous stems, and a few old stumps—set us up for the day. One of the finds of the Foray was picked up here, the beautiful, deep blue agaric *Leptonia euchroa.* The place yielded several other agarics and numerous micro-species. This was one of the best bits. On comparing notes with the other party, which we rejoined at the lodge refreshment room, we learned they had also done fairly well. On going over the day's finds at the end of the Foray it was found that no fewer than 123 species had been collected, 12 being additions to the Mid. W. Division, including *Omphalia bullata, Pleuteus spilopus, L. euchroa,* and *Pholiota tuberculosa.*

On Tuesday we had a nice variation. The nine miles to and from Buckden was done on the public motor car. It was a very pleasant ride each way, only one could not help wondering what would be the result if an axle broke or a wheel came off while the car was careering down a hill. A full description of the delightful Buckden Wood would occupy too much space. As soon as we put foot into it there was an exclamation, ' Ah ! this is the right kind of place for us.' Although we were not there very long, and the season unpropitious, 105 species were picked up. Among them were three new county records— *Mycena excisa* Lasch., *Omphalia camptophylla,* and *Entoloma bulbigenum,* also sixteen additions to the Mid. W. Division. The motor conveyance returned at two o'clock. On the way back it was pulled up a few minutes to give the passengers an opportunity of calling up the echo across a certain point in the valley. This was interesting enough in itself, but a few of us were anxious to get back to the work-tables before too much daylight had fled.

Grass Woods is known to the majority of Yorkshire botanists for its luxuriant undergrowth. Its stock of flowering plants is large and varied. Many uncommon species have been, and are, found within its boundaries. This led us to expect a variety of plant diseases caused by parasitic fungi, nor were we altogether disappointed. Fifteen species were seen; their names, along with those of their host plants, are given in the list of species which will be found in the 'Transactions of the Yorkshire Naturalists Union,' No. 33, shortly to appear. The great abundance of *Xenodochus carbonarius* on *Poterium officinale* was most striking. A far more lamentable feature was the sycamore leaf blotch. Nearly all the sycamore trees in Grass Woods were affected by this unsightly 'parasite' *Rhytisma acerinum*. We never saw the disease so prevalent anywhere. It prevails, more or less, in all the districts visited. It is even in some of the forest-tree nurseries, where one would least expect to find it. Here, at any rate, some attempt should be made to check it; there may have been, but if so, unsuccessful. Seeing that the propagation spores of the disease do not ripen and germinate on the fallen leaves until spring, there is ample time to collect and burn all that are shed in the nursery area. When every leaf of a tree is attacked, besides making the tree look so repulsive, its timber producing powers are very much reduced; it is utterly impossible for the leaves to perform their functions to the full when one fourth, or one third of their substance factories are choked up and destroyed by the thick, black scabs of this disease.

The Mid. W. Division has been previously so well worked for its Uredinaceous plant diseases by Mr. West and the late Mr. Soppitt that, though we found no fewer than 26 species in the three localities, there are no additions to the Mid. W. Fungus Flora in this branch.

Quite a number of small fry was found at the several places on decaying sticks and herbaceous plant remains. These are as interesting as, and even more so than, the conspicuous kinds, but require more searching for. No fewer than 36 species of Discomycetes were met with; and 15 Myxomycetes, including *Cribraria rufa*, a new county record. While agarics were found in fairly great variety, considering past conditions, they were very sparing in quantity; just ones and twos here and there. The commonest and most abundant agaric was the strong smelling *Hebeloma crustuliniforme*. Of the genus *Mycena* 15 were found, Cortinarius 11, Russula 12, and Hygrophorus 10.

Two or three of the members revisited Grass Woods on the

Wednesday morning and brought in several additions. The plentiful unworked-out material remaining on the tables overnight induced the remaining three to stay in and tackle it. A short stroll in the hotel garden was indulged in as a break in the work, when two *Coprinii* were noticed on the manure heap, *C. plicatilis* on the lawn, *Uromyces fabæ* on the bean leaves, *Sphærotheca castagnei* on the peas, and *Cladosporium herbarum* on an empty decaying pea-swad thrown on the ground.

Miss Johnstone, on descending into the Elbolton cave near Thorpe, found *Polyporus nidulæus* growing on a pole in the cave bottom.

While each one worked hard to make the meeting a success, Mr. Malone, of the Bradford Society, excelled in collecting and bringing in specimens.

It was acknowledged by all sufficiently experienced to be able to give an opinion that many uncommon species had been found. The weather was all that could be desired for collecting purposes.

Papers and discussions were the order of the evenings. On the Saturday evening Mr. Clarke read a copy of a lecture by Dr. M. C. Cooke on 'How to Study Fungi,' which was very instructive; also a copy of a characteristic address given by the same talented author many years ago to one of the London Natural History Societies. This was very much enjoyed.

On another evening Mr. Harold Wager, F.R.S., spoke at some length on the 'Life History of a Parasitic Fungus'—*Polyphagus euglenæ.* Mr. Wager has devoted much time and attention to this parasite, and appears to have successfully observed its complete life cycle. The details of its development, and the methods this particular parasite adopts to find and secure its prey were clearly given. The lecture, which was illustrated by diagrams drawn by the lecturer as he went along, was listened to with rapt attention by a most appreciative audience.

On another evening Mr. Gibbs dealt with 'Some of the Smaller Coprophilus Coprini,' Mr. Gibbs has given much attention to this group for several years past. At least one species new to science has been the result. From his interesting remarks, illustrated by diagrams, one gathered that species considered to be of rare occurrence will prove to be common if properly looked for.

During the course of the evening, discussions on the best methods of carrying on the Mycological work of the county came under consideration. Mr. Wager's suggestion that the work, should be definitely shared out among the Members of the Committee was readily adopted. The idea being that each member take one or more of the various groups of fungi under his special charge. Mr. A. Clarke took the Gastromycetes and the Polyporaceæ; Mr. T. Gibbs part of the white spored Agaricaceæ and the whole of the Melanosporæ; Mr. H. C. Hawley, Ochrosporæ and Hyphomycetes; Mr. C. H. Broadhead, Rhodosporæ; Mr. R. H. Phillip, Uredinaceæ and Ustilagineæ; Mr. W. N. Cheesman, Hydnaceæ and Clavariaceæ; Mr. H. Wager, Phycomycetes; Mr. C. Crossland, Ascomycetes and ·other unalloted groups; and Mr. J. W. H. Johnson, Myxomycetes.

Stereo photos and coloured drawings of fungi, by Messrs. Clarke, Gibbs, and Crossland, were at the service of those who desired to look them over. All the necessary books and microscopes were available.

The doctor and one or two other Grassington friends interested themselves in the proceedings. Altogether it was a most enjoyable Foray. One of the greatest drawbacks was, the experienced hands could not devote as much time as they would have liked in assisting the less experienced members to determine the species they collected. This was felt, but could not well be avoided. To add to this, there was not sufficient table room to allow of the named specimens being laid out for the study of those members to whom they were unfamiliar.

About 200 species were collected in and about Grassington, 123 at Bolton Woods, and 105 at Buckden; the net number being 320; six being new to Yorkshire and 53 to Mid. W. These will all be particularised in the list.

I am indebted to Mr. A. Lister, F.R.S., for kindly undertaking the work of determining the Myxomycetes on my behalf.

At the business meeting the Rev. F. H. Woods proposed, and Mr. H. Wager seconded, a vote of thanks to the landowners for permission to explore their domains.

It was decided to recommend Mulgrave Woods as the place for next year's Foray, the dates to be September 19th to 24th. It may be urged that these woods have been investigated by the Yorkshire Mycologists two or three times already; so they have, but the woods are so extensive and varied that were they explored nine times there would certainly be some additions to the Fungus Flora were a tenth visit made.

FIELD NOTES.

GEOLOGY.

Striated rock-surface in Ribblesdale.—An interesting surface of glaciated Silurian Grit is plainly traceable over an area of about fifty square yards on the hillside 320 yards north of Arcow, Horton-in-Ribblesdale, at 875 feet above Ordnance Datum, and the striæ are very distinct over the greater portion. A casual observation shows that most of this steep hillside bears unmistakable evidence of glaciation, due to the Ribblesdale glacier making its way southwards along this valley. The direction of the grooving is practically due north and south, and therefore conforms to the general trend of this portion of the valley.

The grits are well exposed in the wood where this striated surface is seen, which is about a mile-and-a-quarter almost due south of Horton Station. Above this the Mountain Limestone of Moughton Scar unconformably overlies the Silurian rocks. —E. E. GREGORY, Bingley.

—: o :—

MAMMALS.

Bat mobbed by Swallows.—For two days in succession a curious sight was witnessed this week at the Wetwang village pond. The sun was shining brilliantly at 2-30 p.m., when a little bat was seen hunting backwards and forwards over the pond, occasionally pursued by two or three Swallows and House-martins, which seemed to wish to drive it off.—E. MAULE COLE, Wetwang, September 13th, 1907.

—: o :—

BIRDS.

Bird Notes from the Whitby District.—Reading Messrs. Booth and Fortune's report of the Vertebrate Section of the Yorkshire Naturalists' meeting at Robin Hood's Bay last May, in your July number, I am induced to send you a few notes on some of the birds of the district which may interest a portion of your readers.

SPOTTED FLYCATCHER.—It is not surprising that this bird was not seen during the visit of the Union. As in other districts, it was later than usual in putting in an appearance here, the first to be observed being on May 19th. Later it came in even more than usual numbers; so late as September 5th it was observed feeding newly-fledged young.*

---

* The late arrival of Spotted Flycatchers seems to have been general throughout the county this year.—R. F.

RING OUZEL.—This bird is fairly common in the district, and should have been met with by the members of the Union. On May 27th a nest with four young was found near Ramsdale Beck, a short distance from the Whitby and Scarbro' highroad.

WHINCHAT.—Was also very late in arriving in this part of Yorkshire. May 11th appears to be the earliest date recorded. Many did not come to us till much later.

TREE CREEPER.—This is by no means common. Odd birds are to be found in suitable localities.

STONECHAT.—It is so thinly distributed over this locality that it might easily escape observation during a short visit.—THOS. STEPHENSON, Whitby, September 17th, 1907.

An ancient Blackbird.—It may be interesting to record the death of an old blackbird which has just taken place in Harrogate. He was captured by his owner 15 years and 7 months ago, and as he would be at least a month old when caught, he has reached the ripe age of nearly 16 years. For five years he was clothed in the normal plumage of his kind, but after the fifth moult white feathers began to appear, increasing in numbers after every moult, until at the time of his death a considerable portion of his plumage was white. —R. FORTUNE, Harrogate.

—: o :—

FISHES.

Fishes at Whitby.—The following species were obtained during September at Whitby :—

Sept. 25th.—Ballan Wrasse, *Labrus maculatus*, weighing 2½ lbs. was taken off and brought into Whitby.

Sept. 26th.—A large Porbeagle or Beaumaris Shark, *Lamna cornubica*, a male, measuring from end of nose to end of caudal fin 7 ft. 6 ins., and estimated to weigh about 3 cwt., was taken off Saltburn by John Dryden in the herring nets of the coble ' Ann Elizabeth,' W Y 184, and brought into Whitby, where it was exhibited. One net was completely spoiled.

Sept. 26th.—A fine example of Garfish or Sea Pike, *Belone vulgaris*, measuring from end of jaw to end of caudal fin 27 ins., called locally Swordfish.

Sept. 27th.—A large specimen of Great Weever or Sting-bull, *Trachinus draco*, measuring 16 ins. in length and weighing 1 lb., was taken off, and brought into, Whitby.

Sept. 28th.—A small Sunfish, *Orthagoriscus mola*, measuring

31 ins. from tip of dorsal to end of ventral fin, was taken while floating on the surface of the water near the pier end at Whitby by the crew of the Filey coble S H, 299, when about to enter the harbour.—THOS. STEPHENSON, October 2nd, 1907.

Giant Mackerel landed at Grimsby.—A giant mackerel, said to be the largest ever landed on the Grimsby market, was brought to that port recently by the steam trawler Jersey. The fish, which was 24 inches long, with a girth measurement of 15 inches and a weight of $6\frac{3}{4}$ lbs., was caught some 63 miles from Spurn. It was sold for exhibition purposes, and realised 18s. I saw the fish myself.—F. M. BURTON, Gainsborough.

—: o :—

MOLLUSCA.

Petricola pholadiformis **in Lincs.**—I collected this shel in 1900, and since, but have failed to find any published record of its occurrence in Lincolnshire. The empty shells, in varying stages of growth, occur in considerable numbers on the shore about Mablethorpe* and Sutton-on-Sea. Mr. V. Howard, M.A., informs me that about two years ago he found them living abundantly at extreme low tide.

Mr. A. S. Kennard informs me it was obtained some years ago at Burnham-on-Crouch in Essex, then he found it living abundantly at Herne Bay, and Mr. J. E. Cooper found it at Shellness, near Deal, and this year Mr. Kennard found it at Dunwich, Suffolk, and Mr. Mayfield saw it at Lowestoft and Felixstow. In addition to the above-mentioned localities I have a few fine examples which I collected near Warden Point, Isle of Sheppey, in August, 1896.—C. S. CARTER, Louth, October 4th, 1907.

—: o :—

FUNGI.

Lachnea hirto-coccinea **Phil. and Plow.**: An addition to the Yorkshire Fungus Flora.—I found this fungus on Strensall Common on the 6th July, 1907. It was growing on wet sand and rotten wood on the side of a temporary water splash, and was distinct by its scarlet colour and round shape, the size of a shilling. Mr. C. Crossland regards it as an interesting addition to Yorkshire.—W. INGHAM, York, 11th Aug., 1907.

---

* A piece of peat from this locality, with shells in position in their borings, has recently been brought to us, and is now on exhibition at the Museum, Hull.—ED.

REVIEWS AND BOOK NOTICES.

The Use of Life, by Lord Avebury. Macmillan & Co., 1907. 208 pp., price 2/- This well known book, which contains a variety of essays on a variety of subjects, has now been issued for something like the 21st time. The present edition is very handy in size, and is well printed and well bound.

The Protection of Sea Shores from Erosion, by A. E. Carey. Greening & Co., London, 36 pp., price 1/. In this pamphlet Mr. Carey gives a careful summary of the various methods in vogue for protecting the coast from the ravages of the sea, and the illustrations which accompany his remarks enable these methods to be clearly grasped. The pamphlet is apparently a shorthand report of a lecture delivered before the Society of Arts on May 1st last. It is a pity it was not slightly 'edited,' however, before publication. 'I have been asked to speak to you this evening' seems odd to the purchaser of the pamphlet, and 'the lantern slide now to be presented shows' nothing at all, as it is apparently not reproduced.

Guide to the Great Game Animals (Ungulata) in the Department ot Zoology, British Museum (Natural History). London, 93 pp. Price 1/- The Trustees of the British Museum are undoubtedly doing excellent service by issuing from time to time popular handbooks at a reasonable rate. These, whilst primarily being descriptive of the specimens in the Natural History Museum, are also very useful general guides to the subjects dealt with. Before us is the guide to the Ungulata, written by Mr. R. Lydekker. From the careful way this has been prepared, and the number of illustrations, it will prove most useful to sportsman and naturalist alike. It is well bound. Appended is a serviceable 'list of horns, antlers, and tusks,' with measurements.

We have received two valuable publications from the **Liverpool University Institute of Commercial Research in the Tropics.** One is a paper by Viscount Mountmorres on 'The Commercial Possibilities of West Africa,' and the other is the 'Quarterly Journal' for April. Amongst many articles, the following may be cited as typical: 'The Weevelling of Maize in West Africa,' by R. Newstead; 'The Properties of the Fibres separated from Cotton Seeds,' by Dr. E. Drabble; 'Analysis of the Oil from Inoy Kernels (*Poga oleosa*),' by E. S. Edie; 'A Note on some Chemical Properties of Sierra Leone Gum Copal,' by D. Spence and E. S. Edie, etc., etc. There are several plates. Viscount Mountmorres' paper is on sale at 6d., and the 'Quarterly Review at 2/- net.

Catalogue of the Specimeus illustrating the Osteology and Dentition of Vertebrated Animals, recent and extinct, contained in the Museum of the Royal College of Surgeons of London, yb **W. H. Fowler.** Part I., Man. Second edition. Taylor & Francis, 1907. 433 pp., price 10/- net.

Though with the somewhat unattractive title of 'Catalogue,' this excellent work can be particularly recommended to anyone interested in anthropological work. The first edition, prepared by the late Sir William H. Flower, was issued in 1879, and contained 264 pages. The present edition, for which the late Prof. Charles Stewart, was responsible, is nearly twice the size, and includes particulars of many important additions to this well-known collection, including the skulls presented by Sir Havelock Charles, and the specimens purchased in 1895 from the Anthropological Society. In the 'Introduction' particulars are given of the methods of measuring skulls, etc., and in the catalogue itself are details of skulls from almost every part of the world. Of particular value is the list of measurements of British, Roman, and Saxon skulls, many of which are from Yorkshire. There are scores of 'interesting exhibits' in the Royal College of Surgeons! Amongst them may be mentioned the skull of Eugene Aram, and the articulated skeleton of Jonathan Wilde, the famous thief-catcher. The following entry explains itself:—No. 335, 'The Skull of Charles Nichols, the comic lecturer. *Bequeathed by Mr. C. Nichols.*'

Twenty Country Rambles round Leeds, by **John Hornby.** W. Brierley, Leeds, 64 pp., 6d. In this readable pamphlet Mr. Hornby reprints a series of twenty articles which appeared in the *Leeds Mercury Supplement* two years ago. They are written in a chatty style, and each contains full details of the paths to be followed. We don't know whether the author was short of ink, or whether he writes plays, but 'Enter lane.' 'Enter field.' 'Hedge on left.' 'Path ascends.' 'On.' 'On then.' 'Distance lends, etc.' 'Alight.' 'Come then,' etc., seem unnecessarily brief. But the author evidently believes in 'covering the ground.' He takes us to Seven Arches, Eccup, Otley, Seacroft, Kirkstall, Crossgates, Arthington, Saltaire, Ilkley, Apperley Bridge, Methley, Armley, and a dozen other places ; and informs us that these walks 'may all be compassed in a summer's afternoon.' We don't believe it.

Science Progress for October (John Murray, 5/- net), contains several items of interest to readers of the 'Naturalist.' Two of our contributors have papers, viz., Mr. T. Petch on 'Insects and Fungi,' in which he describes the curious fungus-gardens of ants which he has discovered in Ceylon ; and 'Igneous Rock-Magmas as Solutions,' by Mr. Alfred Harker. Mr. R. Lydekker writes on ' American Economic Entomology'; Dr. N. H. Alcock describes ' A Simple Apparatus for Photo-microscopy,' Mr. F. V. Theobald has a paper on 'Economic Ornithology in relation to Agriculture, Horticulture, and Forestry'; and Mr. F. J. Lewis describes 'The Sequence of Plant Remains in the British Peat Mosses.' In an interesting paper on 'The Origin of the "Flower,"' Mr. W. C. Worsdell concludes that 'a "flower" is the result of the extreme modification of a leafy branch or of a portion of the main axis, the axial part of which has been excessively shortened and contracted, and the perianth, leaves, stamens, and carpels reduced and altered from the condition of large fern-like foliage leaves. Thus, in the long run, it is the Ferns that we have to thank for our 'Flowers.'

Manual of British Grasses, by **W. J. Gordon.** London : Simpkin Marshall, Ltd., pp. vi. and 174. 8vo., 6/-
This is a companion volume to 'Our Country's Flowers,' and is illustrated by 33 coloured plates containing figures of 101 species of grasses. Some of these are not very characteristic, *e.g.*, fig. 37 of *Aira flexuosa* shows a plant with flat leaves, but in the text they are described as 'bristle shaped.' The arrangement followed is that of Bentham, the reason for adopting it being that there is an accessible collection in the Natural History Museum at South Kensington, and this is arranged in 'Bentham's way.' He follows the zoologists in the matter of omitting capitals in specific names, and erroniously states that the use of capitals in certain specific names is ' merely a printers custom,' yet we notice he prints all the specific names with a capital in his index of species ; doubtless this is a printers custom. There is an interesting chapter on British cereals, also chapters on the characters of tribes, genera, and species, the two latter are freely illustrated by drawings of the florets and spikelets respectively ; finally there is a useful tabular view of the species. The habitats given for the species are often inadequate and sometimes misleading, and the extra British distribution is very incomplete. Of the 174 pages of text we find 9 pages are devoted to a list of species (really a list of the plates), notwithstanding the fact that two pages are previously given to such a list. Six pages are filled with what are called 'customary names'; they repeat the 'customary' names in the previous list, together with many which look anything but customary. These are again repeated in a seperate index of seven pages. An index of species occupies eleven pages, an index of genera two pages, and an extra three pages are taken up with a list of illustrations in the text; that is forty pages of this small book are devoted to these lists. Curiously enough, in spite of so much space being occupied in indexing, there are no references to the pages where the species are described, all the numbers refer to plates and figures ; while in the index of genera we find many numbers that refer neither to text, figures, nor plates, and after a long search for their meaning we gave it up in dispair.

NORTHERN NEWS.

Mr. Tennyson Macaulay Robinson has discovered that he is 'almost a poet,' and gives a sample of his 'blank verse' in the 'Country Side.' It *is* a—blank—verse.

In the 'Irish Naturalist' for August Mr. J. W. Taylor figures and describes the specimen of *Vitrina elongata* referred to in our 'Notes and Comments' column for August.

In the August 'Quarterly Journal of the Geological Society,' Mr. H. H. Arnold-Bemrose has a lengthy paper on 'The Toadstones of Derbyshire: Their Field Relations and Petrography.'

We regret to record the death of Mr. W. S. Parrish, of Hull, at the age of 43. For several years he was Treasurer of the Hull Geological Society, and was exceedingly useful to Yorkshire Geologists in taking photographs of geological sections.

Mr. P. G. Ralfe records a common Buzzard from the Calf of Man in June last ('Zoologist,' August). This appears to be the second record for the Island. The same writer records an increase in the numbers of the Puffins and Kittiwakes on the Island in recent years.

We regret to record the death of Mr. J. R. Boyle, who has done much to advance the study of Archæology in Yorkshire and Durham. His 'Lost Towns of the Humber,' 'Guide to Durham,' and 'History of Hedon,' are well-known works. Just before his death he was engaged in preparing a History of Hull.

Mr. A. E. Relph, in a recent number of the 'Antiquary,' suggests that some of the neolithic triangular flint 'spear-heads,' such as are frequently found on the Yorkshire Wolds and elsewhere, were really lateral barbs fitted into shafts of harpoons. A restoration is given, showing the way in which Mr. Relph thinks they might have been used.

The Rev. G. A. Crawshay has a paper on 'The Life History of *Tetropium gabrieli*' in the 'Transactions of the Entomological Society of London' (1907, Part II.). By an ingenious method of placing the wood, upon which the larvæ feed, between two pieces of glass, or in glass tubes, the author has been able to make many useful observations in connection with this species. Several excellent illustrations accompany his notes.

Amongst the grants made by the British Association are 'Fossiliferous Drift Deposits, £11 12s. 9d.; Fauna and Flora of British Trias, £10; Faunal Succession in the Carboniferous Limestone, £10; Erratic Blocks, £17 16s. 6d.; Exact Significance of Local Terms, £10; Index Animalium, £75; Excavations of Roman Sites, etc., £15; Structure of Fossil Plants, £15; and Succession of Plant Remains, £45.'

Mr. J. Carleton Rea, of Worcester, in an address to a Conference of the representatives of the Corresponding Societies of the British Association held at Leicester, suggested the study of fungi as a suitable subject for the local societies to take up. In this respect Yorkshire was shewn to have taken the lead. It is the only county posessing a published Fungus Flora of its own, and the pages of this journal frequently bear witness to the zeal of Yorkshire Mycologists.

We have received the 'Report of the Council of the Natural History Society of Northumberland, Durham, and Newcastle-on-Tyne,' which was presented at the Society's meeting on the 9th October. This is a record of a useful year's work, and also contains a list of additions to the Society's Museum. We notice that an assistant curator has recently been appointed, which will enable Mr. Gill to cope with the arrears of work. It seems a pity to learn that the heating of the Museum is unsatisfactory, and must remain so for the present, because an outlay of £50 will be necessary to put it in order.

'The Zoo in your own home' (Advert in the 'Weekly Wisdom').

Uncle Westell knows of a chicken that '*taps at the back door with its foot* every morning to let the lady of the house know that it is breakfast time.' What an interesting photograph this 'knocker-up' would make!

Amongst the 'new arrivals' in an American dealers list we notice '100 fine specimens of *Helix*' from Europe. He also has an 'enormous stock in marine univalves and bivalves.' We'll gurantee many of our brother conchologists at Billingsgate can equal his 'enormous stock.'

'To the majority of people the cosmopolitan sparrow, etc., constitute practically the whole of London's feathered folk, but *to those possessing the seeing eye* a greatly increased avi-fauna is presented. . . . *I have been myself* amazed at the variety of feathered folk to be met with.' Uncle Westell in 'The Animal World.'

Mr. H. S. Toms, of the Brighton Museum, sends us a reprint of an interesting paper on 'Pigmy Flint Implements found near Brighton.' We were relieved to find that Mr. Toms applies the word 'Pigmy' to the implements, and not to the people who made them. As is usual in these 'pigmy' papers, the illustrations are given in a reduced scale, which makes the pigmies appear more pigmy still.

We have received from Messrs. Marion & Co., 22 and 23 Soho Square, W., an exhaustive 'Catatogue of Photographic Apparatus and Materials,' to which we have pleasure in drawing the attention of our readers who are interested in photography. It contains nearly 200 pages, enumerates some thousands of appliances, etc., and is well illustrated. Amateur and professional alike will find much of value in this list.

In the 'Mineralogical Magazine' (No. 67), Mr. L. J. Spencer, of the British Museum, gives 'a (fourth) list of new mineral names.' The first in the list is 'Aegirine-hedenbergite,' and the last 'Zeyringite,' and these are fair samples of the names which occupy some twenty pages, Amongst them are Argento-algodonite, Chlormanganokalite, Hibschite, Kassiterolamprit, Metachalcophyllite, Silicomagnesiofluorite, Titaneisenglimmer, and Tschernichewite. The mineralogists have our sympathy. We hope they do not try to discuss these new names after an anniversary dinner.

We thought there were enough of the 'talky-talky' books on birds. But we learn from a recent writer that 'the great public is awaiting the publication of a book on British Birds,' 'written in plain language,' which will be 'pleasant to read,' and enable 'all readers to identify for themselves all the birds.' He is therefore undertaking this 'very difficult though congenial task.' During the past two or three years we have noticed some dozens of books, each of which proposed to have all the characters for which 'the great public' is said to be in such need, and each of which is about the same value as that now being issued in penny numbers by Mr. Audubon Cuvier Robinson.

Under the heading 'An unrecorded British Mammal'? a contemporary has a note on some 'mole-tailed rats,' which, 'about twenty years ago,' overran Sunk Island, at the mouth of the Humber. 'The ground was undermined by them and farm buildings had to be abandoned.' The editor, who is oblivious of the fact that someone may be 'pulling his leg,' and is ever ready with an explanation for anything under the sun, suggests that the account given might very well refer to the lemming, an animal which is sometimes disastrously abundant in Norway, whence it is conceivable that it might have been brought on shipboard to the mouth of the Humber.' There are no B.E.N.A.'s on Sunk Island, but a little local enquiry would have informed the editor that the animals in question were kangaroos, which reached Sunk Island by clinging to the propellor of a steamer bringing frozen rabbits to Hull. They were eventually exterminated by poisoned tobacco being strewn about on the ground. The kangaroos collected this, put it into their pouches, and died. As proof of this there is a skeleton of a kangaroo in the Hull Museum.

DECEMBER 1907.

No. 611
(No. 389 of current series)

THE NATURALIST.

A MONTHLY ILLUSTRATED JOURNAL OF

NATURAL HISTORY FOR THE NORTH OF ENGLAND.

EDITED BY

T. SHEPPARD, F.G.S.,

THE MUSEUM, HULL;

AND

T. W. WOODHEAD, Ph.D., F.L.S.,

TECHNICAL COLLEGE, HUDDERSFIELD.

WITH THE ASSISTANCE AS REFEREES IN SPECIAL DEPARTMENTS OF

J. GILBERT BAKER, F.R.S. F.L.S., GEO. T. PORRITT, F.L.S., F.E.S.,
Prof. P. F. KENDALL, M.Sc., F.G.S., JOHN W. TAYLOR,
T. H. NELSON, M.B.O.U., WILLIAM WEST, F.L.S.

Contents :—

LONDON:

A. BROWN & SONS, LIMITED, 5, FARRINGDON AVENUE, E.C.
And at HULL AND YORK.
Printers and Publishers to the Y.N.U.

PRICE **6d.** NET. BY POST **7d.** NET.

199066

PUBLICATIONS OF
The Yorkshire Naturalists' Union.

BOTANICAL TRANSACTIONS OF THE YORKSHIRE NATURALISTS' UNION, Volume I.

8vo, Cloth, 292 pp. (a few copies only left), price **5/-** *net.*

ontains various reports, papers, and addresses on the Flowering Plants, Mosses, and Fungi of the county

Complete, 8vo, Cloth, with Coloured Map, published at **One Guinea.** *Only a few copies left,* **10/6** *net.*

THE FLORA OF WEST YORKSHIRE.   By FREDERIC ARNOLD LEES, M.R.C.S., &c.

This, which forms the 2nd Volume of the Botanical Series of the Transactions, is perhaps the mos complete work of the kind ever issued for any district, including detailed and full records of 1044 Phanero. gains and Vascular Cryptogams, 11 Characeæ, 348 Mosses, 108 Hepatics, 258 Lichens, 1009 Fungi, and 39% Freshwater Algæ, making a total of 8160 species.

680 pp., Coloured Geological, Lithological, &c. Maps, suitably Bound, in Cloth.   Price **15/-** *net.*

NORTH YORKSHIRE: Studies of its Botany, Geology, Climate, and Physical Geography.
By JOHN GILBERT BAKER, F.R.S., F.L.S., M.R.I.A., V.M.H.

And a Chapter on the Mosses and Hepatics of the Riding, by MATTHEW B. SLATER, F.L.S. This Volume forms the 3rd of the Botanical Series.

396 pp., Complete, 8vo., Cloth.   Price **10/6** *net.*

THE FUNGUS FLORA OF YORKSHIRE. By G. MASSEE, F.L.S., F.R.H.S., & C. CROSSLAND, F.L.S

This is the 4th Volume of the Botanical Series of the Transactions; and contains a complete annotated list of all the known Fungi of the county, comprising 2626 species.

Complete, 8vo, Cloth.   Price **6/-** *post free.*

THE ALGA-FLORA OF YORKSHIRE. By W. WEST, F.L.S., & GEO. S. WEST, B.A., A.R.C.S., F.L.S

This work, which forms the 5th Volume of the Botanical Series of the Transactions, enumerates 1044 species, with full details of localities and numerous critical remarks on their affinities and distribution.

Complete, 8vo, Cloth.   Second Edition.   Price **6/6** *net.*

LIST OF YORKSHIRE LEPIDOPTERA. By G. T. PORRITT, F.L.S., F.E.S.

The First Edition of this work was published in 1883, and contained particulars of 1340 species o Macro- and Micro-Lepidoptera known to inhabit the county of York. The Second Edition, with Supplement contains much new information which has been accumulated by the author, including over 50 additiona species, together with copious notes on variation (particularly melanism), &c.

In progress, issued in Annual Parts, 8vo.

TRANSACTIONS OF THE YORKSHIRE NATURALISTS' UNION.

The Transactions include papers in all departments of the Yorkshire Fauna and Flora, and are issued i separately-paged series, devoted each to a special subject The Parts already published are sold to the publi as follows (Members are entitled to 25 per cent. discount) : Part 1 (1877), 2/3 ; 2 (1878), 1/9 ; 3 (1878), 1/6 ; 4 (1879) 2/- ; 5 (1880), 2/- ; 6 (1881), 2,- ; 7 (1882), 2/6 ; 8 (1883), 2/6 ; 9 (1884), 2/9 ; 10 (1885), 1/6 ; 11 (1885), 2,6 ; 12 (1886), 2/6 13 (1887), 2/6 ; 14 (1888), 1/9 ; 15 (1889), 2/6 ; 16 (1890), 2/6 ; 17 (1891), 2/6 ; 18 (1892), 1/9 ; 19 (1893), 9d. ; 20 (1894), 5/- 21 (1895), 1/- ; 22 (1896), 1/3 ; 23 (1897), 1/8 ; 24 (1898), 1/- ; 25 (1899), 1/9 ; 26 (1900), 5/- ; 27 (1901), 2/- ; 28 (1902), 1/3 29 (1902), 1/- ; 30 (1903), 2/6 ; 31 (1904), 1/- ; 32 (1905), 7/6 ; 33 (1906), 5.-.

THE BIRDS OF YORKSHIRE. By T. H. NELSON, M.B.O.U., WILLIAM EAGLE CLARKE, F.L.S. M.B.O.U., and F. BOYES. 2 Vols., Demy 8vo 25/- net. ; Demy 4to 42/- net.

Annotated List of the LAND and FRESHWATER MOLLUSCA KNOWN TO INHABIT YORK SHIRE. By JOHN W. TAYLOR, F.L.S., and others. Also in course of publication in the Trans actions.

THE YORKSHIRE CARBONIFEROUS FLORA. By ROBERT KIDSTON, F.R.S.E., F.G.S. Parts 14 18, 19, 21, &c., of Transactions.

LIST OF YORKSHIRE COLEOPTERA. By REV. W. C. HEY, M.A.

THE NATURALIST. A Monthly Illustrated Journal of Natural History for the North of England. Edite by T. SHEPPARD, F.G.S., Museum, Hull; and T. W. WOODHEAD, F.L.S., Technical College Huddersfield ; with the assistance as referees in Special Departments of J. GILBERT BAKER, F.R.S. F.L.S., PROF. PERCY F. KENDALL, M.Sc., F.G.S., T H. NELSON, M.B.O.U., GEO. T. PORRITT F.L.S., F.E.S., JOHN W. TAYLOR, and WILLIAM WEST, F.L.S. (Annual Subscription, payabl in advance, **6/6** post free).

MEMBERSHIP in the Yorkshire Naturalists' Union, **10/6** per annum, includes subscription to *The Naturalist* and entitles the member to receive the current Transactions, and all the other privileges of the Union A donation of **Seven Guineas** constitutes a life-membership, and entitles the member to a set of th completed volumes issued by the Union.
Members are entitled to buy all back numbers and other publications of the Union at a **discount of 2 per cent.** off the prices quoted above.
All communications should be addressed to the Hon. Secretary,
T. SHEPPARD, F.G.S., The Museum, Hull.

NOTES AND COMMENTS.

LINNÆUS.

We have received two pamphlets having reference to· Linnæus, the bicentenary of whose birth has recently been fitly celebrated in various parts of the world. The first is· 'Special Guide, No. 3, to the British Museum (Natural History),· and is entitled ' Memorials of Linnæus, a collection of Portraits, Manuscripts, Specimens, and Books exhibited to commemorate the bicentenary of his birth.'* This gives an excellent idea of the wealth of relics of the botanist in our national collection. Accompanying this guide are two excellent plates—one a portrait of Linnæus in Lapland dress. The second pamphlet contains an excellent account of 'Linnæus, 1707-1778,' by Mr. W. Hillhouse, and is issued by the Birmingham Natural History and Philosophical Society.†

EARLY IRON-AGE BURIALS.

From the pen of the Rev. Canon Greenwell has recently· appeared a paper, on which he has been engaged for some time, dealing with 'Early Iron-Age Burials in Yorkshire.'‡ In this he has gathered together probably all that is known of the Late Celtic and Early Iron-Age burials that have been found in the· county. In addition to isolated burials that have occurred, he gives full descriptions of the cemeteries of Arras, Hessieskew, and the so-called Danes' graves near Driffield. They are thought to contain remains of the Parisi, who, according to· Ptolemy, occupied this area. Canon Greenwell calls attention to the interesting fact that although a fair number of chariot burials occur in these late Celtic graves in Yorkshire, only two other such burials appear to be known in the whole of the rest of Britain, notwithstanding the circumstance that the chariots were fairly abundant throughout the country. From a bibliographical point of view, the paper presents items of interest. It was 'Read 23rd March, 1905'; it bears the date 1906 on the· cover and title, and includes an account of the chariot burial found at Hunmanby in 1907!

---

* London : 'Brit. Mus. (Nat. Hist.),' 1907. 16 pp., 3d.

† Proceedings, Vol. XII., pt. 2, 1907. 20 pp. Published at 55, Newhall Street, Birmingham, 1/- net.

‡ Archæologia, Vol. 60, pp. 251-322, plates.

HISTORY OF 'THE NATURALIST.'

Having at last got together what is possibly the only complete set of 'The Naturalist' extant, it is perhaps of value to place on record particulars of the various series. The first consisted of sixteen monthly parts, issued between January 1833, and April 1834, under the name of 'The Field Naturalist,' and was edited by Prof. James Rennie. The second ran into five volumes, between 1836 and 1839, the last four being printed at Doncaster. The first of these volumes was edited by B. Maund and W. Holl, and the remainder by Neville Wood. The third series consisted of eight volumes (1851-1858), the first five being edited by Beverley R. Morris and the remainder by F. O. Morris. In 1864 commenced 'The Naturalist, the journal of the West Riding Consolidated Naturalists' Society,' which went on for nearly three years. This was edited by George H. Parke and C. P. Hobkirk, and published by George Tindall. Next followed the 'Yorkshire Naturalists' Recorder,' issued by the same society, under the editorship of Joseph Wainwright. This was in 1872-3, and lasted a year. The 'Naturalist' followed, and between 1875 and 1884 nine volumes were issued, C. P. Hobkirk and G. T. Porritt being the editors. Without a break W. Denison Roebuck and W. Eagle Clarke conducted the journal until 1888, from which date the former also edited it until 1902, when the journal was published at Hull, having previously been printed at Wakefield, Huddersfield, and Leeds.

RESTORATION OF YORK MINSTER.

In the tenth 'occasional paper' on the restoration of the Minster, the Rev. the Dean of York announces the completion of a work which has been proceeding for eight years. Every care has been taken in the present restoration of the building to preserve as far as possible everything evidently original, and where insecure to replace it in a more permanent manner. Whatever, from partial decay, could not be retained with safety to the public, has been carefully removed and placed in the grounds behind the deanery. The disintegration of the stonework has been rapid, doubtless due to the action of sulphurous smoke upon the Magnesian limestone. The restoration has been carried out with Ketton stone, which, being Oolitic, will, it is hoped, prove invulnerable to this influence. As the number of smoke-emitting chimneys throughout York has increased, it is evident that unless some measures are adopted to mitigate the smoke the disintegration will be even more rapid. We understand, however, that some of the larger firms are doing all they can in the desired direction.

NEW PRESIDENT OF THE YORKSHIRE NATURALISTS' UNION.

At the unanimous invitation of the Executive Committee of the Yorkshire Naturalists' Union, Dr. Wheelton Hind, F.G.S., F.R.C.S., has accepted the presidency of the Union for the forthcoming year. Dr. Hind is well-known throughout the country for his successful work amongst Carboniferous rocks, and in Yorkshire he has been unusually successful in identifying and tracing various zones in the Carboniferous limestone. This valuable Yorkshire work has been supplemented by researches amongst rocks of the same age in other parts of Britain, resulting in Dr. Hind being placed in the front rank of authorities on this interesting subject. He has also paid much attention to the detailed study of the neglected series of fossil bivalves occurring in coal measures, and he is the author of monographs on Carboniferous mollusca published by the Palæontographical Society. In 1902 he was the recipient of a reward from the Lyall Geological Fund at the hands of the Geological Society of London, of which he has been a fellow for many years. Dr. Wheelton Hind's excellent work in Yorkshire makes this selection of him as President of the county society most appropriate, and will doubtless result in even greater attention being paid to the geological problems of the Carboniferous period by the members of the Uhion.

———◆◆———

The 'Hull Literary Club Magazine,' recently published, contains abstracts of papers on ' The Migration of Birds,' by Thomas Audas, and ' The Roman, Angle, and Dane in East Yorkshire,' by T. Sheppard.

'The Yorkshire Archæological Journal' (part 75), recently issued, is almost entirely devoted to an exhaustive and well-illustrated paper on ' Anglian and Anglo-Danish Sculpture in the North Riding of Yorkshire,' by W. G. Collingwood, F.S.A.

' Fenland Notes and Queries ' (No. 75) contains an analysis of the Oxford Clay at Whittlesey. There is also a note on the composition of virgin fen soil from Methwold Fen. This contains, moisture 72.8%, organic vegetable matter 24.3%, and mineral matter 2.9%.

A ' Note on two rare forms of *Actinocamax* from the English Upper Chalk,' by Dr. G. C. Crick, appears in the September 'Geological Magazine.' One of the species. which Dr. Crick names, *Actinocamax Blackmorei*, is from near Salisbury, and the other is from Gravesend.

The New Mexico College of Agriculture and Mechanic Arts still continues to issue its useful Bulletins. No. 64 has recently been received, and deals with the Tuna (Prickly Pear) as a food for man. It is written by Messrs. R. F. Hare and D. Griffiths, and is well illustrated.

In Memoriam.

JOHN FARRAH, F.L.S., F.R.Met.S.
(1849-1907).

(PLATE XLII.)

BUT two short months ago we placed on record the decease of John William Farrah, of Harrogate, the son and constant companion of John Farrah, a well-known figure in Yorkshire natural history circles. And now, in the middle of November, it is our painful duty to notify the death of the father. He has been for some time suffering from a painful illness, and insomnia, and there is little doubt that his son's sad end, at so young an age, which was a great blow to Mr. Farrah, hastened his own death.

'John Farrah,' as he insisted on being addressed on all occasions (Mr.'s and Esq.'s he abominated), was honest and straightforward to a degreee that is rarely seen now-a-days. He hated deceit and shams of every description, and rarely was he so bluff and out-spoken as when roused by some mean action, or by what he considered a high-handed or improper procedure. Proud of his county and of the characteristics of Yorkshiremen, he strongly upheld these, no matter in whose company. Anything approaching to cant met with his disapproval, and caused him to express his feelings very forcibly. He was also very severe with those who collected rare plants or eggs, or anything which might in any way have a harmful effect upon the flora or fauna of the county. More than one over-enthusiastic collector has received such a 'dressing-down' from John Farrah, that they will remember it to the end of their days.

With these qualities it will be apparent that he was often misunderstood, and those who did not meet him frequently were liable to misjudge his character. To know him intimately was to admire his sterling worth, his straightforwardness, and his ready wit. He was ever willing to help a good cause, and on many occasions has assisted the work of the Yorkshire Naturalists' Union by substantially contributing to its funds. 'Thorough' was his watchword, and rather than a report should not be printed as it ought to be, or a fund not receive the support he thought it should have, he would give the writer of these notes instructions to have the work done *well*, and he would pay the balance. In this way have Yorkshire Naturalists received many benefits from his generosity, without knowing it, as he would not allow his name to appear, nor would he permit the committee to record their thanks on the minutes. On such occasions he characteristically remarked that he gave the money to help the society's work, not to advertise John Farrah.

Whilst he was a good 'all-round' naturalist and antiquary,' botany seemed to have the greatest charm for him. On the memorable and exceedingly successful excursion to Bowes in 1903, Mr. Farrah took a prominent part, and wrote the account of 'The Flowering Plants of Bowes,' * one of the few occasions on which he could be persuaded to put pen on paper, and even then he would not allow 'his stuff' to be any cost to the Union. The first paragraph is so typical of the man that we quote it : 'It is not my intention to write a string of dry scientific names in the body of this article ; if these appear at all it will be at the end, in a list to themselves, where they will stand in stern forbiddingness, the bug-bear of many a would-be botanist. Bowes is delightfully quiet and peaceful, and I pray God that it will for ever remain so. The motor car—the latest curse inflicted upon the country—is comparatively rare. I used to have a contempt for cyclists ; now I am beginning to respect them. They glide along noiseless and stinkless, and comparatively dustless, and the tinkle of their bells is heavenly music compared with the horn of the motor.' †

Mr. Farrah took a leading part in connection with the Botanical Section of the Yorkshire Naturalists' Union, of which he held the office of President. He also took a practical interest in the 'Committee of Suggestions for Research,' and acted for several years on the Executive Committee of the Union. He attended the annual Fungus Foray of the Mycological Committee, and on the Terrace at Rievaulx Abbey in 1903, he found a large indigo-blue fungus, new to science, to which the name *Entoloma Farrahi* has been given. ‡

Mr. Farrah was at one time Honorary Meteorological Recorder for Harrogate, and in the capacity of Secretary and President took an active interest in the old Harrogate Naturalists' Society.

Perhaps next to plants the birds were his favourites, and, in company with Mr. Riley Fortune, he was frequently in the field by 3 a.m. in the spring and early summer months, in order to

* 'Nat.,' Sep. 1903, pp. 359-369. He also wrote the ' Flora of Nidderdale' for Harry Speight's work on that valley.

† Accompanying the general report of the Bowes excursion, appearing in the same issue of the 'Naturalist,' is a plate from a photograph of 'Four F.'s L.S.,' one of whom is John Farrah.

‡ See 'Naturalist,' January, 1904, where a coloured representation of the specimen is given.

study the birds before his day's work started. He was also a good antiquary, and had a thorough knowledge of the past and present history of the Harrogate district. For several years he was the constant companion of the late William Grange, whose 'History of the Forest of Knaresborough' is well known. In connection with this work Mr. Farrah helped a good deal.

His 'collecting' was restricted to books, and of these he had a very fine library. He purchased most books of any importance, and if, from the nature of the contents, or the way they were produced, they particularly appealed to him, he would obtain two copies 'in order to encourage their publication.'

Since he retired from business some time ago he had not enjoyed the best of health. He recently purchased two farms at Felliscliffe, and has since spent most of his time there. Upon these he has expended considerable sums of money, and his work there accounts for his absence from the excursions of the Yorkshire Naturalists' Union during the past two years. In letters to the undersigned, he has frequently expressed a hope that the farms would soon be as they should be, when he would again join in the rambles. But this is not to be.

He was married three times, and leaves behind a son and a daughter.

Many a Yorkshire Naturalist has cause to regret the death of John Farrah. T. S.

The Proceedings of the Liverpool Geological Society for 1906-7 (Vol. X., pt. 3) have been received. In addition to the list of members, etc., there are three useful papers, viz., 'The Storeton Find of 1906,' by H. C. Beasley; 'Desert Conditions and the Origin of the British Trias,' by J. Lomas; and 'Analyses of Ludlow Rocks,' by T. Mellard Reade and Philip Holland. The first is Mr. Beasley's presidential address, in which the recent exposure in the foot-print bed at Storeton is fully described. Mr. Lomas's admirable paper compares recent desert conditions with those obtaining in Triassic times. In the last paper attention is drawn to the intimate relations between the chemical constitution of rocks, their external forms, the soils resulting therefrom, and the vegetation they support.

Gowan's Nature Books, Nos. 6-15. Gowan and Cray, Glasgow. 6d. each net. This excellent series of illustrated hand-books, which was referred to in our columns for November last (p. 378), has been added to by volumes dealing with 'Fresh-water Fishes'; 'Toadstools at Home'; 'Our Trees and how to know them'; 'Wild Flowers at Home' (3rd series); 'Life in the Antarctic'; Reptile Life'; 'Sea-shore Life'; 'Birds at the Zoo'; 'Animals at the Zoo'; and 'Some Moths and Butterflies and their Eggs.' Each of these contains some dozens of carefully selected photographs, beautifully reproduced, and accompanied by suitable letterpress. Whilst each contains many charming photographs, perhaps the one which appeals to us most is No. 10, dealing with 'Life in the Antarctic.' In this are reproduced many of the photographs of birds which were shown by Mr. Eagle Clarke at the York Meeting of the Yorkshire Naturalists' Union. The low price at which these books are sold, at once indicates that there must be a large demand for them. They certainy warrant it.

THE HAIRY-ARMED BAT.

(*Vesperugo leisleri*).

ARTHUR WHITAKER,
Worsbrough Bridge.

(Continued from page 388).

Several nights I allowed my two Leisler's bats to fly in a room about twenty feet long and fourteen feet wide, and I found they could fly far better in such a confined space than a Noctule can. They made a considerable whirring sound when flying. They did not attempt to turn in the air when they reached the end of the room, but would pitch lightly up against the cornice or wall, turning and dropping off again instantly ; so quickly, in fact, that they could hardly be said to have settled in doing so ; it was simply a case of touch and off again. Sometimes they would select one particular spot on the cornice and pitch on to it time after time in quick succession, taking a circular sweep round between each touch. They would not continue flying for more than five minutes at a time, and generally alighted on the cornice to rest, hanging head downwards.

When I had had my Leisler's bats for about a week, I decided one evening to allow all my bats to exercise themselves at once in this room. At the time I had three Long-Eared, two Pipistrelles, and two Lesser Horseshoes, as well as the two Hairy-Armed bats. I allowed all these to fly about together or rest as they pleased for two or three hours, whilst I sat reading, and I observed that they seemed to take very little notice of one another.

When I came to catch them and put them back into their several cages I found that one of the Leisler's was missing. I turned the room topsy-turvy, and searched high and low, but to no purpose ; it seemed to have vanished completely, and I never found it again, though I kept the door of the room closed as much as possible for several days. It may perhaps have succeeded in getting up the chimney, as there was no fire in the grate at the time.

Its companion I continued to keep without difficulty, and it made several journeys, going first to Capt. Barrett Hamilton, at Arthurstown, in Ireland, then to Dr. Wilson, who was in Scotland at the time, and desired to make some drawings of it, and subsequently to Mr. Riley Fortune, at Harrogate, who took a number of photographs of it, six of which he has very

kindly supplied for reproduction here (see Plate XLI.).
From each of these journeys the bat returned in the best of
health, and, if anything, I think, with an increased appetite!
I decided to try to keep it through the winter in a natural
temperature, in order to make a few observations on it during
hybernation, and accordingly, on September 27th, I put it into
a smaller cage some eighteen inches square inside, with
perforated zinc panels in the top and two sides. In one
corner of this, supported on a couple of nails, I placed a large.
piece of semi-decayed wood, behind which the bat always
retired when sleeping. I placed a small thermometer in the
cage so that I could readily note the temperature. By this
time the bat had become so accustomed to artificial conditions,
and to mealworms as an article of diet, that if a liberal supply
of these were thrown into the bottom of the cage it would
manage to secure a fair proportion of them, usually about two-
thirds. It was never anything like so expeditious in finding
them as are some species of bats, notably the Long-eared, an
individual of which species I have known to secure every one of
thirty mealworms put into a large cage, and this on the very
first evening of captivity.

Unfortunately, it was quite impossible for me to examine
my Leisler's bat *every* evening, and as on most evenings when
I did so it was simply sleeping peacefully, I contented myself
with taking occasional notes, and the most interesting of these,
extracted from my diary, are as follows. The temperature
given in Fahrenheit degrees (as 58 F.) was the actual temperature *in* the bat's cage at dusk :—

Sept. 27th (58 F.)—Bat awake and fed well.
Sept. 28th (54 F))—Bat did not wake up.
Sept. 29th (54 F.).   ditto
Oct. 1st (57 F.). Bat awake, and ate seventy mealworms.
Oct. 2nd (60 F.).   ditto
Oct. 3rd (60 F.). Bat awake, and crawling about its cage almost before
  dusk. Had disposed of nearly 100 mealworms since dusk of the previous
  day, when I had thrown them into its cage. I gave it a further supply,
  which it ate ravenously. The same evening I noted an unusual number
  of Noctules flying about at dusk.
Oct. 4th to 12th. Temp. at dusk averaging about 54 F. I do not think my
  bat was awake at all on any of these evenings.
Oct. 12th (58 F.). Bat awoke at dusk and ate about 65 mealworms. I put
  a further supply of about 130 into its cage.
Oct. 25th (54 F.). (Higher temperature than any evening since the 12th).
  I noticed a Noctule flying over Worsbro' reservoir at 3-40 p.m. in full
  daylight, though the afternoon was a dull one. On getting home about
  4 o'clock I went straight to look at my Leisler's bat, to see if the same
  circumstances which induced the Noctule to fly had aroused it, but I
  found it sleeping. About 5 o'clock it awoke, however, and ate about
  50 mealworms.

Nov. 22nd (58 F.). Very much milder than any evening since the 25th of Oct. During the last four weeks I have looked at my bat every evening when it has been mild, but I do not think it has been awake at all during the time. To-night, however, it awoke in good time, and rapidly disposed of about 64 mealworms, after which it at once retired to its sleeping corner and dozed off. I noticed Long-eared bats feeding in Elmhirst Park late the same night.

Nov. 27th (51 F.). Milder than it has been for several days, but the bat did not wake up.

Dec. 12th (36 F.). Examined the bat in the daytime, and found it hanging somewhat limply, with a more flaccid appearance than normal, the legs being more stretched. The fur and skin of the shoulders seemed quite puffed out, making the neck and shoulders look disproportionally thick. Breathing hardly apparent at all.

Dec. 17th (48 F.). The evening being much milder than any since the 27th of Nov., I looked at my bat just after dusk, but it had not moved. When I gently lifted the piece of bark behind which it was resting, I found its eyes were open and it was breathing quite rapidly. Upon my holding a mealworm about half-an-inch from its nose, it promptly seized it, and subsequently disposed of some 45 more.

Jan. 11th (42 F.). In the morning I examined my bat, and found it hanging as usual, but I could not detect the breathing. As I feared it might possibly be dead, I touched one of its hind claws very lightly, when it instantly squeaked and moved the foot a little. Upon lifting up the piece of bark to which it was clinging, I found it had opened its eyes, and had the lips drawn back, especially the upper. It commenced to draw unsteady shuddering breaths, and to move its head a little from side to side, and also 'muscled up' a time or two, *i.e.*, draw up its whole body by contracting the leg mucles, a common action of many bats when hanging. I replaced the bat in its cage, and in about ten minutes time it had relapsed into its usual torpor.

Mar. 16th (51 F.). The bat was not awake at dusk, but as I had seen one or two Noctules flying in the open, I put about 100 mealworms into its cage.

Mar. 17th (49 F.). Bat asleep, but most of the mealworms placed in the cage last night were missing. Strong presumptive evidence that it must have been awake sometime during last night!

Mar. 30th (51 F.). Bat awake at dusk, and ravenously ate about 70 mealworms. I think it may very probably have been awake last night also, when it was fairly mild, but I was prevented from getting a look at it.

April 2nd (54 F.). Bat awake at dusk and fed well.
April 3rd (55 F.). Bat awake and ate some 4 dozen mealworms.
April 4th (59 F.). Bat awake and ate about 70 mealworms.
April 5th (54 F.). Bat awake and ate about 30 mealworms.
April 6th (50 F.). Bat not awake.
April 8th (51 F.). Bat not awake.
April 9th (48 F.). Bat not awake.
April 15th (55 F.). Bat awoke and ate about 40 mealworms.
April 16th (58 F.). Bat awoke and ate 63 mealworms.

After the middle of April I did not trouble to keep daily record of the temperature or notes as to whether the bat was awake or not. The cold and wet spring and early summer doubtless caused it to sleep on a much greater number of evenings than would ordinarily have been the case. And even when June had set in there were as many nights on which my bat did not wake as when it did. This, I fancy, was the case equally with bats in a perfectly natural condition, for I never remember a summer when they were so little in evidence.

Time after time I have visited their customary haunts at dusk, on what seemed to me favourable evenings, and scarcely seen a bat. If the Noctule, which has been fairly well in evidence, be put out of count, I think I might safely say that for every bat to be seen this summer twenty might have been counted last. This leads one to fear that the cold, wet spell of weather, which was so protracted at the time when the bats should have been feeding well to make up for their long winter fast, took very heavy toll of their numbers. A Long-eared bat (*P. auritus*) obtained by Mr. Armitage in Edlington Wood, near Doncaster, early in June, was found to be in a most emaciated and starved condition, and died within a short time of its capture, apparently from weakness. I am strongly inclined to think that this only gives an indication of the fate of very many bats this season in the Barnsley district, and that a good many seasons more favourable to them than this has been must elapse before they will again become as abundant as they were in 1906. It would be difficult to estimate the relative extent to which the several species occurring here have suffered, but were I to hazard a guess, I should say the Noctules (and probably Leisler's Bats also) have been least affected, and the Long-eared Bats have stood the season next best. The Pipistrelle (*P. pipistrellus*) appears to have suffered greatly, and three is the largest number I have seen on the wing together during the whole summer, and that occurred where scores were flitting about at one time in previous seasons—and in many places where one could have relied on finding a few any fine evening in 1906, not a sign of them could be found in 1907.

The Whiskered Bat (*M. mysticinus*) has certainly been scarce, but how much more so than usual I dare not guess, as it is almost indistinguishable from the last mentioned species when seen on the wing, while of Daubenton's Bat (*M. baubentoni*) I have not seen a single specimen all the summer.

It is a great pity that these useful little creatures have suffered such diminution in point of numbers, and one can only hope that the next few seasons may be sufficiently favourable to them to allow them to become as abundant as they were previously. Possibly the bat population in other parts of the country may not have suffered so much, but bats are very stay-at-home sorts of animals, migrating very little, if at all, and no one could well judge of their increase or decrease in any place. who has not been constantly in the habit of observing them on the spot for several seasons

EAST YORKSHIRE BIRD NOTES, 1907.

E. W. WADE, M.B.O.U.

Hull.

The season of 1907 is one that will long be memorable amongst ornithologists for its consistently unfavourable weather.

Commencing at Christmas 1906 with heavy snow storms and hard frost, we had a long and rigorous winter, followed by a cold, wet, and stormy spring and summer, with a few bright intervals only, conspicuous amongst which was the bright sunny weather during the second half of March. The effects upon bird life were disastrous. In winter, Redwings died by thousands all over the country, being driven by hunger to eat not only the Holly berries and Hips, but Mistletoe berries, which, as a rule, no bird but the Mistle Thrush will touch. The winter migrants were unusually plentiful in East Yorks., but the return migration was considerably delayed, even the hardiest species commencing their nesting duties much later than usual. Peewits, especially on the higher ground, could not face the frosts in March. Rooks were at least a week later than usual on the average in laying, and many nests were blown out of the trees by the March gales. Even the Heron, usually the earliest of nesting birds, which sometimes sits on its eggs through snow and frost, had, in one colony at least, not commenced to lay on 24th March. Carrion Crows in many instances were sitting on clutches of three incubated eggs on 20th April. Jackdaws, which usually have full clutches by the first week in May, had as many empty nests as full at that date; whilst the spring migrants, warblers, &c., were conspicuous by their absence, and when they did come, were almost silent. Early clutches of Garden Warbler were abnormally small—three and four instead of the usual five eggs—and many of these birds were singing and laying in late summer, as if to make up for lost time. At the end of May, Tree Pipits were still in full song. Greenfinches and Brown Linnets had, as a rule, only just laid full clutches by the fourth week in May, and the latter continued their breeding operations till August, a clutch of fresh eggs being found at Hornsea on 1st August; whilst Mr. F. Boyes ("Field," 4th September) reported that on 1st September a pair was feeding a young Cuckoo in the nest. Mr. H. R. Jackson reports that in middle June he saw a Reed Warbler's nest which, owing to the cold and stormy weather, had broken its supporting reeds and fallen into the water. The Swifts, instead of arriving in small parties, came in flocks of hundreds (F. Boyes). The Pink Footed Geese, whose arrival on the Wolds on 19th September is considered an event as fixed as the seasons themselves, this year made its appearance on 21st September.

In spite of the inclement season, we have to record unusual appearances of some birds, which we should hardly have expected to see so far north this year, viz., a Nightingale sang regularly at Sutton, another at Waghen, and a third at Elloughton this spring, and must have been breeding birds. On 28th May the Grasshopper Warbler, the scarcest of visitors to Holderness, was singing between Sutton and Waghen. On 20th May Mr. R. Haworth Booth saw a male Red-breasted Flycatcher at Hull Bank House ; and on 4th June Mr. H. R. Jackson saw male and female of the same birds at Thearne, about a mile distant. The Turtle Dove has nested at Burton Constable this year, the first time on record, and is undoubtedly extending its range in Holderness, where ten years ago it was practically unknown as a breeding species.

We have one or two interesting observations to record from this district, viz :—

The noticeable scarcity of the Grey Crow in autumn and winter, as compared with previous seasons. Is it possible that some process of extermination is being carried out in their Scandinavian homes, say, by placing a capitation grant upon their destruction ?

The extended breeding range of the Red-legged Partridge, which has become a well-known species on the Wolds and in Holderness in recent years. It would seem that this is a partially migratory species, contrary to what one would expect of so skulking a bird, for in April 1907, twelve were picked up on the beach at Hornsea in so exhausted a condition that the finder knocked them on the head and sold them. There were two occurrences of their being caught alive in back-yards, near Pearson Park, in Hull, this spring, and the birds were presented to the Park collections. Mr. H. R. Jackson shot one at Riston on 5th October, 1907, and reports that he has shot over the same ground twenty years, and never seen the bird there before.

On 5th June a Quail was killed on the telegraph wires at Buckton ; and Mr. M. J. Stephenson shot one on 16th September and one on 20th September in the highest part of Arras Wold, one a bird of the year, and another an old bird.

The gradual disappearance of the Corncrake from the East Riding is a melancholy fact. In Holderness, where it used to be common, it is scarcely ever heard or seen now. The cause for this is difficult to understand, as the conditions as to food, &c., have not changed.

On 9th May a Starling's nest with eight eggs, all of one type, was found in Burton Constable woods in a decayed tree-stump far from any other breeding place of the species. This is the third year is succession of the birds' laying the same number.

On 11th May a common Buzzard was seen at Burton Constable.

A pair of Lesser Spotted Woodpeckers has been at a certain locality in Holderness all this season.

In early May a large flock of Dotterel was seen on the low lands between Easington and the Humber, and several were shot. The Act for the Protection of Wild Birds seems to be generally a dead letter, owing to the supineness of local authorities. On May 11th-18th a flock of about 100 Dotterel frequented the wolds near Bempton, and on 2nd June 30 or 40 of these birds were seen at Bempton Cliffs. Possibly the heavy snowstorms in the north prevented their going to the nesting grounds.

For game birds, with the exception of the Grouse, which did well in North Yorks., the present season has been most disastrous. Partridges hatched first clutches well, but the chicks died almost immediately, the unusually long grass in wet cold weather being too much for them. The birds, however, in some cases reared second broods, for cheepers were seen up to the middle of September. The game bags show that not more than 10 per cent. of young birds have been shot this year. Pheasants in the wild state, on the other hand, hatched badly, many eggs being left in the nests, but the chicks actually hatched did fairly well. Among hand reared chicks the mortality was great, the season being reported the worst since 1863.

At Spurn Point on July 14th, two pairs of Ringed Plover still had eggs, viz., three just hatching and four fresh, and the number of birds about had visibly increased. The Lesser Tern on the same date were carrying fish, about 100 of them being in the air together, but the young, which would be in hiding on the beach, were invisible ; three odd eggs were seen. The common or Arctic Tern, which haunted the district all spring, was not visible.

The Sheld Duck is said to have reared four or five broods here this season, but in any case this bird shows a great increase in the Humber district of late years, probably owing to the destruction of its former feeding haunts at Crosby Warren, Lincs.

HORNSEA MERE.—The keeper, J. Taylor, reports that the

Herons have decreased, but the Pochards show a slight increase over recent years. Only one pair of Shovellers bred here, against seven pairs in 1906, probably owing to the greater amount of water having flooded out the nesting locality. A pair of Tufted Ducks have again been on the Mere all this season. Only three pair of Great-crested Grebe remain, and of these only one pair reared young, two in number. The increase of boating is undoubtedly responsible for driving away these birds. The Goldfinch has not been seen here this year, as has also been the case in some other of its usual haunts, but as it has appeared in greater numbers thnn usual in some localities, doubtless its appearance is merely a question of the local food supply.

BEMPTON.—We have again to congratulate ourselves upon the breeding of the Peregrine Falcons, which reared two young for the second year in succession.

The year 1907 has been the most disastrously wet season at the cliffs within living memory; an unfortunate factor for climbers, but fortunate for the birds. In consequence of the interruption in gathering eggs at the right time, many became stale before taken, and the third laying was thus postponed until the climbing season had terminated, as the interval for relaying, after sitting some days, was naturally greater than if the eggs had been taken when quite fresh. Most of the eggs of this third laying were hatched off in consequence.

COLEOPTERA.

Abundance of *Coccinellidæ*.—One result of the swarms of Aphides which were so obnoxious last month may now be seen in the numbers of Ladybirds maturing.

Yesterday, while walking from Wath-on-Dearne to Swinton, the palings bordering the road-side were literally swarming with ladybirds. So far as I could see, there were only two species represented, *Adalia bipunctata* and *Coccinella 10-punctata*, the latter in its usually very varied markings.

A few larvæ were also to be seen, one of which, much too large for either of the above species, was probably *Coccinella 7-punctata*.

Since the autumn of 1884, when, as I well remember, the railings and walls in and around West and South Parades, Wakefield, were literally alive with these same two species, I have never seen so many ladybirds congregated.—E. G. BAYFORD, October 18th, 1907.

SYLVAN VEGETATION OF FYLINGDALES, N.E. YORKS.

J. W. BARRY.
Fylingdales.

IN this, as in other of the highland parts of Yorkshire, the natural woods are confined to the ravines and to the slopes of the hills. The summits of the ridges, or of the tableland, out of which the dales have been carved, is either moorland, or moorland reclaimed, and shows only a few artificial plantations. The landscape has therefore a double aspect. From some points of view the whole impression is of bareness. From others the succession of woodland scenes unending.

Woods and Moors.—This subject will be dealt with in detail elsewhere.

Woods and Ravines.—In our ravines woods play a most important part. They are the great safeguard against the slipping away of the sides; that is to say, in the glaciated part of the township, or in the main dale, as opposed to the basin of the Derwent. The ravines in the area thus indicated are largely filled with glacial deposits : clays, sands, and morainic debris, the wasting away of which is a great source of trouble. The *fons et origo* of most of the landscapes is the alternation of clays with bands of sand. With clay alone, or with sand alone, the question would be less serious ; but when beds of sand are sandwiched in between the clays, the introduction of the least amount of water (which is usually affected by nothing greater than a mole) may cause a catastrophe. The only hope lies in a constant covering of trees. When woods are felled the land goes. Thus, soon after succeeding to the property, I yielded to professional advice, and cut down an old wood in the ravine, within the boundary of the original park. Landslips followed at once, and to such an extent that in one place what had been a prominent ridge became a deep hollow.

Natural Woods.—As regards natural wood, there are two or three patches in the Jugger Howe Gill (the upper ravine of the Derwent), which seem to be in a ' virgin ' or ' primæval ' condition. At the Peak again (though this is not within the bounds of Fylingdales parish) there is a considerable extent of wild vegetation in the cliffs and the ' coombs.' The latter is the old local and Celtic name for what the Hotel Company now call the ' Undercliff.' This vegetation, more of a 'macchia' than a wood, has been but little interfered with.

... There are also fragments of woodland in various parts of the parish which, though they have been cut, have evidently not been planted by man.

INDIGENOUS TREES OF THE FIRST CLASS.

From an inspection of these fragments of natural wood, as well as of planted woods sufficiently old to have reached a semi-natural condition, it is evident that the indigenous timber trees of the first dimensions are only three in number ; viz., the Oak (*Quercus pedunculata*), the Ash (*Fraxinus excelsior*), and that species of Elm which is called in the south the 'Wych,' but here the 'Wild' or the 'Rock' Elm ; botanically, of course, *Ulmus montana.*

To allocate to each of these trees their respective shares in the aboriginal woodland is a somewhat difficult task. In regard, however, to

The Elm (*Ulmus montana* Stokes).—There can be no doubt that in the ravines and narrower dales it once occupied a more conspicuous position than it does to day. It must have contended along the sides of the becks with the Ash and the Alder, and monopolised in many places the lateral runlets ; and it must have again come to the front in rocky places. It must have been found, in fact, scattered everywhere at moderate altitudes. When the moorland region, however, is entered it suddenly fails.

The Ash (*Fraxinus excelsior* L.).—The same failure at the moor-edge occurs with the Ash. Lower down the valleys, and near running water, the Ash must, as has been seen, have been a competitor with the Elm, whilst on the north exposures of the ravines it may perhaps have been the prevailing tree of the woodland.

The Oak (*Quercus pedunculata* Ehrh.).—The great extension of the Oak in this Dale is no doubt due, in large part, to artificial planting. Even the old parts of the Ramsdale Woods, where the Oaks seem to be not less than two hundred and fifty years old, prove, on close examination, to be no exception, and were, doubtless, planted by Sir Hugh Cholmley in the reign of Charles I.

At the same time the Oak was evidently plentiful by nature. Indeed, an examination of the old natural woods, of the hedge-rows, and of the Dale generally, shows that, in whatever proportion it stood, the Oak must have been absolutely ubiquitous ;

and that after the Elm and the Ash had been left behind in the upper part of the gills, it must have been almost without a competitor.

In strange contrast, however, with the Oak's ubiquity is the shyness of its fruiting. Such a thing as a regular acorn year, as one sees frequently in the Midlands and occasionally in the inland parts of Yorkshire, I have never witnessed here, and never expect to witness. Indeed, down to 1893, the sight of a ripe acorn was a very rare one, and Fylingdales acorns were indelibly associated with green fruit in their cups blown off by autumn gales. Since then, and since our cycle of dry summers, experiences in regard to acorns have been different; but even in the best years it has been only on the outer and sunnier edges of the Oak woods, or on hedgerow Oaks in sheltered situations, that I have observed acorns in any quantity.

Oaks and the Caterpillar Plague.—This scarcity of mast may possibly be connected, at least during the last twenty-five years, with the ravages of the caterpillar of the Oak leaf Roller Moth (*Tortrix viridana*). On an average of perhaps three seasons out of four, during the period just mentioned (prior to that there was nothing of the kind), the Oak woods have been almost entirely stripped from this cause; the general appearance presented in the beginning of July being about the same as that in the beginning of May, *i.e.*, a few trees (mostly Elms) in leaf, and a few signs of green on the rest. By the end of July the foliage recovers, and the Oaks do not appear to suffer, as far as I can see, in general health. I believe, however, though I have not properly proved it, that those most affected fail to bear fruit.

Last year, and for the first time, at any rate to any great extent, the foliage of the Hazel undergrowth was likewise devoured. This, on the other hand, unlike that of the Oak, made no recovery during the rest of the year, nor has it done so this present season. The consequence is that great numbers of the Hazel bushes, and particularly the younger ones, are now in a dying condition, and in some cases dead.

When the caterpillars are at their height there is a sound throughout the woods as of a light but continuous shower of rain. This is, of course, from the fall of the excrement. At the same time the insects themselves are noticeable, from their habit of letting themselves up and down by long silken threads. In making your way, therefore, through the coverts, you can scarcely avoid coming in contact with them, as they hang about

1907 December 1.

2 E

or sway in the breeze.   Large flocks of Jackdaws and Starlings invariably pay their attention to the caterpillars ; and, indeed, generally give one the first notice of the plague having begun. The effect, however, produced by them in the main woods is quite insignificant.   Near the house, on the other hand, where Oaks are intermixed with other trees, and where birds of all kinds are constantly present, but little damage is, as a rule, to be perceived.

The years in which the Oaks escape the enemy seem to be : (1) Those in which there are heavy and constant falls of rain during the caterpillers season.   (2) Early years, when the leaf gets a start of the caterpillar, and hardens.   At any rate, it is noticeable that in the plague years those Oaks which are exceptionally forward are immune ; whilst in plague-free, or nearly plague-free, years the late-leafing trees, as those on the north edges of the woods, are attacked.

INDIGENOUS TREES OF THE SECOND CLASS.

The Birch (*Betula verrucosa* Ehrh.) takes a subordinate position in this Dale as compared with that taken by it in other dales of Yorkshire.   With the exception of a few scattered standards, and these chiefly near the moor edge, the only woods, or remains of woods, of birch are in the Derwent section of the township.

As to the causes of this feature, no doubt some part has been played by man.   In the main dale, which has always been the centre of population, and must always have been more under the woodman's eye, Birch may possibly have been largely cut out in mixed woods, in order to make room for more valuable timber, and not being a species which is tolerant of shade, has failed to reappear, as other trees would have done.

At the same time, I think that there may have been natural causes as well.   And I should be inclined to specify under this head :—(1) A soil unsuited to the Birch in the cold, stiff glacial clay, which fills the ravines and much of the hillsides in the area last noticed.   (2) Exposure to sea winds.

In regard to the latter circumstance, my attention was first drawn to it by losses after planting in such situations.   Indeed, after the sudden loss, about eighteen years ago, of nearly all the then-to thriving Birch in a mixed plantation, some ten years of age, and some three acres in extent, which was on a steep slope, with a loamy soil, but exposed to the east and north-east, I was careful not to plant this species of tree wherever the sea winds could penetrate. Then, after a long interval, I thought that I might have been mistaken. At any rate, I determined to try again on a soil that was an ideal soil for the Birch.

The site was an old freestone quarry that had grown a crop of Larch and stunted Oak, succeeded by a luxuriant jungle of Bracken. The quarry was under the brow of the moorland plateau, 700 feet above the level of the sea, with a fair amount of shelter from the west and south-west, but with exposure, though not complete exposure, to the east and north ; there being sufficient remnants of the old crop to break the extreme bitterness of the winds. The result, however, has been completely disastrous. Though all other sorts of trees have survived, the birch has succumbed, as in previous cases, and in a nearly similar manner : namely, by losing all their foliage in the early summer after a long continuance of our British Etesians. They have not, however, gone all at once, as in the previous instance, but some one year and some another, at different stages of their growth.

These details I have thought proper to give, since I have not come on any botanical or any sylvicultural author (excepting one, to whom I furnished the information myself) who has so much as hinted at the Birch being a sea-shunning tree, and since, therefore, a general statement might fail to convince. At the same time I may mention that I had long noticed that the Birch did not occur within a mile or two of our coast where it would be fully exposed to the north and north-east. I then observed that even under half exposure to these points of the compass, and at this distance from the sea, no Birch, or trace of Birch, was to be found in our natural woods or coppice. Finally, on looking over my journal of Norwegian travel, I saw that at the outer extremities of such fiords as I visited, and amongst the islands, the Birch, which all along the slopes of the fiords themselves had been the characteristic tree, was replaced by Scots Pine.

With all these facts, therefore, to go upon, I think that the conclusion which has been drawn may be fairly justified. If, accordingly, it be accepted, then the parts of the main dale of Fyling adapted to the Birch are limited from considerations of aspect alone, whilst they are still further reduced by considerations of soil. Hence, in combination with human handiwork, the scarcity of the tree in this, the seaward part of the township. The Derwent basin, on the other hand, has less of these adverse circumstances.   It is at once the inland section of the township, and also free, or nearly so, from glacial clays.   Hence the presence in its slacks of birchwoods still existing and unmistakable evidences that in times gone by these birchwoods were of considerable extent.

The Alder (*Alnus glutinosa* Gaert.) calls for no remark.

The Aspen (*Populus tremula* L.).—Though rare, is nevertheless found here.   The slacks of the moors or the moor edges are places in which I have seen it, so that it appears as essentially a highland tree ; and this, as far as I have been able to judge personally, is its character in Northern Europe as well as in Southern.   Again, I have only seen it on the slopes of these slacks ; never at the bottom by the becksides.   Indeed the only specimens in the Ramsdale woods are on freestone rock with a south aspect, the situation being a clearing within a hundred yards of the moor, and the Aspen having sprung up since the clearing was made.   This, again, agrees with what I have observed elsewhere.   For though the authorities usually write of the Aspen as being essentially a wet soil tree, yet its constant occurrence in dry and rocky places (its stature there is naturally small) is one of the facts that has always struck me in the mountains of Norway, Scotland, and Corsica.   On the slope facing east of Jugger Howe Gill the little woods or groves of primæval aspect are, I think (for I have never climbed up to them), composed of this tree.   There, however, I should say, the soil was humid.

INDIGENOUS TREES OF THE THIRD CLASS.

The Rowan (*Pyrus aucuparia* Ehrh.), in contrast to the Birch, is one of the most ubiquitous and one of the most impressible of trees.   It does not seem particular as to either situation or soil.   It springs up even on the sea cliffs of the Peak, and is part of the main undergrowth in the Oak woods of the dales.   It accomodates itself to clays and to sour moorland, as well as to its favourite rocks and rubble.   Whenever a moor edge plantation is made, and has become tall enough to harbour birds, the Rowan is sure to appear, and in increasing numbers, owing to the voiding of its seed by wood pigeons and smaller birds.   On pastured ground it is kept down, of course, by browsing ; but in rough enclosures, where there are neglected whin bushes, it often manages to make a show, thanks to the protection of those thorny thickets.   In this way a natural wood of Rowan is forming slowly and by irregular steps, according to the degrees of inattention of the successive tenants, on the slopes of the little moor 'intake' near Swallow Head.   In the lowest part adjoining the Ramsdale Woods some of the trees are already old, and close enough together to keep down all undergrowth.

NATURALISED TREES.

The Sycamore (*Acer pseudo-platanus* L.), known locally as 'The Plane,' is now around homesteads perhaps the most characteristic tree of the Dale.   The choice is a wise one, since though liable from its early leafing to be cut by spring storms, it usually recovers itself before the season is out, and is, of all trees of first-rate dimensions, the one that keeps its shape best under exposure to the sea winds.   At the same time, like its scriptural namesake,* it is less liable to be uprooted than perhaps any other tree, the proverbial Oak not excepted.

Of course the Sycamore is an exotic, and the evidence of its exotic character can been seen here, as elsewhere, in the fact that, whenever planted and allowed to mature, it rapidly spreads itself.   Had it therefore been indigenous it must necessarily have been ubiquitous.   That, however, is not the case.   It does not occur in natural woods, and in no woods except those at Fyling Old Hall do I know any specimens a hundred years old.   Seedlings, in fact, are seen only where they can be accounted for by the presence of standards from which the seed can have blown.   And the seed will generally, I think, be found to have blown in an easterly direction, since the prevailing winds at the time of leaf-shedding are the westerly.   Thus, in the Ramsdale woods, the seedlings and saplings of Sycamore which now, to the owner's annoyance, abound, and which cannot be extirpated, can be traced to some standards of this tree planted near the edge of the woods on a moor-edge farm that was sold out of the estate about a century-and-a-half ago.   The seedlings and saplings have been gradually spreading in a down-valley or easterly direction, so that they may now be traced that way for about a mile.   Up the valley, or in a westerly direction, only one or two plants are to be found.   These, moreover, are at not more than a hundred-and-fifty yards from the presumed parent tree, and have shown themselves only in the last few years.

---

* Luke xvii. 6.

As to the introduction of the Sycamore into these parts, I cannot doubt that here, as in so many other places, the original introducers were the monks. The abbot and monks of the Whitby Monastery were the possessors of the greater part of Fylingdales from the time of William II. to that of Henry VIII., and the oldest Sycamore trees in the township are around the two homesteads with which the monks seem to have been brought in special connection, viz., Fyling Old Hall, the ancient 'Manerium de Fyling,' and Fyling New Hall, formerly Park Gate, and so called from its proximity to the ancient Park.

The Beech (*Fagus sylvatica* L.) has been introduced here much later than the Sycamore. In fact, as I have not seen a Beech tree which can be a hundred years old, I do not think that there can have been one in the Dale in 1819, when my great grandfather bought this estate and began to plant. As it is, there are but a few odd specimens away from the plantations which adjoin this house.

At the same time the Beech reproduces itself even in this climate ; not, of course, as freely as the Sycamore, but at least as freely as the indigenous Oak.   This circumstance is again peculiar, because it is a rare thing with us to find Beech mast which has any kernel.   The crops of it are both more abundant and more frequent than those of acorns, but the nut almost invariably proves to be 'deaf.'   And yet there is scarcely a standard of over sixty years of age near which seedlings cannot be found.   The seedlings and saplings seen in Ramsdale Woods by the side of the beck can be traced to two Beech trees about eighty years old that have been planted at the upper and outer edge of the coverts.

The Beech here, as in many other parts of Britain, is infested by a species of Scale, *Cryptococcus fagi,* some trees being already in a dying condition.   Those which suffer most seem to be saplings which are shaded by taller timber over them.

The Scots Pine (*Pinus sylvestris* L.) is excellently adapted to the slacks and more sheltered parts of our moorlands.   On the open plateau it requires to be massed in great bodies in order to resist the winter gales, and even then is liable to be blown over where the subsoil is impervious or the surface moist. But in the situations that have been mentioned it is quite at home, and in a climate so congenial to it that matured timber from the Foulsyke plantations has been found, according to the late I. Taylor, to be as good as any imported from the Baltic. At the same time it seeds itself freely. Indeed, were it not for the moor sheep, the turf paring, and the annual moor burning, or 'swiddening,' the seed blown from plantations of this tree would, everywhere where there is heather, give birth to natural woods. This may be seen on the Sneaton Moor Intake between the two Red Gates, round which there are, or have been, Scots plantations formed some eighty or so years ago. In one place seedlings of various sizes (some being already a few feet) stand thick on the ground for a considerable distance ; so that, if fenced, they would soon make a covert. On Silpho Brow, towards Hackness, where the soil is sandy and the climate less marine, a still better example may be noted ; for there nearly the whole moorland brow is becoming, and already has become, a scattered, picturesque, forest-like wood, the pines being in every stage of growth from small seedlings to middle-aged trees.

Yet with all this the pine is clearly an exotic as far as regards this part of the country. There is no trace of woods of it that have not been planted, or, at least, sprung from artificial plantations, nor of any isolated specimens or seedlings at any distance from such plantations. Neither yet can we suppose that it has been extirpated by man. Had it existed here since the glacial period it must, with its facility for reproduction, have gained too firm a hold in the lower uplands to have been banished utterly from every corner and crag. As a matter of fact there are no remains of it even in the peat bogs.* The upper timber bed in the bogs between the two peat layers, where we should expect to find it, as we find it in Scotland, Ireland, and Norway, since the Scots pine surpasses perhaps any other tree in its adaptability to growing on pure peat, is occupied here by the remains of Birch woods.

During the summer of 1906 the Pine Beetle (*Hylurgus piniperda*) committed great ravages among young pines, some of the trees being killed outright.

————◆◆————

Mr. C. B. Crampton has a paper on 'Fossils and Conditions of Deposit, and Theory of Coal Formation' in the 'Transactions of the Edinburgh Geological Society' (Vol. 9, Pt. I., 1907).

Amongst the many interesting papers bearing upon the natural history and archæology of Hertfordshire, which appear in the 'Transactions of the Hertfordshire Natural History and Field Club' (Vol. 13, Pt. I., 1907), are 'On a recent Palæolithic discovery near Rickmansworth,' by Sir John Evans ; 'Witches' Brooms,' by J. Saunders ; and 'Ostracoda and Mollusca from the Alluvial Deposits at the Watford Gas Works,' by the Editor of the Proceedings, Mr. John Hopkinson.

---

* It occurs in the peat of South-east Yorks.—ED.

FIELD NOTES.

ARACHNIDA.

A Pseudo-scorpion new to Yorkshire.—*Chernes rufeolus* Sim. To the list of the counties (*Vide* 'Nat.' for Oct.) in which this pseudo-scorpion has occurred, Yorkshire may now be added. On Oct. 12th I secured two examples in the cracks between the flagstones which formed the floor of a barn at Broak Oak, Linthwaite, Colne Valley. The full list for the county embraces five species, *Chthonius rayi* L. Koch., *Obisium muscorum* Leach., *Chelifer latreillei* Leach., *Chernes nodosus* Schr., and *Chernes rufeolus* Sim.—WM. FALCONER, Slaithwaite, October 21st, 1907.

A Spider new to Northumberland.—*Tigellinus furcillatus* Menge. In the 'Naturalist' for May, 1901, p. 160, the Wessenden Valley, near Huddersfield, is noted as being the most northerly point at which this very rare spider has been found. This is no longer correct; an adult female, which I took near Staward Peel in the above county, Aug., 1907, places its limit of distribution much further north. The specimen has been kindly compared by Dr. A. Randell Jackson with females obtained by him in Delamere Forest, Cheshire, July, 1906. —WM. FALCONER, Slaithwaite, October 21st, 1907.

—:o:—

BIRDS.

Woodcock in Littondale.—It may interest those members of the Yorkshire Naturalists' Union who have been investigating the fauna of Littondale to know that in 1905, on three occasious in June and July, I flushed a Woodcock in 'Gildersbank,' near the Cave. I could not find a second bird or a nest, but it is very probable that there would be a second bird if I had only been fortunate enough to see it.

A few are always shot in Littondale later in the season, but these will no doubt be the usual autumn visitors.—J. W. DALTON, 7, Manningham Lane, Bradford.

Winter Migrants at Sedbergh.—The following are particulars of the arrival of our winter migrants in this district :— Bramblefinch, Oct. 7th; Redwing, Oct. 10th; Siskin, Oct. 15th; Fieldfare, Oct. 15th. I saw two Swallows on the 21st Oct., rather late for our district.—WM. MORRIS, Sedbergh.

Red-backed Shrike nesting near York in 1881.— Referring to Mr. Oxley Grabham's note on the nesting of the Red-backed Shrike near Pickering this summer (*ante* p. 325), it may be of interest to record that during my schooldays in York I found a nest of this species at Acaster, not far from Naburn Locks, in the year 1881, and I have still one of the eggs in my possession.—HARRY B. BOOTH, Shipley.

The Waxwing (*Ampelis garralus* L.) in Fylingdales.— At the recent excursion of the Yorkshire Naturalists' Union to Robin Hood's Bay, I saw in the cottage of Mr. J. W. Barry's gardener (formerly his gamekeeper) a fine specimen of the Waxwing, which he had shot in his garden some seven or eight years previously, during the winter months. As the district of Fylingdales has lately received considerabe attention from a natural history point of view (thanks to its courteous owner), I think it well to send in this additional note.—HARRY B. BOOTH, Shipley.

Migration of Swifts, etc.—During part of August I had occasion to cycle several times daily along the road between Leeds and Menston by way of Horsforth, Rawdon, and Guiseley, and had therefore the opportunity of noting any migration' which was taking place in that district, of the Swallow, House Martin, and Swift. The fact which I noticed most particularly about the migration taking place was that the Swift apparently migrated from this part of Airedale in an opposite direction to the Swallow and House Martin. The Swift flying up the valley (west) and the Swallow and Martin down the valley (east). I have noticed the latter in Lower Wharfedale. On one occasion (see table below) I watched a party of Swifts fly towards Horsforth from a northerly direction and then strike up the valley (W.) to Rawdon. I should be very pleased if readers of the 'Naturalist' would give the benefit of their experience with reference to the migration routes out of the Yorkshire Dales taken by the above-mentioned birds.

Aug. 11th.—Two Swifts over the moors at Hawksworth, near Menston.
Aug. 12th.—Two Swifts over Menston.
Aug. 13th—18-20 Swifts over Menston in the morning, none in the evening.
Aug. 14th.—One Swift over Menston.
Aug. 16th.—One Swift over moors at Hawksworth.
Aug. 17th.—One Swift over Burley in Wharfedale.
Aug. 18th.—House Martin flying down Airedale (E.), none at Guiseley.
Aug. 20th.—Swallows flying down the Aire Valley (W.). House Martins at Rawson, none here on the 19th.
Aug. 20th.—Five Swifts flying up the Aire Valley (W.) towards Rawdon.
Aug. 22nd.—18 Swifts at Rawdon, and a party of 12 flew to Horsforth from a northerly direction and then flew up the valley (W.) towards Rawdon.
Aug. 23rd.—Swallows flying down the valley (E.). 7 or 8 Swifts over the east end of Rawdon.

—S. HOLE, Leeds.

REVIEWS AND BOOK NOTICES.

SOME NEW BOOKS.*

WE have recently had the disagreeable duty of perusing and commenting upon such an enormous number of would-be natural history books, that to get hold of really useful publications that will further the study of natural history is indeed a pleasure. The four works referred to in the following notes are so far above their fellows that we are glad to draw special attention to them. We can only express the hope that in their own interests publishers will exercise some care in the selection of works placed upon the market. In recent years there have been many books issued—in some cases by well-known and usually reliable houses—which have been a disgrace to the twentieth century.

'Mammals of the World,' is a well-illustrated volume, printed in good clear type, and is very cheap at six shillings. Obviously for the use of the young naturalist, the descriptions of the various mammals are given in an interesting style, as might be expected from such a well-known writer as Mr. W. F. Kirby. No space is wasted in fine writing. Dr. W. E. Kirby writes a useful 'Introduction,' dealing with the structure of the mammalia, which is rather more technical, and will better appeal to older readers.

Of a somewhat similar type is ' Cassell's Natural History for Young People,' which is also well illustrated, and exceedingly cheap. In addition to the mammals, however, Mr. Bouser deals with birds, amphibia and reptiles, fishes, and the invertebrata. His descriptions are accompanied by characteristic anecdotes, which add to the interest of the work. Either this or the preceeding work can be well recommended as a prize or present to a boy or girl, and we are sure that either volume will be appreciated.

Dealing with another branch of nature knowledge altogether is the charming volume issued by Mr. Heinemann. In this the authoress has depicted in a beautiful series of no fewer than 75 large plates, all the important members of the British flora.

Mammals of the World, by W. F. Kirby. London : Sidney Appleton, 1907. 141 pp., coloured plates. 6/- net.

Cassell's Natural History for Young People (Second edition), by A. E. Bouser, with over a hundred full-page illustrations. London : Cassell & Co., 1907. 280 pp. 5/-

Wild Flowers of the British Isles, by H. Isabel Adams. London : William Heinemann, 1907. 168 pp., 75 coloured plates. 30/- net.

Microscopy : The Construction, Theory, and Use of the Microscope, by Edmund J. Spitta. London : John Murray, 1907. 468 pp., plates. 10/6 net.

The drawings are made with unusual accuracy as regards both colour and form. These are reproduced by the three-colour process, which, with the exception of a slight preponderance of purple, represents the drawings very faithfully. From the decorative point of view, however, the illustrations are unusually valuable, and we can strongly recommend them to art students. From the arrangement and method of portrayal it is evident that the artist-authoress has intended that every advantage shall be given to the beauties of our flora from the point of view of design and decoration. In the seemingly careless arrangement of flowers, and in the apparently accidental twist of the stalks or tendrils, it is clear that every effort has been made to secure the greatest artistic effect from the subjects drawn. But beyond their success as pictures, the plates are unusually useful to the botanical student for the ease with which they enable the different species to be recognised. Towards this end the plants are classified, and at once show the resemblances to and the differences between the various species. Occasional drawings of parts of the flower, etc., facilitate this. With regard to the letterpress there is an exceedingly modest preface ; then some ' descriptions of botanical terms,' which are followed by clear and accurate descriptions of the various plants,

In the fourth volume before us is a work alike useful to student of animals or plants. It is a scholarly treatise, and as might be expected from the President of the Quekett Microscopical Club, is written by one who has made a special study of the subject with which he deals. In his previous works dealing with Photomicrography and Bacteriology respectively, Mr. Spitta gave evidence of the thoroughness of his work. In ' Microscopy,' however, he can safely be said to have excelled himself in his regard for treatment of even the smallest details of microscopical research. His knowledge of the various lenses and parts of the microscope is unrivalled, and of this the full benefit is given to readers of ' Microscopy.' Every possible detail of microscopic work is described in a way which at once shows that the author is giving the information from practical experience, and not from second-hand and antiquated sources. The volume is illustrated by no fewer than 241 blocks in the text, and by 41 half-tone reproductions from original negatives. These latter largely deal with the structures of diatoms as revealed by different powers and appliances. It goes without saying that the publisher, Mr. John Murray, has done his share of the work well.

1907 December 1.

CHILDRENS BOOKS.

The 'Look-about-you' Nature Book, by **T. W. Hoare** and others. T. C. & E. C. Jack, Edinburgh, 1907, 5/- net. In this thick volume are bound together seven of Messrs. Jack's well known shilling 'Look-about-you' nature study books, which are specially adapted for young readers. They are in simple language, in large clear type, and beautifully illustrated. They deal with all manner of subjects, and will certainly prove an attraction to juveniles.

The Fairy-Land of Living Things, by **R. Kearton.** Cassell & Co. 1907. 182 pp., 3/6. In this volume the brothers Kearton, by pen and picture, have produced a volume specially for young readers. It is well

Dandelion Heads.

printed, and contains nearly 200 illustrations, all of which are good. It is very suitable for boys in the upper standards at school, and would make a useful prize. One of the illustrations we are kindly permitted to reproduce.

The Fairyland of Nature, by **Wood Smith.** London: S. W. Partridge & Co. 126 pp., Cloth 1/- This book contains a number of essays, written in rather school-boy phraseology, upon a variety of subjects, such as 'The Mammoth Cave,' 'Flesh Eating-Plants,' 'A Piece of Sponge,' etc. Each appears to have been written by the author immediately after he had read something upon the subject elsewhere. As a sample of the author's style, we read (p. 64), 'From six to twelve trucks, each containing usually half-a-ton of coal, at a time are taken by a boy called a "driver," with horse, into a part of the mine called the "station," where he exchanges the empties for full trucks.' There are several illustrations, mostly ancient, one (p. 67) is upside down, and another (p. 62) shows 'a stem of a *palm-tree* found in coal seams.' But the book is remarkably cheap at one shilling.

Cremation: The planning of Crematoria and Columbaria is the title of a pamphlet (A. C. Freeman, 72, Finsbury Pavement, E.C., 20 pp., 1/-). The address was originally read to the Society of Architects, and contains much useful information, as well as some interesting designs. It is surprising how few Crematoria there are in Britain to-day—the ancient Britains, Romans, and Saxons all taught us the proper way to dispose of the dead; but in this present century of prejudices we are too slow to learn. We notice from Mr. Freeman's pamphlet that the Hull Crematorium is of the early perpendicular style, *freely treated.* It's a way architects have in Hull.

THE FIRST RECORDED BRITISH EXAMPLE OF THE WHITE-SPOTTED BLUETHROAT.

T. H. NELSON, M.B.O.U.

HAVING recently had an opportunity of examining the Scarborough specimen of this form of the Bluethroat, which I exhibited at the British Ornithologists' Club on 16th October last, I am pleased to be able to state that its identity is fully established.

It is in every way typical of the white-spotted form (*Cyanecula wolfi*), the white in the centre of its dark blue throat being most distinct, and about half-an-inch in diameter. The plumage, even after the lapse of thirty years, still retains its deep intense hue.

On questioning the present owner of the specimen as to the facts of the occurrence, he fully corroborated the original statements of the Rev. J. G. Tuck and Mr. Eagle Clarke; though, as the females of the two forms of Bluethroat cannot be distinguished, it is somewhat unfortunate that a misleading statement was made as to the sex of this bird, which accounts for its rejection by the authors of recent ornithological works. My informant remembers his father bringing home the bird, telling him he had found it under the telegraph wires, and pointing out where it had been damaged by coming in contact with them.

For further particulars see ' Birds of Yorkshire,' pp. 38-39.

Mr. J. W. Harrison has a ' Note on *Oporobia (Larentia) autumnata*, in Cleveland,' in the November ' Entomologist.'

We have received the ' Transactions of the Eastbourne Natural History Society' (Vol. IV., Pt. 2). It contains papers dealing with France, Portugal, Switzerland, etc. Some of the items are of distinct interest to Eastbourne naturalists, however.

An interesting phase of the marine biological work which is being carried on at various parts of the coasts of the British Isles is illustrated by the capture, which has just been effected, of a marked crab which has travelled from Scarborough to Montrose—a distance of about 155 miles—in 689 days. Not very long ago a crab was marked and returned to the sea at Flamborough, was caught at Beadnell in Northumberland, having walked a distance of 108 miles in 114 days. It is an interesting fact also that one of the crabs liberated by Professor Meek, of the Marine Laboratory at Cullercoats, twelve miles north of Beadnell, was recaptured after an interval of nearly four months at Portlethen, in Kincardineshire, a distance of eighty miles. In all, Professor Meek, who acted for the Northumberland Sea Fisheries Committee, marked 145 crabs, whilst Dr. H. C. Williamson, of the Marine Laboratory at Aberdeen, a few years ago marked over 1,500 crabs ; but of the small percentage subsequently recovered none appeared to have been retaken far from the point at which they were liberated in the first instance.

NEW PLANT RECORDS FOR NORTH LANCASHIRE.

S. L. PETTY.
Ulverston.

ON behalf of the recently formed North Lonsdale Field Club,
I give below a few new records.

* *Monotropa hypopetys* L.   Grange, 1907, W. Duckworth.
    New to the Lancashire portion of V.-C. 69; and the
    Westmorland record of many years ago has not been
    confirmed, I believe.   Is recorded for W. Lancs. recently.

* *Epipactis atro-rubens* Schultz.   Grange, 1907, W. Duckworth.
    New to Lancashire portion of V.-C. 69.

Habenaria viridis R.Br.   Urswick, 1907.   J. Dobson.
    New to the Furness area.   On record for Cartmel,
    but no recent confirmation.

Allium scorodoprasum L.   Grange, 1907.   W. Duckworth.
    New to Grange area.

Colchicum autumnale L.   Finsthwaite, 1907.   E. T. Baldwin.
    Recorded 1805, Bot. Guide by Jas. Woods, Jnr.   A
    little below Newby Bridge ; on the left hand side of the
    road to Ulverston.   Never confirmed.

The first three species have been confirmed by Mr. Arthur
Bennett, to whom my thanks are due.

By Seashore, Wood, and Moorland, by **Edward Step** (3rd edition).
London : S. W. Partridge & Co.   320 pp., price 2/6.   The fact that this
volume has reached a third edition speaks well for its popularity.   It is very
suitable for a Sunday School prize, and is written so as to be easily understood by children.   The chapters deal with all manner of subjects from
jelly-fishes to mermaids and sea cows, squirrels to cuckoos, Nature's water-
pots to parrots, butterflies to bats, grasshoppers, beetles and fishes.   More
or less suitable references occur to the Band of Hope, *The Childrens' Friend*,
etc.   The style of the book, 'Oh, Mr. Weston, do tell me something about
Crabs' (p. 64), might almost have given Uncle Westell the idea for his book,
noticed in our columns for October.   There are several illustrations, most of
which have an ancient look about them.   At half-a-crown the book is cheap.

British Country Life in Spring and Summer, edited by **Edward
Thomas.**   Hodder and Stoughton, 1907.   239 pp., Price 8/6 net.   If the
increased interest in 'Nature Study,' which has been manifest during the
past few years, had no other result than the publication of such books as the
one before us, there would be cause for gratification.   In 'British Country
Life' are bound together, in attractive form, the first six parts of 'The
Book of the Open Air,' referred to in these columns for July (p. 261).   The
book throughout is of the character of the two parts then noticed.   Having
regard to its size, the excellence alike of paper, type, and matter, and the
nature of the illustrations, 'British Country Life' is a long way the cheapest
book we have handled for some years.   Besides the chapters referred to
in the previous notice, there are articles by A. H. Patterson, W. H. Hudson,
Gerald H. Leighton, Richard South, and many others.

NORTHERN NEWS

'Answers to correspondents.—G. Fish. Of course the mussel belongs to the animal kingdom, *as all other fish* do. Editor, *Hull Daily News.*'

M. Gustave-F. Dollfus has an interesting article, 'La Géologie il y a cent ane, en Engleterre,' in 'La Feuille Des Jeunes Naturalistes' for November.

According to Dr. F. Ameghino, of the La Plata Museum, South America was the birth-place of the human race, and he traces man back to Miocene opossums !

We are pleased to learn, from a contemporary, that the illustrations in the 'Birds of Yorkshire' 'evidence wonderful patience in the field and laboratory skill.'

From a Hull Newspaper. 'Time was when the Annual Conversazione of the Hull Literary and Philosophical Society' was held 'amid the bones of whales and other amphibious monsters.'

A train carrying a travelling menagerie recently was wrecked, and caught fire. The elephants filled their trunks at a neighbouring stream, and ran backwards and forwards and eventually extinguished the flames. It was in America.

Two men were experimenting with a Chameleon ; they placed it on surfaces of blue and white and green, with the normal results. At last the diabolical idea occurred to them of placing it upon a tartan shawl, with the result that the poor thing died in ten minutes of nervous exhaustion !

The 'Lancashire Naturalist' (No. 5), under the head of 'Lancashire Naturalists of Note,' has an account of Mr. W. A. Parker, F.G.S., who has done so much amongst the fossils from the Coal-Measures of Sparth Bottoms. We learn that 'his motto has always been "Love of the work with hard and persistent effort."'

Some Museums seem to have all the luck. In the will of the late Miss A. Mason is the following :—'I also give to the said Corporation of Nottingham for the said Art Museum the last Mule Canary that travelled from Nottingham to the Cross Keys, Wood Street, Cheapside, on the (almost) last "old times coach," driven by "Old Joe Pearson" (as the next visit I made to Loughborough, Leicester, was by railway). Also the two gadflies which were also in the same glass case as the Canaries, and were caught at Prince Albert's Exhibition in 1851 at Hyde Park.'

Under 'Notes on Flowering Plants' a Lancashire contemporary has the following information :—'Adder's Tongue consists of a single leaf with a slender spike of seeds rising from its bottom, which is supposed to resemble the tongue of the serpent.' The Moonwort.—'The stalk is round, firm, and thick ; it is naked in the middle, and there grows the leaf, which is composed, as it were, of several pairs of small ones, or rather, is a whole and single leaf indented deeply, so as to resemble a number of smaller ; these are rounded and hollowed, and thence came its name of Moonwort.' Evidently the moon has something to do with it !

The October 'Bradford Scientific Journal' is a particularly good number. Of local interest are 'The Bradford Botanical Garden,' by W. P. Winter, and 'The Status of the Polecat in the Bradford District,' by H. B. Booth, and there are several shorter notes. The Polecat is evidently scarce in the Bradford area ; and one 'probably true Polecat' was 'captured *and killed,*' and whilst the men were inspecting its retreat it got up and *slowly dragged itself away !* There are some so-called 'Science Notes, Past, Present, Future,' which are certainly not worth the space they occupy. In referring to the spelling of a certain moor near Ilkley a correspondent informs us that 'Whichis' the correct spelling. It may be so, but we should not have thought it.

A specimen of a false-scorpion, *Chelifer cancroides* Linn., from ¡Manchester, is figured in the October 'Zoologist.'

A photograph of a nest of a common Heron on the ground in a wood near Scarborough is reproduced in 'British Birds' for October.

In the October 'Geological Magazine' Dr. F. A. Bather has a useful note on 'Nathorst's use of Collodion Imprints in the Study of Fossil Plants.'

We are sorry to record the death of Mr. Howard Saunders, at the age of seventy-two. His admirable work in the ornithological world is well known.

We regret to have to record the death of Mr. I. Chalkley Gould, F.S.A., who has done so much in connection with the study and preservation of ancient earthworks and other monuments of the past. He was in his sixty-fourth year.

We regret to record the death of Prof. Charles Stewart, LL.D., F.R.S., Conservator for the past twenty-three years to the Royal College of Surgeons, London. His work at the College of Surgeons' Museum has been almost superhuman, and on the shelves of that Institution, rather than by his writings, will Prof. Stewart's work be remembered in the future.

Several remains of Hyæna, Bear, Rhinoceros, Lion, and Elephant, from the Hoe Grange Cavern, near Brassington, Derbyshire, have been added to the palæontological collection in the Natural History Museum, South Kensington.

In 'Man' for October Mr. S. H. Warren has a 'Note on some Palæolithic and Neolithic Implements from East Lincolnshire.' Judging from the drawings, however, the evidence of, at any rate, the Palæolithic implements, is by no means convincing.

A mammoth tusk is recorded, at a depth of $15\frac{1}{2}$ feet, 'in the upper layer of the Keuper Marl,' near Water Orton, in the Midlands. Another 46 inches in length, and weighing 70 lbs., is recorded from the boulder-clay at Hornsea. It is 'free from all shelling.'

In 'A few notes on nature's year,' published in a contemporary, we learn that in October the woods smell 'leafy,' in November the rabbits gambol, and that December 25th is Christmas Day! Did not Wordsworth say, 'To the solid ground of Nature trusts the mind which builds for aye'?

A 'TAIL-PIECE.'

The Elephant (very disgusted). 'DASH THAT SHORTSIGHTED FOOL OF A KEEPER. THAT'S THE SECOND TIME HE'S PUT MY GRUB AT THE WRONG END!'—Reproduced from 'Punch' by permission.

CLASSIFIED INDEX.

COMPILED BY W. E. L. WATTAM.

It is not an index in the strict sense of that term, but it is a classified summary of the contents of the volume, arranged so as to be of assistance to active scientific investigators, the actual titles of papers not being regarded so much as the substantial nature of their contents.

CONTRIBUTORS.

2 F

BOOK NOTICES.

ASTRONOMY.

BOOK NOTICES—*continued.*

BOOK NOTICES—*continued.*

BOOK NOTICES—*continued.*

ILLUSTRATIONS.

1907 December 1.

ILLUSTRATIONS—*continued.*

SPECIES AND VARIETIES NEW TO BRITAIN BROUGHT FORWARD IN THIS VOLUME.

ARACHNIDA.

Chernese cyrneus, with illustration, 129

FUNGI.

Geaster triplex Jungh from Hebden Bridge, C. Crossland, 99; Hebeloma subsaponaceum Karst., from near Pocklington, C. Crossland, 99; Cantharellus hypnorum from Ferrymoor, C. Crossland, 99; Lentinus suffrutescens Fr., from Milnsbridge, C. Crossland, 52, 99; Lachnea gilva (Boud.) from Hebden Bridge, C. Crossland, 99; Zygodesmus fulvus Sacc., Var. olivascens Sacc., from Selby, C. Crossland, 99; Graphium xanthocephalum from Masham, C. Crossland, 100.

MOLLUSCA.

Vitrina elongata Dp., first record for Britain, found in Co. Louth, 265, 266, 407

ORTHOPTERA.

Capture of Phoraspis leucogramme at the Liverpool Docks, noted, 186

SPECIES AND VARIETIES NEW TO SCIENCE DESCRIBED IN THIS VOLUME.

CRUSTACEANS.

Pygocephalus parkeri, with illustration, found in Coal Measures at Sparth, Rochdale, by Mr. W. A. Parker, 333

FUNGI.

Clavaria gigaspora Cotton n. sp., C. Crossland, 97; Verticicladium Chees-manii Crossl. n. sp., with illustration (Plate IX.), C. Crossland, 98

RHYNCHOLOPHIDÆ.

Description and illustration of Ritteria mantonensis, found at Manton, Lincs., C. F. George, 357

NOTES AND COMMENTS.

January.—Liverpool Biologists— Liverpool Geologists—Leeds Geologists—Nature Photographs.

February.—The Lincoln Boring— Educational Museums—Local Museums—Selby Museum—Liassic Dentaliidæ—Coast Erosion,

March.—Roseberry Topping—British Eggs of Pallas' Sand-Grouse—Distorted Strata in the Little Don Valley —Nottingham Naturalists—Bradford Naturalists—Recorders' Reports— Fossil Mushrooms—British Tunicata —The Slaughter of Kingfishers.

April.—Chernese cyrneus, a new False Scorpion—Malton Museum—Mammal v. Animal—Naturalist Associations — Naturalist Magazines — Deformed Belemnites — Griffithides Barkeri.

May.—York Report of the British Association—Local v. General Museums — Selby Museum — International Ornithological Congress — Flamingoes—Yorkshire Wild Birds and Eggs Protection Committee— Derbyshire Naturalists — Speeton Ammonites—A New Magazine [The Lancashire Naturalist]—Marine Biology — Lancashire and Cheshire Entomologists.

June.—Bradford Scientific Journal— Bradford Museum—Inclosure Acts in Lincolnshire and East Yorkshire —Lakeland Ravens—Plant Associations and Golf—A Darlington 'Find' [Lower jaw of Whale]—Scarborough Museum—Scarborough Naturalists— Model of Ebbing and Flowing Well.

NOTES AND COMMENTS—*continued.*

July.—A New Magazine ['British Birds']—Booming Bempton—Wade's 'Birds of Bempton Cliffs'—'Birds of Yorkshire'—Stonham's 'Birds of the British Islands'—Dentalium giganteum—Sir H. H. Howorth and Glacial Nightmares—Mr. Lamplugh's British Association Address—Bird Migration

August.—York Philosophical Society—New British Land-Shell [Vitrina elongata Dp.[.

September.—The British Association at Leicester — Attendance — Handbooks—President's Address—Value of a Standard—Growth of Botanical Science—History of the Earth—Shape of the Farth—Cephalopods—Natural Monuments — Other Addresses—Plankton Investigations—Mimicry in Insects—Viking Relics at York—Holderness Gravels—Iron Ore

Supplies—Marine Peat—Yorkshire Fossil Plants.

October.—The Pacific Eider—Food of the Black-headed Gull—The 'Grey Wethers'—Lamarck's Evening Primrose—Footprints in the Sands of Time—Norfolk Naturalists—Nests of Coot and Crested Grebe—A New Crustacean [Pygocephalus parkeri].

November.—The Geological Society's Centenary—The Father of English Geology [William Smith] — Toads Embedded in Rocks—The Leicestershire Coalfield—Sycamore Leaf Blotch —Walton Bone Cave, near Clevedon The Lion and the Mouse—Country Side 'Natural History.'

December.—Linnæus—Early Iron-age Burials—History of 'The Naturalist'—Restoration of York Minster—New President of the Yorkshire Naturalists' Union.

CHESHIRE.

Birds: Occurrence of Spotted Crake ♂ at Corwen, A. Newstead, 189

Lepidoptera: Note on a Melanic Race of Agrotis ashworthii, 23

Societies' Reports: Chester Society of Natural Science, &c., Annual Report and Proceedings of, 376; Lancashire and Cheshire Entomological Society, Annual Report of, 166

CUMBERLAND.

(SEE LAKE COUNTIES).

DERBYSHIRE.

Birds: Ornithological Notes from Derbyshire, 1905, published in Derbyshire Archæological and Natural History Society Journal, 1906, 64

Geology: Comment on Mr. C. Fox-Strangways' 'The Geology of the

Leicestershire and South Derbyshire Coalfield,' 379-380

Societies' Reports: Derbyshire Archæological and Natural History Society, Trans. of, Vol. XXIX, 164

DURHAM.

Coleoptera: Recent additions to Coleoptera of Northumberland and Durham District, 1906, 64

Hymenoptera: Notes on the Solitary Wasp (Adynerus parietum Linn.), W. M. Egglestone, 38-39

Lepidoptera: List of species captured in Cleveland District in 1906, T. A. Lofthouse, 188

Personal Notices: Death of Mr. J. E. Robson, 155

Societies' Reports: Cleveland Naturalists' Field Club, Proceedings of, 1905-6, 360; Durham and Newcastle-on-Tyne Natural History Society, Trans. of, 233, 407

ISLE OF MAN.

.LANCASHIRE.

(SEE LAKE COUNTIES).

LAKE COUNTIES.

(CUMBERLAND, LANCASHIRE, AND WESTMORLAND).

LINCOLNSHIRE.

NORTHUMBERLAND.

NOTTINGHAMSHIRE.

WESTMORLAND.
(SEE LAKE COUNTIES).

YORKSHIRE.

YORKSHIRE—*continued.*

Crustacea: Giffithides barkeri, with figure from Anagram, Nidderdale, 133

Diatoms: Distribution of Diatoma hiemale in East Yorkshire, etc., with illustration of species, R. H. Philip, 312-313

Fishes: Notes on the capture of a Basking Shark (Selache maxima), with portrait of same, at Redcar in 1906, T. Sheppard, 10 ; Short-finned Tunny (Orcynus thynnus) exhibited at Bradford, H. B. Booth, 29-30 ; Large Trout, with illustration, captured near Harrogate, R. Fortune, 134 ; Capture of a Fox-Shark at Whitby, T. Stephenson, 273 ; Fish Remains from the Kimeridge Clay, T. Sheppard, 279; List of Species found on visit of Y.N.U. to Litton-dale, H. B. Booth and R. Fortune, 346 ; Sunfish captured near Saltburn, T. Stephenson, 375; List of uncommon Fish captured in the vicinity of Whitby during September, T. Stephenson, 403-404

Flowering Plants: The montane form of Myosotis sylvatica from Gordale Scar, P. F. Lee, 73 ; Cornus suecica, distribution of, in Yorkshire, H. J. Burkill, 135-136 ; Plant distribution in the driftless area of North-East Yorkshire, F. Elgee, 137-143 ; Plants noted on visit of Y.N.U. to Robin Hood's Bay, J. Hartshorn, 200; Habenaria bifolia near Tib-thorpe, L. F. Piercy, 256 ; Plants noted on visit of Y.N.U. to South Cave, J. F. Robinson, 287-288 ; Plants noted on visit of Y.N.U. to Thorne Waste, Bellerby, H. H. Corbett and E. A. Woodruffe. Peacock, 320-322 ; Potamogeton alpinus, Balb., at Sandal Beat, near Doncaster, H. H. Corbett, 327, 375 ; Lists of Plants noted on visit of Y.N.U. to Litton-dale, T. W. Woodhead, 348-350 ; Botanical features of Littondale, W. A. Shuffrey, 354-356 ; Plants of Allerthorpe Common, J. J. Marshall, 356 ; Sagina Reuteri Boiss. on Skip-with Common, W. Ingham, 383 ; Sylvan Vegetation of Fyliugdales, N.E. Yorks., J. W. Barry, 423-431

Fungi: Y.N.U. Fungus Foray at Farnley Tyas, in 1906, with lists of all species collected, not previously recorded for Huddersfield District, including Lentinus suffrutescens Fr.

new to Britain, and Hebeloma nudipes Fr., Galera tenera (Schæff), var. pilosella Pers., Tubaria cupularis (Bull), Psilocybe canobrunnea Fr.. Russula furcata, Pers., var. ochroviridis Cke, Cantharellus Friesii Q, and Clavaria incarnata, Weissm, new to Yorkshire, C. Crossland, 50-57 ; List of two species new to Science, seven species new to Britain, and forty-six species and three vars. new to Yorkshire, with illustration of Verticicladium Cheesmanii Crossl., C. Crossland, 97-105 ; Fungi found during visit of Y.N.U. to Robin Hood's Bay, including Triposporium elegans new to Yorkshire, and five first records for N.E. Yorks., C. Crossland, 253-255, 285-285 ; Volvaria parvula from Wooldale, C. Crossland, 257 ; List of Species found during visit of Y.N.U. to South Cave, C. Crossland, 288-289 ; List of Species noted on visit of Y.N.U. to Thorne Waste, including Clitopilus popinalis and Inocybe scabella new to Yorkshire, and one undetermined species, C. Crossland, 322-323; List of Species found during visit of Y.N.U. at Littondale, comprising Ascophanus cinereus and Trichia Karstenii (= Hemitrichia Karstenii (Rost.) Lister), C. Crossland, 351-353 ; List of Fungi found at Horton-in-Ribbles-dale, including Lycoperdon velatum Vitt., T. Gibbs, 395-396 ; Y.N.U. Fungus Foray at Grassington, Bolton Woods, and Buckden, C. Crossland, 397-401 ; Lachnea hirto-coccinea Phil. and Plow. on Strensall Common, first county record, W. Ingham, 404

Geology and Palæontology: Life Zones in British Carboniferous Rocks, Part II., with lists of fossils of the Millstone Grits and Pendleside Series, and plate, W. Hind, 17-23, 90-96 ; Note on exposure of New Red Sand-stone at Middlesborough, W. Y. Veitch, 31-32 ; Liassic Dentaliidæ, with illustration of Dentalium gigan-teum, Phillips, 35-36 ; Spondylus latus from the Lincolnshire Chalk at Barton, North Lincs., with illustration of same, 40 ; Note on the Geological formation of Roseberry Topping, Guisborough, with illustration, 65 ; Distorted strata in the Little Don Valley, with illustration, 66 ;

YORKSHIRE—*continued.*

Prominent Yorkshire Workers:
Baker, John Gilbert, F.R.S., F.G.S.,
etc., with portrait, T. Sheppard, 5-8.
Cole, Rev. E. Maule, M.A., F.G.S.,
with portrait, T. Sheppard, 267-269,
309

Reptilia: Species noted on visit of
Y.N.U., to Robin Hood's Bay, H. B.
Booth and R. Fortune, 252: Species
noted on visit of Y.N.U. to Littondale, H. B. Booth and R. Fortune,
346

Scientific History: List of papers
and monographs written by John
Gilbert Baker, F.R.S., F.L.S., etc.,
6-7; Bradford Scientific Journal, 68,
193, 311, 439; Victoria History of the
Counties of England—Yorkshire, 181-
186, 217-218; List of papers and
monographs written by Rev. E. Maule
Cole, M.A., F.G.S., 268-269, 309

Societies: Bradford Natural History
and Microscopical Society, Recorders'
Reports, 1906, 68-69; Cleveland Naturalists' Field Club, Proceedings of,
1905-6, 360: Hull Scientific and Field
Naturalists' Club, Transactions of,
1906, Vol. III., Part IV., 63-64; Leeds
Geological Association, Transactions

of, 3-4; Scarborough Philosophical
Society, Annual Report, 1906, 195;
Yorkshire Geological Society, Pro-
ceedings of, Vol. XVI., Part 1, 158-
159; Yorkshire Philosophical Society,
Annual Report, 1906, 265

Tardigrada: Macrobiotus papillifer,
var. of, found at Penyghent, with
illustrations, G. S. West, 72-73

Yorkshire Naturalists' Union:
Annual Gathering at York, Decem-
ber, 1906, T. Sheppard, 26-28; Fungus
Foray at Farnley Tyas in 1906, C.
Crossland, 50-57; Officers of the
Y.N.U., 1907, 58-59; Y.N.U. at
Robin Hood's Bay, T. Sheppard, 198-
202, 250-255, 284-285; Appointment
of Mr. H. Culpin as Treasurer of the
Y.N.U., 224; Y.N.U. at South Cave,
T. Sheppard, 286-289; 'Birds of York-
shire,' with plate of Young Grey
Wagtails, 292-293; Visit of Y.N.U.
to Thorne Waste, T. Sheppard, 316-
324; Visit of Y.N.U. at Littondale,
Yorks., T. Sheppard, 342-353; Fungus
Foray at Grassington, Bolton Woods,
and Buckden, C. Crossland, 397-401;
History of 'The Naturalist,' 410;
Election of Dr. Wheelton Hind, F.G.S.,
F.R.C.S., as President for 1908, 411

MISCELLANEA.

Anthropology: Anthropology at the
British Association meeting at Leicester, G. A. Auden, 366-369

Arachnida: A curious faculty in
Spiders of the family Salticidæ, O.
Pickard-Cambridge, 9-10; Note on
paper published by O. Pickard-Cambridge 'On some New and Rare
British Arachnida, 37; Peculiarities
in Attis Spiders, W. W. Strickland,
147-148

Birds: White Lesser Black-backed
Gull, and reported meeting of the
Ivory Gull, on the Farne Islands in
1906, H. B. Booth, 172; Birds of the
Farne Islands, with many illustrations, R. Fortune, 234-238; Notes on
the Lapwing, F. Stubbs, 310-311;
Pacific Eider, note on occurrence of
in Britain, 329; Black-headed Gull,
report of Cumberland County Conucil

on food of, 329; Coot and Crested
Grebe, notes on and plates of nests
of each species, 332-333; List of
Bones of Birds found in Walton
Bone Cave, Clevedon, 381

Diptera: A Net-Building Chironomus
larva, with notes on C. pusio, and
illustrations, A. T. Mundy, 389-391

Evolution: Theories of Evolution,
Agnes Robertson, 167-171, 209-215,
241-249

Ferns: Paisley Fern (Allosorus crispus), the Chemistry of, P. Q. Keegan,
25

Flowering Plants: Orchis mascula,
Narthecium ossifragum, and Polygonum bistorta, P. Q. Keegan, 153-
155; Birch Tree, P. Q. Keegan, 205-
208; The Ancestors of the Angiosperms, M. A. Johnstone, 338-341;

MISCELLANEA—*continued.*

CORRIGENDA.

Page 42, *Delete word* "water" *under illustrations.*
,, 44, *Delete word* "water" *under illustrations.*
,, 225, line 15, *for* "excess" *read* "access."
,, 355, line 28, *for* "heather" *read* "Halton."